Bridge Design and Evaluation

Bridge Design and Evaluation

LRFD and LRFR

Gongkang Fu

WILEY

John Wiley & Sons, Inc.

Library of Congress Cataloging-in-Publication Data:
Fu, Gongkang.
 Bridge design and evaluation : LRFD and LRFR / Gongkang Fu.
   p. cm.
 Includes bibliographical references and index.
  ISBN 978-0-470-42225-0 (cloth); ISBN 978-1-118-32993-1 (ebk); ISBN 978-1-118-33220-7 (ebk);
ISBN 978-1-118-33268-9 (ebk); ISBN 978-1-118-40255-9 (ebk); ISBN 978-1-118-40256-6 (ebk);
ISBN 978-1-118-40257-3 (ebk)
 1. Bridges–Design and construction–Textbooks. 2. Bridges–Evaluation–Textbooks.
3. Bridges–Maintenance and repair–Textbooks. 4. Load factor design. I. Title.
 TG300.F82 2013
 624.2′52–dc23
                                                          2012026192

To my father who taught me algebra when I did not have a high school to attend,
and to my mother who taught us love when we did not have enough food.

Contents

Preface

Since the first edition of the American Association of State Highway and Transportation Officials (AASHTO) *LRFD Bridge Design Specifications* in 1994, U.S. highway bridge design has been moving towards load and resistance factor design (LRFD). Prior to that, for more than a half of century, highway bridge design practice has been using the load factor design (LFD) and service load design (SLD), according to the AASHTO Standard Specification for Highway Bridges. Furthermore, in 2003, the first edition of *AASHTO Manual for Condition Evaluation and Load and Resistance Factor Rating (LRFR) of Highway Bridges* was issued for bridge evaluation and load rating. In 2007, AASHTO mandated the LRFD specifications in the United States after more than a decade of preparation, including trial use and software development. With the Federal Highway Administration's (FHWA's) requirement and also for consistency, a number of states have started the practice of LRFR for bridges designed according to LRFD. This introductory book covers the LRFD and LRFR methods for highway bridges as a textbook for a first undergraduate and/or graduate course on highway bridge design and/or evaluation.

The sixth edition of the AASHTO *LRFD Bridge Design Specifications* issued in 2012 was the latest design code when this book was prepared. Therefore, in this book, this edition of specifications is referred to as the AASHTO specifications, AASHTO design specifications, or AASHTO LRFD specifications, depending on what is being emphasized in the context. For bridge evaluation and load rating, the second edition of the *AASHTO Manual for Bridge Evaluation* dated 2011 was the latest when this book was prepared, which is referred to as the AASHTO manual or AASHTO evaluation specifications, depending on context. This set of specifications is also sometimes referred to as the AASHTO specifications collectively along with the LRFD specifications.

While some information provided here may be used for preliminary design of highway bridges, this book mainly covers the detailed design of short- to medium-span highway bridges according to the AASHTO specifications. In addition, structural design is the main focus in detailed design. In other words, detailed design covered in this book proceeds with given or already optimized structure type, span arrangement, span length, and so on. This situation is assumed so that the student will not have to be concerned with whether the being-designed bridge needs to be optimized, with respect to span type, length, width, geometric parameters, and so on, which would unlock everything and leave no direction for the student to follow. On the other hand, the student needs to understand that determination of these factors may require additional information and/or knowledge not completely covered in this book, such as cost effectiveness of alternative span lengths and/or arrangements associated with different materials and climate conditions, and so on.

Furthermore, this book has a focus on short- to medium-span highway bridges for detailed design. The short- and medium-span lengths here are defined as those in the range of 20 to about 200 ft, within the application range of the specifications. This book also covers those bridges' evaluation (particularly load rating) according to the AASHTO manual. For the given span length range, wind is usually not the major governing load. However, longer highway bridge spans whose design often is controlled by wind load effect still need to meet the requirements of the AASHTO specifications presented here. Therefore, understanding the requirements covered in this book may be viewed as a prerequisite to studying long-span bridge design and evaluation.

This book has seven chapters. Chapter 1 is an introduction to bridge engineering, including design and evaluation. Chapter 2 covers both the general and specific requirements in the AASHTO specifications for designing highway bridges. The concepts of structural reliability are also presented, which were used for the calibration of both sets of AASHTO specifications focused on here. Chapter 3 presents further detailed specific requirements of the specifications for loads, load effects, and their combinations for highway bridge components. Chapter 4 of this book covers the superstructure part of bridge design, Chapter 5 the bearings, and Chapter 6 the substructure. Chapter 7 shifts focus from design to evaluation (load rating) of the same bridges covered in the book.

This book is designed as a textbook for first courses of undergraduate and/or graduate studies on highway bridge design and/or evaluation according to current AASHTO specifications. For an undergraduate course of three credit hours, Chapters 1, 3, and 4 (or along with 5) are recommended to be covered. If the undergraduate course is designed for four credit hours, Chapter 5 (or 6) may be added. For a graduate course, Chapters 1 through 6 may be covered for three credit hours, and Chapter 7 can be added for four credit hours. Structural analysis will be a prerequisite for using this textbook, and steel and concrete designs are preferred to be prerequisites but may be allowed as corequisites. Successful completion of the course will enable the student to perform duties of an entry-level engineer in bridge design and evaluation, according to the current AASHTO specifications.

Another alternative way of using this book is to teach only its bridge analysis and the related examples or portions of the examples. It can be part of a structural analysis course or an independent course at either undergraduate or graduate level.

The examples included in this book can be used without referring to the text. Such use can be particularly convenient for review after a level of understanding of the relevant text material is established. Therefore, they may also serve as a helpful reference for junior engineers with limited familiarity with bridge design and/or evaluation and those who are preparing for a professional engineer license exam, particularly where bridge design is a required subject.

A complete highway bridge will be visible to the student when several of these examples are integrated together, including its load rating of primary members. For instance, the following examples can make a typical highway commonly seen in one of today's bridge design offices: the reinforced concrete deck designed in Examples 4.1, 4.3, and 4.4 plus the steel plate girders designed in Examples 4.9 to 4.11 plus the shear studs designed in Example 4.8 plus the abutment designed in Examples 6.1 to 6.4. Examples 7.1 and 7.2 then provide the bridge's superstructure member load rating. Of course, the deck design in Example 4.1 may be replaced by another deck design in Example 4.2. Furthermore, the superstructure steel beams in Examples 4.9 to 4.11 may be replaced by the prestressed concrete beams in Examples 4.12 to 4.14 or another steel beam superstructure in Examples 4.5 to 4.7. To minimize inconvenient cross referencing between different examples, a few steps of calculation in these examples may have been repeated in other examples, so that such cross referencing will involve only nearby pages if at all.

The student is not required to have the AASHTO LRFD or LRFR specifications while taking a course using this book. However, reference to the AASHTO specifications provisions are provided in the text so that the instructor and/or student can easily find more information in the specifications if needed. It is also worth noting that AASHTO offers U.S. faculty three free AASHTO publications a year, while the faculty needs to pay for shipping and handling.

This book is based on the author's experience of teaching bridge design and evaluation at Wayne State University and Illinois Institute of Technology for 16 years. Many students of the author have assisted in preparing the manuscript. He would like to express gratitude to them: Drs. Dinesh Devaraj and Pang-jo Chun and Messrs. Tapan Bhatt, Jason Dimaria, Alexander Lamb, and Justin Sikorski. Bridge design plans and/or calculations also have been received from state bridge owners of California, Georgia, Idaho, Illinois, Michigan, New York, and New Jersey. The author is very grateful to the state agencies for this set of information related to current practice of bridge design and evaluation. It certainly has influenced the design and preparation of this book.

The following unit abbreviations have been used:

in. = inch

ft = foot

k = 1000 pounds

kin = 1000 pound inch

kft = 1000 pound foot

k/ft = 1000 pounds per foot

k/in^2 = 1000 pounds per square inch or ksi

Bridge design and evaluation is a widely spread subject, and many engineers in the world have spent their entire career on it. An introductory text book as this may easily miss some relevant and/or important information or contain errors on a topic. The author would be grateful if the reader could identify them to him. He can be reached at gfu2@iit.edu.

Bridge Design and Evaluation

1 Introduction

1.1 Bridge Engineering and Highway Bridge Network

Human beings have been constructing bridges for about four thousand years. The oldest and still existing bridge in the world is perhaps the Zhaozhou Bridge in Hebei Province in China, originally constructed approximately in A.D. 600. However, bridge design and construction then may not be considered bridge engineering practice by today's definition. Instead, work was done based more on experience as opposed to quantitative planning as done now. Bridge engineering today uses calculus-based analysis and detailed planning.

Materials used in bridge construction have also changed noticeably through a good number of years, from mainly natural materials such as stones and wood then to mainly man-made materials such as steel and Portland cement concrete today. Due to great improvement in the strength and production quality control of these materials, bridge components have become smaller, thinner, skinnier, and lighter to reduce self-weight and be more economical.

1

In 1866 Wayss and Koenen in Germany conducted a series of tests on reinforced concrete beams (Heins and Lawrie, 1984), which started the era of concrete for bridge construction. More tests and research work were done in the following decades. The first bridge using reinforced concrete in the world was credited to Monier in 1867 (Heins and Lawrie, 1984). The first bridges using steel are believed to be constructed in the United Kingdom and United States in the 1880s. These pioneering projects began what is known today as modern bridge engineering.

Another important aspect characterizing modern bridge engineering is the tools used to perform quantitative modeling and planning. They include calculus and calculus-based mechanics, acknowledged as the foundation of modern bridge engineering as practiced today. This knowledge was established in the seventeenth century. With the new materials and advanced analysis tools, fast development of modern bridge engineering had its technical strength.

The fuel for substantial developments of bridge engineering was the need or desire for economic development. For example, today's highway bridge technology in the United States is largely a result of rapid development of the interstate highway system in the 1950s and 1960s after World War II. As a product, a comprehensive highway system has been established consisting of about 50,000 miles of roadways and about 600,000 bridges.

It is also interesting to mention that a number of developing countries are currently experiencing a similar "boom" in their surface transportation systems. This has become the driving force for bridge engineering development in those parts of the world.

1.2 Types of Highway Bridges

In highway systems bridges maintain the continuation of the roadway, for the traffic of vehicles and/or pedestrians as needed. The American Association of State and Highway Transportation Officials (AASHTO) design specifications include the following definition for highway bridges: any structure having an opening not less than 20 ft that forms part of a highway or that is located over or under a highway. Note that this definition actually covers a number of large culverts with a span longer than 20 ft, although structurally different from the highway bridges covered in this book because soil interaction is involved in the load-carrying behavior and performance.

Based on the superstructure type, highway bridges may be recognized as slab bridges, beam or girder bridges, arch bridges, truss bridges, cable-stayed bridges, and suspension bridges. These names directly refer to the main spanning structure of the bridges. Namely, slab bridges use a slab to span the opening, beam or girder bridges use beams, arch and truss bridges use arches and trusses, cable-stayed bridges use cable-stayed main girders, and suspension bridges use suspension-supported main girders.

Figures 1.2-1 to 1.2-7 illustrate some of these bridge types. More details of these bridge types will be presented in Chapter 4 on highway bridge superstructure.

It also should be mentioned that, by number of bridges and by number of spans, beam or girder bridges represent by far the most popular highway bridges in the United States and the world. Their span lengths are usually within 300 ft to be cost effective. For longer spans, arch, truss, cable-stayed, and suspension spans may become more preferred due to cost consideration. Some examples for these spans are shown in Figures 1.2-6

Figure 1.2-1
Two simple steel beam spans sharing a pier.

Figure 1.2-2
Continuous steel beam bridge.

Figure 1.2-3
Curved concrete box
beam bridge.

Figure 1.2-4
Prestresed concrete
I-beam bridge of single
span.

to 1.2-9. Typically more than 90% of highway bridges are beam or girder
bridges in the United States. This percentage is larger for number of bridge
spans because very long bridges often consist of many short or medium
beam spans. Therefore, this introductory book on highway bridge design
and evaluation focuses on beam or girder bridges of short and medium
spans covering a vast majority of current highway systems.

Figure 1.2-5
Prestressed concrete box beam bridge.

Figure 1.2-6
Steel truss bridge.

It is also very common to classify bridges into different groups according to their superstructure type. For example, classification according to superstructure material is a common one. Thus, there are steel bridges, prestressed concrete bridges, timber bridges, aluminum bridges, and so on, referring to the major material used to construct the superstructure. Another common classification is based on superstructure configuration, such as continuous bridges and simple span bridges, referring to spans made of continuous beams and simply supported beams, respectively. Sometimes they are used to refer to truss spans as well because trusses can

Figure 1.2-7
Steel arch bridge.

Figure 1.2-8
Cable-stayed bridge.

Figure 1.2-9
Suspension bridge.

also be easily made continuous over a support. Segmental bridge refers to the construction method used: namely, these bridges are constructed using segments and assembled into a system. Usually segmental bridges have long spans that require the spans to be constructed by connecting small pieces (segments) to form the spans. It is thus important to ensure the connections in segmental bridges to be positive and reliable for many years to come. A typical approach to this type of construction is to use prestressing: pretensioning each segment to reduce the self-weight and posttensioning each segment to the part already in place during erection.

Note also that it is important to differentiate a bridge from a bridge span. A "bridge" refers to all the spans of the bridge. A "bridge span" refers to one of the possibly many spans of the bridge. Of course, when a bridge has only one span, that span is the bridge. Nevertheless, the word *bridge* sometimes also includes the approach spans at the two ends of the bridge as well as the substructure system near and below the ground. In that case, the word *span* often only means the superstructure and does not include other parts of the bridge system.

1.3 Bridge Construction and Its Relation to Design

It is essential for a bridge designer to understand the planned construction process of the bridge being designed. It is very important that the designer take into account the construction procedure in the design process because the construction procedure may subject bridge components to loading conditions not experienced in service and these conditions require special design. For example, lifting a prestressed concrete beam as a primary component for a beam bridge using a crane may cause a significant negative moment that would never be induced in the service condition. Thus this negative moment needs to be carefully designed and detailed for. Otherwise, the beam may fail or be damaged during construction. This check in design is referred to as a constructability check in the AASHTO design specifications. Figures 1.3-1 to 1.3-3 show a few steps of highway bridge construction.

It is also worth mentioning that many bridge failure incidents occur when the bridge is under construction. Therefore, construction load is an important load to consider and to cover in design.

Figure 1.3-1
Concrete deck placement using a concrete pump.

Figure 1.3-2
Steel piles placed as the foundation for an abutment to be constructed.

Figure 1.3-3
A steel shear stud being welded
to the top flange of a steel beam.

1.4 AASHTO Specifications and Design and Evaluation Methods

Load factor design (LFD) and service load design (SLD) were the two traditional design methods used in the United States for highway bridge design
over many decades and prior to the current AASHTO specifications. Sometimes SLD is referred to as allowable stress design (ASD) in the literature.
In 1994, AASHTO adopted the first edition of the *LRFD Bridge Design Specifications* with the load and resistance factor design (LRFD) method. As mentioned in the preface, the latest sixth edition (AASHTO, 2012) is exclusively
referred to in this book. This set of design specifications is also referred to
hereafter as the AASHTO specifications, AASHTO design specifications, or
AASHTO LRFD specifications.

   For highway bridge evaluation, AASHTO has another set of specifications, the *Manual for Bridge Evaluation*, which references the design
specifications for consistency. The latest second edition of the evaluation
specifications (AASHTO, 2011) is referred to in this book as the AASHTO
manual, evaluation manual, or evaluation specifications. Nevertheless,
sometimes "AASHTO specifications" is also used to refer to both sets of
specifications when the context is clear. Figures 1.4-1 and 1.4-2 show the
cover pages of the electronic version of these specifications.

   The articles of the two sets of AASHTO specifications are referred to in
this book using italic Futura Condensed font, with an additional letter *M*
for the AASHTO manual. For example, *1.3.1* means Article 1.3.1 in the
AASHTO design specifications, and *M6A4.2.1* refers to Article 6A4.2.1 in
the AASHTO manual.

Figure 1.4-1
Cover page of *AASHTO LRFD Bridge Design Specifications*
(electronic version). Used by permission.

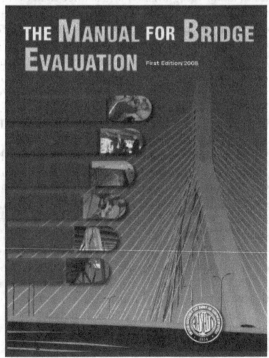

Figure 1.4-2
Cover page of *AASHTO Manual for Bridge Evaluation*
(electronic version). Used by permission.

1.5 Goals for Bridge Design and Evaluation

The AASHTO specifications set forth the following goals for highway bridge design in the United States: constructability, safety, serviceability, inspectability, economy, and aesthetics. Traditionally, the requirement on inspectability has not been adequately emphasized and is therefore worth special attention here. Apparently including this item as one of the design objectives in the design code is based on experience learned from maintaining the extensive network of bridges in the past several decades after its establishment. This has been seen as an issue because of the inadequate attention received.

It is important that the bridge designer focus on the above goals in the design process instead of merely trying to satisfy the provisions and/or equations in the specifications. Obviously the specifications cannot cover every situation that can possibly be encountered. These goals, however, should be targeted at all times during the design.

The AASHTO manual also serves as a standard and provides uniformity in the procedures and policies for bridge evaluation. Such evaluation is charged to determine the physical condition, maintenance needs, and load capacity of U.S. highway bridges. Chapter 7 addresses the determination of highway bridge load capacity or load rating as referred to in the community and in the specifications. While the topics of bridge inspection and scoping for renewing bridge condition are briefly mentioned there, their in-depth coverage is beyond the scope of this introductory book.

1.6 Preliminary Design versus Detailed Design

Highway bridge structures are part of a roadway for surface transportation. Accordingly, bridge structures need to meet the requirements for the roadway. These requirements may include but are not limited to roadway width, horizontal or vertical curvature profiles, elevation, cost effectiveness, maximized life span, and so on. In the first of the two stages of highway bridge design, or the preliminary design stage, these factors or issues are considered and covered. The factors accordingly determined are width, length, how many spans, types of the spans (suspension, cable stayed, truss, arch, or beams; simply supported or continuous), material type (steel, concrete, timber, etc.), construction method, and so on.

After the preliminary design is completed, the second design stage, or the detailed design stage, will begin. This stage includes determination of each member's dimensions, location, connection with other members, quantity of steel reinforcement if any required, possible maintenance approaches, and so on. The detailed design works toward the goal of a set of plans for the contractor to build the bridge. Occasionally the construction may not take place, although the plans have been completed. For example, alternative designs for a bridge may be required and performed using different materials to compare the costs (such as steel superstructure vs. concrete superstructure). Then the lower cost option is accepted and the higher cost one is eliminated, although both designs have been completed to the same level of detail so that construction could have been carried out for either option.

This book focuses on the detailed design of highway bridges in accordance with the AASHTO design specifications, while preliminary design is only briefly discussed when necessary. In addition, structural design is the main focus here since it is a major consideration in detailed design. In other words, detailed design covered in this book proceeds with given or already optimized structure type, span arrangement, span length, and so on. This situation is assumed so that the student will not have to be concerned with whether the being-designed bridge span has been optimized with respect to type, length, width, geometric parameters, and so on, which would unlock everything and leave no direction for the student to follow. On the other hand, the student needs to understand that determination of these factors may require additional information and/or knowledge not completely covered in this book, such as cost effectiveness of alternative span lengths and/or arrangements associated with different materials and climate conditions, and so on.

1.7 Organization of This Book

There are six more chapters beyond this introductory chapter.

Chapter 2 covers the general and specific requirements in the AASHTO specifications for designing highway bridges. Some background information on the derivation of some of these requirements is also presented in this chapter, such as the structural reliability theory. This theory along with reliability analysis methods were used to derive or calibrate the load and resistance factors in the load combination formulas for both design and evaluation of highway bridges in both sets of AASHTO specifications.

Chapter 3 presents further detailed specific requirements of the AASHTO specifications for loads, load effects, and their combinations for highway bridge components. These items represent a major advancement carried by the LRFD and load and resistance factor rating (LRFR) approaches in the specifications, as opposed to the LFD and SLD methods. The LRFD and LRFR concepts are distinctively based on probabilistic coverage of the risk involved in the entire life of the bridge. While the AASHTO specifications do not require the design or evaluation engineer to explicitly use the theory of probability and statistics in practice, it is definitely helpful to understand the background material in Chapter 2 to fully understand the concepts. This understanding would also be required when the engineer needs to extrapolate the concepts to cases not clearly covered in the current specifications, such as long bridge spans.

Not every load discussed in Chapter 3 may be critical or even present for every component that may be designed. However, these loads represent the most commonly observed ones that need to be considered in bridge design and evaluation. Note also that there may be loads that have not been explicitly covered in the specifications. The engineer needs to identify and estimate, spending the fullest effort, all possible governing loads at all stages of construction and operation of the bridge and have them covered in design. This book is unable to exhaustively identify all the loads given in the specifications. One example of such cases is the load that may be applied to bridge components during construction. Due to the wide variety of construction procedures that may be used, construction load may vary significantly from one bridge to another. It is the responsibility of the design engineer to exhaustively identify and quantify these loads and cover them in the design with an adequate safety margin. This concept is included in the specifications and covered in this book.

A bridge structure system may be divided into three major parts: the superstructure, the bearings (or connections between the superstructure and substructure), and the substructure. A more common classification includes only two parts: the superstructure and the substructure with the bearings included in the substructure. This book separates the bearings from the other two parts for an explicit understanding of the bearings. Chapter 4 of this book covers the superstructure part of bridge design, Chapter 5 the bearings, and Chapter 6 the substructure. Of course, these parts need to be integrated into a bridge structure system to function and last. Therefore, these three chapters are interrelated and will need to be so used during the course. It should also be noted that the order of these three chapters also represents a typical order of bridge design practice: from the top to the bottom of the structure system.

Chapter 7 shifts focus from design to evaluation (load rating) of the same bridges covered in the book. Since bridge load rating uses many provisions in the AASHTO design specifications, this topic is readily covered after these provisions are presented and learned in the previous chapters. It also should be noted that U.S. bridge engineers today often spend more of their time on bridge evaluation and maintenance than design in the United States. Therefore, this chapter should receive adequate attention in using this book.

The examples included in this book may be used without referring to the text. Such use can be particularly convenient for review after a level of understanding of the relevant text is established. Therefore, the examples may also serve as a helpful reference for junior engineers with limited familiarity with bridge design and/or evaluation and those who are preparing for a professional engineer license exam, particularly where bridge design is a required subject. A complete highway bridge will be visible to the student when several of these examples are integrated together, including its load rating of primary members. For instance, the following examples may make a typical highway commonly seen in one of today's bridge design offices:

the reinforced concrete deck designed in Examples 4.1, 4.3, and 4.4 plus the steel plate girders designed in Examples 4.9 to 4.11 plus the shear studs designed in Example 4.8 plus the abutment designed in Example 6.1 to 6.4. Examples 7.1 and 7.2 then provide the bridge's superstructure member load rating. Of course, the deck design in Example 4.1 may be replaced by another deck design in Example 4.2. Furthermore, the superstructure steel beams in Examples 4.9 to 4.11 may be replaced by the prestressed concrete beams in Examples 4.12 to 4.14 or another steel beam superstructure in Examples 4.5 to 4.7. To minimize inconvenient cross reference between different examples, a few steps of calculation in these examples may have been repeated in other examples.

References

American Association of State and Highway Transportation Officials, AASHTO (2012), *LRFD Bridge Design Specifications*, 6th Ed., AASHTO, Washington, DC.

American Association of State and Highway Transportation Officials, AASHTO (2011), *Manual for Bridge Evaluation*, 2nd Ed., AASHTO, Washington, DC.

Heins, C. P. and Lawrie, R. A. (1984), *Design of Modern Concrete Highway Bridges*, Wiley, New York.

Tang, M.-C. (2008), "*Evolution of Bridge Technology*," T.Y. Lin International, paper presented by T. Ho at ASCE/SEI Workshop, Washington DC, Feb. 2008.

Baker, R. M and Puckett, J. A. (1997), *Design of Highway Bridges Based on AASHTO LRFD Bridge Design Specifications*, Wiley, New York.

2 Requirements for Bridge Design and Evaluation

2.1 General Requirements

The AASHTO specifications require highway bridges be designed for constructability, safety, and serviceability, with due regard to issues of inspectability, economy, and aesthetics. These requirements are highlighted in this chapter.

The National Bridge Inspection Standards (NBIS) has established the national requirements for bridge inspection. The AASHTO manual accordingly specifies the inspection procedures and evaluation practices to meet these requirements. While constructability, inspectability, and aesthetics are often not to be altered in bridge evaluation, the objectives of safety, serviceability, and economy are still relevant in bridge evaluation practice.

2.2 Limit States

In the AASHTO specifications many aspects of the general requirements identified above are addressed in the form of limit states. For example, the safety requirement is met by assuring cross-section capacity limits are not exceeded, collapse mechanism limit states are prevented, and so on. The concept of limit state is focused on in this section and the following sections will discuss each of the general requirements. Keeping the limit state concept presented in this section in mind, it will hopefully be easier to understand the general requirements.

The limit states specified in the AASHTO codes are intended to provide a buildable and serviceable bridge capable of safely carrying the design loads for the specified life span of 75 years. It also should be noted, though, that some of the requirements are more difficult to quantify and accordingly formulate. An example is the esthetics requirement. The AASHTO design specifications thus can only provide descriptive guidelines for those requirements.

In the AASHTO specifications, four groups of limit states are specified to be used in the design of bridge elements: service, fatigue and fracture, strength, and extreme-event limit states.

The Service Limit State is taken as restrictions on stress, deformation, and crack width under regular service conditions. Therefore the load factors in the Service Limit State are typically 1.0, representing the service condition, although there are exceptions where the design load is considered as not representing the routine service condition. This group of limit states also provides certain experience-related requirements that cannot always be derived solely from strength or statistical considerations. More details of this load limit state are given in Chapter 3 and design application examples are presented in the following chapters for specific bridge components.

The Fatigue and Fracture Limit State represents restrictions on the stress range due to a design truck with respect to fatigue and/or fracture failures seen as material cracking in bridge components. This limit state in the specifications is to limit the probability of cracking occurrence under repetitive loads and/or limit crack growth under such loads to prevent fracture during the design life of the bridge. More details on the mechanism of such failure are discussed in Chapter 3 and design application examples for steel bridge components are presented in Chapter 4 for bridge superstructure design.

Some structural engineers and/or structural engineering students may be more familiar with the Strength Limit State, since it is the most basic and the first one discussed in our education curriculum. It may also be most intuitive to us. This Strength Limit State in the AASHTO specifications is taken to ensure that strength and stability are provided to resist the specified load combinations that a bridge component or system is expected to experience in the design life. Note that the strength and stability of concern may be local or global. For example, the design of a beam for moment often needs to ensure that not only global buckling (such as lateral torsional buckling) but also local buckling (such as that involving a flange or the web of the cross section) will not occur with the specified margin of safety reflected in the load and resistance factors. More details of this limit state are discussed in Chapter 3 and design applications are presented in the following chapters.

The Extreme Event Limit State refers to the requirements in the AASHTO specifications for structural survival of the bridge component or system during a rare event. Such an event may be an earthquake, significant flood, vessel collision, truck collision, ice flow, or scour condition.

The AASHTO specifications prescribe a general format of load combination as follows:

2.2.1 General Formulation of Limit State Load Combination

1.3.2.1
$$Q_n = \sum \eta_i \gamma_i Q_i \tag{2.2-1}$$

where Q_n = total design load effect (nominal value)
Q_i = load effects to be discussed in Chapter 3, such as those due to self-weight, trucks, wind, and so on
γ_i = load factors for respective Q_i, to be defined in Chapter 3
η_i = load modifier associated with: η_D relating to ductility, η_R relating to redundancy, and η_I relating to operational classification

The load factors γ_i in the AASHTO design specifications are allowed to vary depending on the situation to ensure safety. For example, the self-weight load factor is larger if the self-weight acts as a load generating stresses to be superimposed with other load stresses. However, the self-weight load factor is smaller is the self-weight acts as a strength or resistance. An example for that situation is in a stability check for a bridge pier. These different treatments are to account for uncertainties in estimating the self-weight under these different situations.

For loads for which a maximum value of γ_i is appropriate,

1.3.2.1
$$\eta_i = \eta_D \eta_R \eta_I \geq 0.95 \tag{2.2-2}$$

and for loads for which a minimum value of γ_i is appropriate,

1.3.2.1
$$\eta_i = \frac{1}{\eta_D \eta_R \eta_I} \leq 1.0 \tag{2.2-3}$$

The values of η_D, η_R, and η_I are given in the AASHTO specifications, as shown in Tables 2.2-1 to 2.2-3.

In general, the concepts of enhancing safety through redundancy and ductility are emphasized. Examples of redundancy include using more than

Table 2.2-1
Load modifier relating to ductility, η_D

For the strength limit state:	
≥ 1.05	For nonductile components and connections
1.00	For conventional designs and details complying with these specifications
≥ 0.95	For components and connections for which additional ductility-enhancing measures have been specified beyond those required by the specifications
For all other limit states: 1.00	

Table 2.2-2
Load modifier relating to redundancy, η_R

For the strength limit state:	
≥ 1.05	For nonredundant members
1.00	For conventional levels of redundancy, foundation elements where φ already accounts for redundancy as specified
≥ 0.95	For exceptional levels of redundancy beyond girder continuity and a torsionally closed cross section
For all other limit states: 1.00	

Table 2.2-3
Load modifier relating to operational classification, η_I

For the strength limit state:	
≥ 1.05	For critical or essential bridges
1.00	For typical bridges
≥ 0.95	For relatively less important bridges
For all other limit states: 1.00	

three parallel beams in the cross section, multiple columns in the pier or bent, pier cap on piles to ensure pile group effect, and so on. Ductility refers to the capability of structural systems or components to sustain the first failure of the material used without system failure. For instance, reinforced concrete columns should be designed to be able to prevent the main vertical steel reinforcement from buckling after the concrete is crushed under a seismic event by designing confinement to retain the crushed concrete in place to support the steel reinforcement. Furthermore, composite beam sections with concrete and steel interacting with each other should be designed to have steel yielding as the ultimate limit state not concrete crushing for a more ductile behavior.

Strength limit state refers to the limit for the strength and stability of a bridge component, bridge subsystem, or bridge system. For example, a bridge bearing should be designed to sustain the limit strength for the load combinations transmitted from the member it supports. A bridge truss system with multiple members needs to be designed to have adequate strength for the designated load combinations.

Strength limit states may refer to local or global behaviors. An example of the former is the shear capacity controlled by the stability of the web in a steel I beam. On the other hand, lateral torsional buckling of the beam is viewed as an example of the global stability limit state. The corresponding loads specified in the AASHTO design and evaluation specifications are statistically significant load combinations that a bridge is expected to experience in its design life. More details and quantitative combinations of the loads are given in Chapter 3.

2.2.2 Strength Limit State

This limit state is relevant to those loads that occur rarely or relatively infrequently, such as those induced by earthquakes, significant floods, collision of trucks, vessels, or ice flows. The extreme-event limit state is to be designed for to ensure the structural survival of the bridge during such an event or a combination of them. While survival may be defined at different levels, such as with severe damage, minor damage, or limited inelastic behavior, this limit state may be applied differently in design for different members depending on their functions in the system. For example, connections between the superstructure and the abutment are required to be designed more conservatively than bents with multiple columns as a relatively redundant substructure system. This is because a superstructure–substructure connection failure may cause a superstructure to fall, making the entire road discontinuous for service, while failure of one of the columns in a multicolumn bent would not cause system failure and roadway close. More details of this limit state and its applications are given in Chapter 3 and the following chapters.

2.2.3 Extreme-Event Limit State

The service limit state is given in the AASHTO specifications as restrictions on stress, deformation, and crack width under regular service conditions. It also provides certain experience-related provisions that cannot always be derived solely from strength or statistical considerations. For example, reinforcement spacing in concrete beams is controlled as a service limit state requirement in addition to the strength limit state. Chapter 3 and example applications in other chapters will have more details for the concept and specific applications of this limit state.

2.2.4 Service Limit State

Fatigue refers to appearance of cracking in the material as a result of repetitive application of stress below or way below the material's static strength. Fracture is the sudden increase of the crack size induced by fatigue. Depending on the location of such failure and the redundancy of the bridge system,

2.2.5 Fatigue and Fracture Limit State

fatigue and fracture failure may or may not be a cause of system failure or collapse. Bridge members whose fracture may cause system failure or collapse are referred to in U.S. bridge engineering practice as fracture critical members.

Existence of material discontinuity at the microscopic scale may be very difficult to eliminate, depending on the process and procedure used to fabricate the material. However, some may more likely induce discontinuity than others. For example, the continuous hot-rolling process of steel shapes produces relatively more uniform material than welding as a process of locally treating and fusing different steel materials. As a result, welding more likely introduces discontinuity resulting from inadequate fusion of the base metal and the weld material (electrode). Therefore, weld details have more often experienced fatigue and fracture failure.

Accordingly, the fatigue limit state in the AASHTO specifications is required to be taken as a restriction on stress ranges. The applied load is specified in the specifications as a single design truck with the number of expected stress range cycles based on an estimation of how many times over the life span of 75 years such trucks may cross the bridge. Depending on the type of weld detail (i.e., how it is completed, such as in the field or in the shop) and the number of truck load repetitive applications, the allowable stress range given in the AASHTO specifications is different. The fracture limit state is taken as a set of material toughness requirements of the AASHTO Materials Specifications.

A typical fatigue-prone bridge component detail is the so called cover plate weld to attach it to a steel I-beam bottom flange for increased moment capacity. Such a weld is subjected to significant bending stress. If the weld is also subjected to a large number of stress cycles applied by trucks of large volume, fatigue cracking has been observed many times. Although such cover plate welding has been significantly reduced or eliminated in new bridge design as a result of this observation and subsequent research on the phenomenon and failure mechanism, many existing bridges still have such details and they need to be evaluated repeatedly to estimate the bridge beam's remaining life.

2.3 Constructability

Constructability refers to the ability to successfully complete the construction of the bridge being designed. This issue is particularly important simply because it is a prerequisite for the bridge to start its design life by entering the stage of operation. Thus, it is discussed before other general design issues. While it is important, it cannot be exhaustively covered in the specifications, since there are a variety of construction techniques and construction procedures. In general, the strength limit states discussed in the next section on safety are applicable to the constructability check, but, depending on the situation, with the load factors reduced to be close to the service limit state level.

The constructability issues explicitly mentioned in the AASHTO specifications include, but not limited to, deflection, strength of steel and concrete, and stability during critical stages of construction. For instance, if the designer requires steel beams with a concrete deck to compositely support both the dead load (self-weight of concrete) and the live load (truck load), this requirement needs to be specified for construction.

Loads applied to bridge components during construction may be different from those during service. Sometimes, construction stresses can be larger than those under normal service conditions. Bridges should be designed in a manner such that fabrication and erection can be performed without undue difficulty or distress and with locked-in construction force effects within tolerable limits. When the designer has assumed a particular sequence of construction, that sequence is required to be defined in the contract documents, such as the plans. If the method selected to construct the bridge structure requires certain strengthening and/or temporary bracing or support during erection, this requirement also needs to be indicated in the contract documents including the plans.

The specifications also identify several other issues that need to be addressed in design. They include, but not limited to, avoiding details that require welding in restricted areas and placement of concrete through congested reinforcing. There also should be adequate considerations to climatic and hydraulic conditions that may affect the construction of the bridge.

2.4 Safety

Conventional structural design and evaluation practices use the allowable stress design (ASD) method and/or the load factor design (LFD) method. The approaches of the load and resistance factor design (LRFD) and the load and resistance factor rating (LRFR), as indicated by the names, have a format of using different factors respectively for the loads and the resistance. The LRFD and LRFR methods allow individual treatment of each load or resistance. It is thus more flexible and with a higher fidelity to handle their different levels of uncertainty. This format of design checking can cover various limit states for design or evaluation, such as bending and shear failures, excessive deflection, cracking potential, seismic load, and wind load.

More significantly, the AASHTO LRFD and LRFR specifications have been calibrated with respect to the risk of failure involved. This risk is quantified as the probability that the real total-load effect exceeds the real resistance. Engineers use nominal values of load effects and resistances to satisfy the design or evaluation requirements. This procedure, however, will not eliminate the possibility of failure since the involved loads and resistances vary randomly, sometimes very significantly.

Examples of such variables are severe earthquake load, the maximum truck load over the 75-year life span, and maximum flood load over the bridge life. Apparently different loads are associated with different levels of uncertainty to model and predict for design or evaluation. The load and resistance factors presented in Chapter 3 are meant to address these uncertainties as the sources of failure risk.

The design and evaluation methods prescribed in the AASHTO LRFD and LRFR specifications are referred to as calibrated because their load and resistance factors are selected along with the associated nominal loads and resistances to maintain the failure risk at an acceptable level. This section will briefly discuss the calibration conducted for the AASHTO LRFD and LRFR specifications. This discussion provides the background information so that the reader will understand the concept and the limitations and correctly apply the specifications in the bridge design and evaluation.

2.4.1 Uncertainty in Design and Evaluation

A typical design or evaluation process involves a significant amount of uncertainty. The sources of uncertainty include imperfect quality control causing variations in the dimensions of designed members, random fluctuation in member strengths, and unpredictably variable loads generating random effects in members being designed. Considering the design cost, these variations are not explicitly addressed in the design process but are covered using conservative nominal values. For example, the strength of a construction material used in bridges is nominally represented using a deterministic value, typically equal to the mean value minus several times the standard deviation of the real strengths physically measured. The wind load, as another example, may be represented using intensity as the maximum with a return period. Bridge evaluation as an engineering practice has similar situations with respect to using nominal values to cover variation or uncertainty.

Furthermore, for evaluating existing structures, the requirement for reducing uncertainty is often higher than for design, because using conservative assumptions may lead to excessive costly repair or replacement and it is more desirable to avoid such overconservative practices. For example, it is overconservative to load rate an existing highway bridge using the design truck load if the bridge is routinely subjected to much lighter truck load than other bridges. Such a practice can unnecessarily require some bridges to be strengthened or replaced when they are found inadequate against the standard design truck load. In other words, these bridges can as safely serve the traveling public as other bridges without strengthening or replacement. Thus, strengthening or replacing those bridges would be overconservative and should not be required.

In calibrating the AASHTO LRFD and LRFR specifications, the uncertainty in design and evaluation practice is covered using the theory of probability not just based on experience and qualitative treatment as done in the past. This section presents some details of the involved modeling, to prepare the reader for the discussion of LRFD and LRFR specification calibration in the next section.

2.4.2 Modeling Uncertainty Using Probability Theory

To cover uncertainty for design and evaluation, random variables are used to describe those quantities with uncertainty and/or random variation. According to the theory of probability, a typical random variable X is described using its probability density function (PDF) $f_X(x)$. Figure 2.4-1 shows an example PDF for a random variable with a normal distribution, and Figure 2.4-2 is an example for a lognormal distribution. One of the major differences between a normally distributed PDF and a lognormally distributed one is that in the former there is the possibility for negative values to occur and the latter eliminates such a possibility. Here the subscript X in $f_X(x)$ identifies the variable and argument x as a realization of X. The PDF is also sometimes referred to as the distribution of the random variable X.

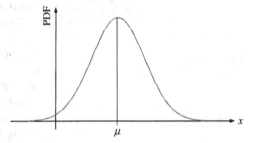

Figure 2.4-1
PDF of a normal random variable.

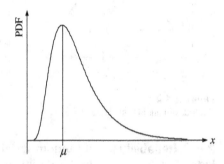

Figure 2.4-2
PDF of a lognormal random variable.

The PDF for a random variable offers important information about the variable. For example, a higher value of PDF for a particular realization x indicates a higher probability or likelihood for that x to be realized. With the PDF available, the mean μ_X and variance σ_X^2 of X are derivable as follows:

$$\mu_X = \int_{-\infty}^{\infty} x f_X(x)\, dx \qquad (2.4\text{-}1)$$

$$\sigma_X^2 = \int_{-\infty}^{\infty} (x - \mu_X)^2 f_X(x)\, dx \qquad (2.4\text{-}2)$$

The mean value is also referred to as the expected value or expectation, meaning that μ_X is the expected or more likely realization of the variable X, and σ_X^2 indicates how variable X may be. The larger σ_X^2, the more variable or unpredictable X is.

In practical applications, the PDF is typically obtained using observations of the variable X to fit an assumed model such as the normal and

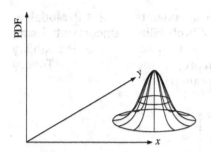

Figure 2.4-3
Joint probability density function.

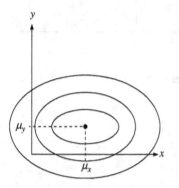

Figure 2.4-4
Contour of joint PDF in Figure 2.4-3.

lognormal distributions shown in Figures 2.4-1 and 2.4-2. This fitting should be evaluated because a poor fitting can result in an unrealistic modeling, thus possibly leading to erroneous estimation of the bridge's reliability or safety. The fitting evaluation is typically performed for several candidate models to finally select a best fitt PDF for the following analysis that uses the PDF.

Note also that sometimes the PDF or distribution of the random variable is not readily available due to a number of reasons. One of them is that a statistically significant number of observations of the random variable are required but not readily available or too costly to acquire. When such a situation occurs, approximation is sometimes advisable by selecting the distribution according to experience and/or limited data.

So far we have discussed the situation involving only one random variable. In practical applications, we often encounter the situation involving two or more random variables. When that is the case, joint PDFs for two or more variables will be needed. Figure 2.4-3 shows a joint PDF for two random variables X and Y, and Figure 2.4-4 shows the PDF's projection on the plane of the two variables X and Y as contours.

2.4.3 Reliability Index for Quantifying Bridge Reliability or Safety

If random variable PDFs are available for uncertain quantities involved in bridge design and evaluation, it becomes possible to quantify associated risk as follows. Assume that there are N such random quantities involved in a specific bridge member design or evaluation. The member's failure risk is quantified as the probability of failure P_f as follows:

$$P_f = \text{probability of failure} = 1 - \text{probability of survival} = 1 - \text{reliability} \tag{2.4-3}$$

For example, the simplest case of probability of failure is defined using a failure indication function or simply failure function g involving two random variables R and Q:

$$g = R - Q \tag{2.4-4}$$

where R represents resistance and Q is the corresponding load effect. For example, R can be a moment resistance and Q is then the sum of the moments due to different loads simultaneously acting on the particular bridge member's cross section. Thus, the probability of failure is

$$P_f = \text{probability}\ (g < 0) = \text{probability}\ (R - Q < 0) \tag{2.4-5}$$

which can be expressed as

$$P_f = \int_{-\infty}^{\infty} \int_{-\infty}^{\infty} I(g < 0) f_{RQ}(r, q)\, dr\, dq \qquad (2.4\text{-}6)$$

where $f_{RQ}(r, q)$ is the joint PDF of random variables R and Q. An example of $f_{RQ}(r, q)$ is displayed in Figure 2.4-3. The function $I(\cdot)$ indicates failure as follows:

$$I(g) = \begin{cases} 1 & \text{if } g < 0 \\ 0 & \text{otherwise} \end{cases} \qquad (2.4\text{-}7)$$

Equation 2.4-6 shows that the probability of failure is defined as the integration of the joint PDF over the domain of failure defined using $g < 0$. In the three-dimensional (3D) space spanned by the two random variables and their probability density, P_f is shown as the volume under the PDF in the region where $I(g) = 1$, also as shown in Figure 2.4-5. For other areas where $I(g) = 0$, the PDF is removed in Figure 2.4-5, as seen by comparing Figures 2.4-3 and 2.4-5. This region has no contribution to the integration defined in Eqs. 2.4-6 and 2.4-7.

Moreover, if R and Q are both normal random variables independent of each other, then the failure function $g = R - Q$ is also a normal random variable with the mean μ_g and standard deviation σ_g related to the means μ_R and μ_R and variances σ_R^2 and σ_Q^2, respectively, for R and Q:

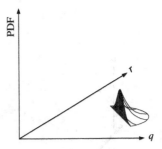

Figure 2.4-5
Probability of failure as volume in space spanned by r and q defined in Eq. (2.4-6).

$$\mu_g = \mu_R - \mu_Q \qquad (2.4\text{-}8)$$

$$\sigma_g = \sqrt{\sigma_R^2 + \sigma_Q^2} \qquad (2.4\text{-}9)$$

and

$$P_f = \int_{-\infty}^{\infty} \int_{-\infty}^{\infty} I(g < 0) f_{RQ}(r, q)\, dr\, dq$$

$$= \int_{-\infty}^{-\mu_g/\sigma_g} f_g(w)\, dw = \int_{-\infty}^{-(\mu_R - \mu_Q)/\sqrt{\sigma_R^2 + \sigma_Q^2}} f_g(w)\, dw$$

$$= \Phi\left(-\frac{\mu_R - \mu_Q}{\sqrt{\sigma_R^2 + \sigma_Q^2}}\right) = \Phi(-\beta) \qquad (2.4\text{-}10)$$

where Φ is the cumulative probability function (CPF) of the standard normal random variable [i.e., function NORMSDIST($\cdot$) in MS Excel] and $f_g(w)$

is its PDF. In Eq. 2.4-10 β is the reliability index defined as

$$\beta = \frac{\mu_R - \mu_Q}{\sqrt{\sigma_R^2 + \sigma_Q^2}} \qquad (2.4\text{-}11)$$

Also note that, in general, R and Q are seldom simultaneously normal variables and the failure function g is possibly not linear as in Eq. 2.4-4. When this situation occurs, Eqs. 2.4-10 and 2.4-11 are not valid. There have been two approaches in defining the reliability index β for those situations.

The first approach is to relate β to the failure probability P_f defined in Eq. 2.4-6:

$$\beta = \Phi^{-1}(1 - P_f) \qquad (2.4\text{-}12)$$

where $\Phi^{-1}(\cdot)$ is the inverse of the CPF of the standard normal random variable. In MS Excel, it is the function NORMSINV$(\cdot)$.

The second approach to treating non-normal variables and the nonlinear failure function g is to define β as

$$\beta = \frac{\mu_g}{\sigma_g} \qquad (2.4\text{-}13)$$

while the failure function is linearized using a first-order approximation (using the linear terms in g's polynomial expansion) at the so called design point. The design point is defined as a point on the failure surface (making $g = 0$) with the shortest distance to the origin in the space of equivalent standard normal variables (Shinozuka, 1983; Rackwitz and Fiessler, 1978). This is shown in Figure 2.4-6. The calibration of the AASHTO LRFD and LRFR specifications used the latter approach (Nowak, 1999; Kulicki et al., 2007).

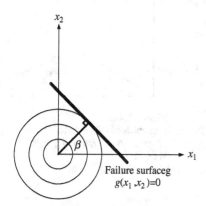

Figure 2.4-6
Space of two standardized normal variables and definition of reliability index β.

2.4.4 Reliability Considerations for Bridge Design and Evaluation (Load Rating)

The structural design of highway bridges addresses the risk of structural failure under a variety of load effects and in a variety of failure modes. For example, a primary load-carrying bridge member of a steel beam is subjected to the risk of failure modes in bending and shear due to vehicular load, seismic load, and so on. A concrete member is subjected to concrete cracking under routine vehicular loads in addition to other failure modes. On the other hand, the target safety or reliability levels against these different risks can be very different due to different consequences of failure and different associated costs. This issue is discussed below.

It is required that the design of new bridges cover the failure risks over their expected life span. The AASHTO specifications have specified this life span to be 75 years. However, in such a long time period, many unpredictable events may take place, changing the risks imposed on bridge structures. For example, the weight limits of truck load may change within the expected life span of 75 years, which may increase the routine loads and the life maximum loads applied to the structures. Moreover, certain political or societal changes may also bring load changes to the structures, such as significant economic development that can cause noticeable increases in truck load. Accordingly, the target reliability is required to be relatively higher for design compared with that for evaluation, because the former needs to cover the 75-year life span and the latter is required to cover up to the next evaluation, typically 2 to 5 years, when the bridge components are observed to have noticeable deterioration or change.

In addition, the cost for increasing the load-carrying capacity for a new bridge to be constructed is relatively low, compared with that for an existing structure. For example, increasing the design live load by 25% for new bridges today has been found to increase the cost by only about 2 to 3%. This is because such an increase will almost not change the labor cost but will only increase the material cost. The labor cost is much more expensive than the material cost, causing the total cost increase due to material cost to be very limited.

On the other hand, increasing the load-carrying capacity of an existing bridge structure can become very expensive. For example, permanently increasing the load-carrying capacity of a concrete beam superstructure is very difficult if not impossible because there have not been acceptable technologies to accomplish that. As a result, such a requirement can only be satisfied by replacing the structure, which can be very costly. This high cost has determined that a different strategy for safety control is needed for existing bridge evaluation. Namely, a lower safety or reliability requirement is needed, compared with the one for design, so that a requirement for replacing the bridge does not occur very frequently. Fortunately, in the United States, there has been a periodic bridge inspection program that monitors the deterioration or change of these structures. This program provides information on the structure's evolution so that only the safety between two inspections needs to be focused on and not a much longer time period such as that for design.

With these considerations, the target safety or reliability level for evaluating existing bridges has been set at a lower level compared with that for new bridge design. Accordingly, many other decisions in evaluation and design can be made based on this concept. Essentially decisions in design should be made considering a relatively remote and thus more uncertain future. In contrast, decisions in evaluation (when to perform maintenance, load posting, etc.) are for short-term effects or for "buying" time. For example, a shallow concrete overlay may be selected to improve a bridge deck's condition and create a smoother riding surface. This will not result in a higher level of structural capacity but it costs less than replacing the bridge.

Some load and resistance factors in the AASHTO design and evaluation specifications have been determined according to this concept of different safety or reliability requirements. On the other hand, it should be noted that current AASHTO design and evaluation specifications have not been developed to such a stage that they completely reflect this concept of a rationally relative relationship in terms of distinct reliability requirements. For example, many calculations in the specifications are recommended to use the same formulas, such as the multilane reduction factors, impact factor, and so on. These parameters are conservatively selected for design to cover a relatively higher risk over the long life span of concern. Therefore, they may not be rational or optimal for the evaluation purposes. However, estimations for these parameters that are more appropriate for the evaluation are not available. Therefore, they are currently "borrowed" from the design specifications for practical reasons.

2.4.5 Calibration for AASHTO LRFD Specifications

For the calibration of the AASHTO LRFD and LRFR specifications, a general safety model or safety margin in Eq. 2.4-4 is used for a typical critical load effect or limit state for a structural component of the bridge, such as the shear at the support of a girder or fatigue limit state for a weld. While the load effect Q is not modeled as a normal variable, the reliability index definition in Eqs. 2.4-12 and 2.4-13 was used in the calibration calculation, according to the method of Rackwitz and Fiessler (1978).

According to Nowak (1999), calibration of the AASHTO LRFD specifications was performed using the following four steps: (1) Selection of typical bridge types and components, (2) establishment of the reliability model for the selected bridge types and components, (3) selection of the target reliability level, and (4) determination of the load and resistance factors to reach the target reliability level. These steps are discussed in more detail next:

1. **Selection of Typical Bridge Types and Components** The following four types of bridge beams were included in the calibration: (a) noncomposite steel beams, (b) composite steel beams, (c) reinforced concrete T beams, and (d) prestressed concrete beams. They represent the most popular highway bridge types in the United States. Their dead-load effects were modeled based on statistical data from the literature. The live load was described using truck weight data from weigh stations. The analysis covered single-lane, two-lane, and multilane bridges of simple and continuous beams. The span length was from 10 to 200 ft. The analysis covered the component's bending and shear limit states, but not deflection, cracking, and so on.

2. **Establishment of Reliability Model for Selected Bridge Types and Components** The model used in this calibration is given in Eq. 2.4-4, which is a linear function of the variables R and Q. However, since not all the variables are normal variables, Eqs. 2.4-10 and 2.4-11 are not valid. Therefore, the first-order approximation method described in Eqs. 2.4-12 and 2.4-13 was used to perform the reliability analysis needed in the calibration.

   In addition, the statistical information for some needed random variables was not available (e.g., the probability of the presence of multiple trucks on the span). Therefore, the Monte Carlo simulation was used to generate pseudorandom samples of truck loads and their locations on the bridge to extract such information. However, it should be noted that such computer-aided simulation is based on a number of assumptions, which need to be validated.

3. **Selection of Target Reliability Index** The target reliability index was defined as the one accepted then, which is further understood as the average reliability index level embedded in the current bridge design specifications then. Using this concept, the reliability indices for the selected typical cases of highway bridge design were computed. An average β value of 3.5 was selected as the target reliability index for calibration. This level has been implemented in the AASHTO LRFD specifications.

4. **Calibration of Specifications** This process was used to determine the load and resistance factors in the specifications. In general, design of a bridge component is required to satisfy

$$\varphi_R R_n = \gamma_D D_n + \gamma_L L_n \qquad (2.4\text{-}14)$$

where φ_R, γ_D, and γ_L are factors for resistance, dead load, and live load, respectively, they were determined in the calibration and R_n, D_n, and L_n are the nominal values of the resistance, dead load, and live load. The subscript n indicates nominal value. For convenience of analysis, the nominal values are related to the mean values as follows through the so-called bias factor:

$$\text{Mean value} = \text{ bias factor} \times \text{nominal value} \qquad (2.4\text{-}15)$$

For the resistance R, for example,

$$\mu_R = \lambda_R R_n \qquad (2.4\text{-}16)$$

where λ_R is the bias factor for resistance. This relation also helps quantify the relation between the process of design using nominal values and the process of reliability index computation using the mean μ and standard variation σ. This relation has made the calibration process possible, which is briefly discussed next.

To explicitly show the process of determining the load and resistance factors, it is assumed that Eqs. 2.4-5 through 2.4-10 are valid. Namely, the failure function is assumed to be a linear function of the involved random variables, which are further assumed to be normal variables. The process of load and resistance factor determination can be shown in a closed form. When these assumptions are not valid, as in the calibration of the AASHTO LRFD specifications, the same concept applies though the process cannot

2.4.6 Determination of Load and Resistance Factors for AASHTO LRFD Specifications

be shown explicitly because iteration is required in determining the so called equivalent normal variables, the design point, and accordingly selecting the linear approximation of the failure function.

Based on the assumptions of linear failure function and normal variables, the reliability index defined in Eq. 2.4-11 is

$$\beta = \frac{\mu_g}{\sigma_g} = \frac{\mu_R - \mu_D - \mu_L}{\sqrt{\sigma_R^2 + \sigma_D^2 + \sigma_L^2}} \qquad (2.4\text{-}17)$$

Substituting the design equation 2.4-14 into the above equation gives

$$\beta = \frac{\mu_g}{\sigma_g} = \frac{\lambda_R[(\gamma_D D_n + \gamma_L L_n)/\phi_R] - \mu_D - \mu_L}{\sqrt{\sigma_R^2 + \sigma_D^2 + \sigma_L^2}} \qquad (2.4\text{-}18)$$

Equation 2.4-18 shows that, when the statistical parameters (mean and standard variation) for R, D, and L are available, the reliability index β is fully controlled by resistance factor ϕ_R and load factors γ_D and γ_L. In other words, one can adjust these factors to reach the predetermined target reliability index 3.5.

Again, calibration of the AASHTO LRFD specifications was more complex, because the assumption of normal random variables was not valid. Therefore the Rackwitz and Fiessler (1978) method was used to compute the reliability index, converting the nonnormal variables to equivalent normal variables. Nevertheless, β is still a function of the resistance and load factors ϕ_R, γ_D, and γ_L, as conceptually shown in Eq. 2.4-18. Of course, the function relation is more complex than that in Eq. 2.4-18 and cannot be expressed in a closed-form formula. Nevertheless, the calibration concept remains the same.

On the other hand, there are three factors to be determined using one equation, Eq. 2.4-18. Therefore, mathematically, there can be many combinations of the three factors. Additional considerations or conditions were used in determining these factors. Accordingly, the resistance factors were taken from investigations of resistance modeling and prediction of the corresponding bridge components. This left two load factors to be determined in the calibration process, γ_D and γ_L. Their combination needs to satisfy Eq. 2.4-14 and maintains a relative constant β at 3.5 over the span range covered. Given the nominal live-load model HL93 in the AASHTO LRFD specifications, $\gamma_D = 1.25$ and $\gamma_L = 1.75$ were found to be able to reach the target reliability of 3.5 for the span range (10 to 200 ft).

2.4.7 Calibration for AASHTO LRFR Specifications

Calibration of the AASHTO LRFR specifications (AASHTO, 2011) used a target β of 2.5 to determine the prescribed load and resistance factors. In other words, a reliability level of $\beta = 2.5$ is ensured relatively uniformly over other parameters in bridge evaluation. This target was determined also by assessing the average reliability level of the previous generation of the AASHTO load-rating specifications [AASHTO (1983) *Manual for Bridge Evaluation* and (2002) *Standard Specifications for Highway Bridges*].

It should be noted that the load-rating step in bridge evaluation often needs to cover overweight vehicles. Such vehicles may travel legally if a permit is applied for and granted. The process of issuing such a permit involves a bridge capacity check to ensure that the bridges on the intended routes do have the capacity to carry the permit load. Calibration for overweight permit checking in the AASHTO LRFR specifications used a relative calibration concept. Namely, the live-load factors for different overweight truck groups were determined with reference to those for the non-overweight trucks and the statistics of the two truck classes (Moses, 2001). Further work of calibration for overweight trucks has been initiated and was in progress when this book was being prepared. Thus new live-load factors for overweight trucks may be included in the AASHTO load-rating specifications in the future.

It should be noted that the calibration of the AASHTO LRFD and LRFR specifications used data collected in Canada for about 10,000 trucks. Truck loads are of major concern in highway bridge design and evaluation, especially for short to medium spans. Therefore the calibration processes focused on the span range of 10 to 200 ft. The Canadian truck load data set was critical for the specifications and thereby governed practice. On the other hand, a large amount of truck data have been collected from thousands of U.S. highway sites since then. Efforts have been and will continue to be spent on updating the specifications to better cover the U.S. situation and realistically reflect U.S. practice.

2.4.8 Future Research Work for Calibration

2.5 Serviceability

Serviceability is understood here as the ability of a bridge to serve the specified functions at an acceptable level over the design life. The following AASHTO specifications are relevant to the serviceability of a bridge:

❑ Clearance
❑ Durability
❑ Maintainability
❑ Rideability
❑ Controlled deformation
❑ Facilitating utilities
❑ Allowance for future widening

These items are discussed next in detail.

2.5.1 Clearance

Clearance refers to the geometric relation of the bridge to the surrounding objects and environment to allow normal traffic flow. The involved traffic may consist of vehicles, vessels, pedestrians, and so on.

In the United States, permits for constructing a bridge over navigable waterways are required from the U.S. Coast Guard and/or other agencies having jurisdiction. Vertical and horizontal navigational clearances need to be established in cooperation with the U.S. Coast Guard. The horizontal clearance may affect the selection of span length and/or span type. The vertical clearance can control the bridge superstructure height. Since the superstructure section height related to the maximum stress in the section is a function of the material strength of the cross section, this may dictate selection of the superstructure material.

When the bridge being designed needs to intersect with another highway, the vertical clearance of the bridge shall be in conformance with the AASHTO (2011b) publication *A Policy on Geometric Design of Highways and Streets* for the functional classification of the highway unless exceptions thereto can be accordingly justified.

Note that possible reduction of vertical clearance over the design life of the bridge needs to be taken into account in the design. For example, this may happen when the overpass bridge has experienced a foundation settlement. If the expected settlement exceeds 1 in., it is required to be added to the specified clearance. Another typical example of vertical clearance reduction is future overlays of the road surface under the bridge. The AASHTO specifications require that the minimum clearance include 6 in. of future overlays. When overlays are not contemplated by the bridge owner, this requirement may be nullified. Figure 2.5-1 shows a case of low clearance that does not meet AASHTO requirement and is warned using a sign on the highway. Taller vehicles will need to detour around such a geometric dimension problem.

The AASHTO specifications also enhance the regular requirement of 16 ft clearance for certain situations. The vertical clearance to sign supports and pedestrian overpasses should be 1 ft greater than the highway structure clearance, and the vertical clearance from the roadway to the overhead cross bracing of through-truss structures should not be less than 17.5 ft. These higher requirements of vertical clearance for sign supports, pedestrian bridges, and overhead cross bracings consider their lesser capacity to possible collision with vehicles under them. In other words, requirements are enhanced for situations where the consequence of possible collision impact is more serious.

The AASHTO also has requirements for bridge width. The width shall not be less than that of the approach roadway section, including shoulders or curbs, gutters, and sidewalks. The usable width of the shoulders should generally be taken as the paved width. No object on or under a bridge, other than a barrier, should be located closer than 4 ft to the edge of a designated traffic lane. The inside face of a barrier should not be closer than 2 ft to

Figure 2.5-1
Inadequate clearance indicated by a highway sign.

either the face of the object or the edge of a designated traffic lane. The specified minimum distances between the edge of the traffic lane and a fixed object are intended to prevent collision with slightly errant vehicles and those carrying wide loads.

When the bridge being designed needs to pass over a railroad, it is required by the AASHTO specifications to be in accordance with standards established and used by the affected railroad in its normal practice. These overpass structures shall comply with applicable federal, state, county, and municipal laws. Regulations, codes, and standards should, at a minimum, meet the specifications and design standards of the American Railway Engineering and Maintenance of Way Association (AREMA), the Association of American Railroads, and the AASHTO.

The contract documents resulting from bridge design are required to call for quality materials and for the application of high standards of fabrication and erection. Structural steel shall be self-protecting or have long life coating systems or cathodic protection. Reinforcing bars and prestressing strands in concrete components, which may be expected to be exposed to airborne or waterborne salts, shall be protected using an appropriate combination of epoxy and/or galvanized coating, concrete cover, density, or chemical composition of concrete, including air entrainment and a nonporous painting of the concrete surface or cathodic protection.

Prestressing strands or tendons in cable ducts in concrete members are required in the AASHTO specifications to be grouted or otherwise protected against corrosion. Attachments and fasteners used in wood

2.5.2 Durability

construction are required to be of stainless steel, malleable iron, aluminum, or galvanized steel, cadmium plated, or otherwise coated. Wood components need to be treated with preservatives. Aluminum products shall be electrically insulated from steel and concrete components to be protected from possible corrosion.

Protection shall be provided to materials susceptible to damage from solar radiation and/or air pollution. Consideration shall be given to the durability of materials in direct contact with soil and/or water.

To prevent water from remaining near the edge of a concrete deck, a continuous drip groove is required along the deck's underside at a distance less than 10 in. from the edge. Where the deck is interrupted by a sealed deck joint, all surfaces of the supporting piers and abutments as applicable, other than bearing seats, shall have a minimum slope of 5% toward their edges. This is to facilitate drainage of water in order to mitigate otherwise possibly caused deterioration. For open deck joints, this required minimum slope is increased to 15%, and the bearings shall be protected against contact with salt and debris. The more significant slope is intended to enable rains to wash away debris and salt.

Wearing surfaces shall be interrupted at the deck joints and shall be provided with a smooth transition to the deck joint device. In the past, wearing surfaces over such joints have been seen to crack, leak, delaminate, and/or disintegrate. Steel formwork, such as steel stay-in-place forms for concrete decks, shall be protected against corrosion in accordance with the specifications of the owner.

2.5.3 Maintainability

Management of operation of the existing roadway system has taught us that highway bridges need adequate maintenance over their design lives. It includes, but is not limited to, washing deicing chemical and other harmful substances off the bridge surface, removal of deteriorated concrete on the top of a reinforced concrete deck slab, repainting steel components every several years, and replacing frozen or rusted steel bearings.

Experience has taught us lessons when bridges are not easy to maintain and their service lives are thereby reduced. Accordingly, the AASHTO specifications require that structural systems whose maintenance is expected to be difficult be avoided.

All highway bridges require a driving surface which is often provided by a deck that is also a structural component to transmit the vehicle wheel load to the supporting components such as trusses, arches, and beams. The deck also serves as a roof for the entire bridge, protecting the components below it, such as beams, bearings, pier cap, pier columns, and so on. Simultaneously subjected to climate loading and vehicular load, the deck also has to resist the deteriorating force of deicing chemicals where they are used. In such a condition, the deck needs to be rehabbed and/or replaced within the design life of the bridge. Namely, the deck cannot serve as long as the bridge system. Often two or more deck lives are equal to one bridge life.

More replacements of the deck may even be needed for those bridges in the northern part of the country where a lot of corroding deicing chemicals are used in the winter season. The AASHTO specifications require that the following considerations be identified in the contract documents:

- ❑ a contemporary or future protective overlay
- ❑ a future deck replacement
- ❑ supplemental structural resistance

Some state highway agencies have established strategies for maintaining reinforced concrete deck slabs and associated design strategies. For example, a design thickness has been specified by some state agencies (although not so in the AASHTO specifications). It is associated with the following stages for the entire life span of the deck slab:

- ❑ Asphalt concrete overlay at about 15 to 20 years of slab life for improved rideability.
- ❑ Removal of asphalt overlay and certain depth of top concrete and then overlay of concrete at about 20 to 30 years of slab life to improve structural performance and rideability.
- ❑ Replacement of slab at about 30 to 40 years of slab in a rehabilitation of the bridge. Thus two such slab life cycles will make one life cycle of the bridge system.

In addition, the AASHTO specifications also identify bearings as a type of bridge component to receive special attention when maintainability is of concern. Bearings often need to be rehabbed or replaced within the expected life span of the bridge. Namely, the bearing's expected life span is much shorter than that of the bridge system. Thus the procedure for removing them for rehabilitation or replacement needs to be taken into account in the design of the bridge. Thus, areas around the bearing seats should be designed to facilitate jacking the component supported on the bearing (such as a truss or beam) and also to facilitate cleaning, repair, and replacement of bearings. The same concepts will apply to areas under a deck joint, since joints need to experience rehabilitation and replacement like bearings and sometimes more frequently. Jacking points for the members supported on the bearings need to be indicated on the plans, and the structure shall be designed for the jacking forces. Inaccessible cavities and corners should be avoided. Cavities that may invite human or animal inhabitants shall either be avoided or made secure.

Maintenance of traffic during rehabilitation or replacement of bridge components or the entire bridge is often required since completely closing the road for such maintenance operations is unacceptable to the traveling public. In general, two options may be available: (1) staging rehabilitation or replacement construction by using partial width of the road and (2) utilization of an adjacent parallel structure. These options need to be

Figure 2.5-2
Replacing left half of the bridge while keeping the right half open.

taken into account when designing the bridge. Figure 2.5-2 shows a case of option 1. The left half of the bridge is being constructed. The right half was completed first and opened to traffic.

Many state departments of transportation or highway have developed and maintained protection systems for improved durability. For example, the epoxy coating of steel reinforcement has become very popular in reinforced concrete bridge members. In reinforced concrete deck slabs, epoxy coating the bottom layer of the two layers of reinforcement was started about 15 to 25 years ago in the United States. Now epoxy coating for both layers is widely practiced. For improved concrete performance, microsilica additives in the deck slab concrete are widely used to increase the density of concrete, reducing the diffusion speed of chloride ions from deicing chemical to reach and start to corrode steel reinforcement. Waterproofing membranes also have been used in concrete deck slabs for improved concrete durability. Many state agencies have specific requirements within the jurisdiction while the AASHTO specifications do not include these details. The design engineer is required to become familiar with these requirements when designing for a specific agency.

2.5.4 Rideability

Rideability is relevant to the bridge deck since the deck provides the driving surface of the bridge. The deck is required to be designed to permit smooth movement of vehicle traffic. On paved roads, a structural transition slab should be located between the approach roadway and the abutment of the bridge. Construction tolerances, with regard to the profile of the finished

Figure 2.5-3
Deck joint as a unit before concrete placement.

deck, are required to be indicated on the plans or in the specifications or special provisions. The number of deck joints (Figure 2.5-3) shall be kept to a practical minimum. These deck joints often leak after some years of service. Continuous spans may be used to reduce the number of these joints as well as integral abutment bridge spans. Edges of joints in concrete decks exposed to traffic should be protected from abrasion and spalling. The plans for prefabricated joints shall specify that the joint assembly be erected as a unit. Where concrete deck slabs without an initial overlay are used, consideration should be given to providing an additional thickness of 0.5 in. to permit future wearing or correction and to compensate for thickness loss due to abrasion. Thus this additional thickness is considered to be sacrificial and not structural.

In general, dead-load deflections if significant should be compensated for using camber. Both steel and concrete superstructures can be cambered during fabrication to compensate for dead-load deflection and vertical alignment. Deflections due to weights of different components should be reported separately. For example, deflections due to steel beams should be included in plans separately from those due to the concrete deck slab. Deflections due to future wearing surfaces or other loads not applied at the time of initial construction are also required to be reported separately.

Structures such as bridges should be designed to avoid undesirable structural or psychological effects due to live load deformations. In the past, live-load deflection was required to be controlled at specified levels with reference to span length L. However, studies have found no evidence of

2.5.5 Deformation Control

serious structural damage attributable to excessive deflection. The current AASHTO specifications thus have made live-load deflection and depth limitations optional, except for orthotropic plate decks. However, any significant deviation from past successful design practice regarding slenderness and deflections should cause attention and thus review of the design to determine that the bridge will perform adequately. Note that deflection may have been addressed in other requirements. For example, the vertical clearance of a highway bridge to the navigable waterway surface or roadway surface needs to cover the possible deflection if significant.

In addition, service load deformations may cause deterioration of wearing surfaces and local cracking in concrete slabs. They also impair serviceability and durability, even if self-limiting and not a potential source of collapse. These concerns should be addressed in design when applicable.

In skewed steel bridges with straight beams or girders and horizontally curved steel girder bridges, there are more significant torsions to the beams and horizontal reactions to the bearings compared with the nonskewed counterpart. The following additional investigations need to be considered as specified in the AASHTO specifications:

❏ Elastic vertical, lateral, and rotational deflections due to applicable load combinations shall be considered to ensure satisfactory service performance of bearings, joints, integral abutments, and piers.

❏ Computed girder rotations at bearings should be accumulated over the intended or assumed construction sequence. Computed rotations at bearings shall not exceed the specified rotational capacity of the bearings for the accumulated factored loads corresponding to the stage investigated.

The AASHTO specifications do offer optional live-load deflection limits. They may be adopted by bridge owners, as summarized in Table 2.5-1.

Nevertheless, these live-load deflection limits are mandatorily applicable for the following cases:

❏ Orthotropic decks

❏ Precast reinforced concrete three-sided structures

Table 2.5-1
Recommended deflection limits for steel, aluminum, and concrete structure

Load and Structure	Deflection Limit
Vehicular load, general	Span/800
Vehicular and pedestrian loads	Span/1000
Vehicular load on cantilever arms	Span/300
Vehicular and pedestrian loads on cantilever arms	Span/375
Vehicular and pedestrian loads on cantilever arms	Span/375

For metal grid decks and other lightweight metal and concrete decks, the following deformation requirements for serviceability apply (*9.5.2*) for live loads including dynamic impact. Deck deformation here refers to local dishing with reference to the deck's supports at wheel loads, not to overall superstructure deformation. The primary objective of curtailing excessive deck deformation is to prevent breakup and loss of the wearing surface:

- ❑ Span/800 for decks with no pedestrian traffic
- ❑ Span/1000 for decks with limited pedestrian traffic
- ❑ Span/1200 for decks with significant pedestrian traffic

If the bridge owner chooses to invoke deflection control, the following principles are given in the AASHTO specifications to be considered for adoption.

- ❑ For the maximum absolute deflection for straight girder systems, all design lanes should be loaded with the multiple presence factor (MPF) applied, and all supporting components should be assumed to deflect equally. The MPF will be explained along with live load in Chapter 3.
- ❑ For curved steel box and I-girder systems, the deflection of each girder should be determined individually based on its response as part of a system.
- ❑ For composite design, the stiffness of the design cross section used for the determination of deflection should include the entire width of the roadway and the structurally continuous portions of the railings, sidewalks, and median barriers, if applicable.
- ❑ For straight girder systems, the composite bending stiffness of an individual girder may be taken as the stiffness determined as specified above divided by the number of girders.
- ❑ For maximum relative displacements, the number and position of loaded lanes should be selected to provide the worst differential effect.
- ❑ The live-load portion of the service I limit state load combination should be used for deformation control design, including the dynamic load allowance IM. The live-load and load combinations are to be introduced in Chapter 3.
- ❑ For skewed bridges, the right cross section, instead of the skewed cross section, may be used.
- ❑ For curved and curved skewed bridges, the radial cross section may be used.

For steel I-shaped beams and girders and for steel box and tub girders, the AASHTO specifications include requirements regarding the control of permanent deflections but through flange stress controls.

Table 2.5-2
Recommended deflection limits for wood structures

Load and Structure	Deflection Limit
Vehicular and pedestrian loads	Span/425
Vehicular load on wood planks and panels	0.10 in.

Figure 2.5-4
Utilities carried by a highway bridge.

In the absence of other criteria, the deflection limits given in Table 2.5-2 may be considered for wood construction.

2.5.6 Utilities When a bridge crosses an obstacle such as a valley, a roadway, a waterway, a railway, and so on, it is often the perfect and safe vehicle to carry needed utilities across. Therefore, where required, the bridge shall be made to support and maintain the conveyance for utilities. Figure 2.5-4 shows an example. This can be viewed as a function requirement for the bridge.

2.5.7 Allowance for Future Widening Some highway bridges have to be widened to accommodate growing traffic demand for more lanes resulting from economic development. The AASHTO specifications thus require that the load-carrying capacity of exterior beams not be less than that of an interior beam unless future widening is virtually inconceivable. This requirement will make it easy to convert an exterior beam to an interior beam.

When future widening can be anticipated, consideration should also be given to designing the substructure for the widened condition. For example, the abutments or foundations may be designed larger than required by the current and unwidened superstructure to carry the conceivable future superstructure.

2.6 Inspectability

It has been learned through maintaining highway bridges that they need to be designed and accordingly constructed to permit adequate inspection, particularly for primary components or primary component systems. For example, large steel and concrete box beams need to have manholes to allow access to the inside for inspecting and detecting possible cracking in the material. The AASHTO specifications also explicitly require inspection ladders, walkways, catwalks, covered access holes, and lighting, if necessary, where other means of inspection are not practical. Where practical, the specifications also require access to permit manual or visual inspection, including adequate headroom in box sections, to the inside of cellular components and to interface areas, where relative movement may occur.

2.7 Economy

Economic consideration is required at every stage and step of bridge design. Starting from the preliminary design to taking into account the location and dimensions of members and the amount of reinforcement in concrete components, cost saving is always a significant factor.

More specifically, the AASHTO specifications require that structural types, span lengths, and types of material be selected with due consideration to projected cost. It should be noted that the cost should include that for operation and renewal. In other words, not only the initial cost but also the future expenditures during the projected service life of the bridge should be considered.

Several factors may influence the cost noticeably. One of them is location. Due to the availability of material and expertise in fabrication and shipping, the cost can be significantly different for the same bridge using the same materials but at two different locations. The site can also make an obvious difference in cost, due to site preparation, erection constraints, maintaining the site for construction, and so on.

In some cases, the construction season can be a factor for cost variation. When relevant information is available, for example for the trends in labor and material cost fluctuation, efforts should be made to minimize the cost by taking into account the effect of such trends. For instance, the construction should be scheduled for a more favorable time period if other factors are not negatively impacted.

Structural type, including the material to be used, is another factor that affects life-cycle cost, sometimes quite dramatically. These alternatives should be compared based on long-range considerations, including inspection, maintenance, repair, and/or replacement. Lowest first cost does not necessarily lead to lowest life-cycle cost.

However, there have been cases where economic studies do not indicate a clear choice. The AASHTO specifications also suggest that the bridge owner may require alternative contract plans for bidding competitively. Of course, alternative plans are required to be of equal safety, serviceability, and aesthetic value. More specifically, movable bridges over navigable waterways should be avoided to the extent feasible, according to the AASHTO specifications. Where movable bridges are proposed, at least one fixed bridge alternative should be included for economic comparisons.

2.8 Aesthetics

Bridges, due to their significant geometric dimensions, become part of the environment or landscape after construction. The design engineer should be conscious about the possible impact of the bridge to the surrounding. As commented on earlier, aesthetics is difficult to quantify, though. The fact that a discussion of aesthetics ends this chapter is not because it is less important but partially because it is not easy to quantify. There are no limit states, per se, associated with aesthetics consideration in the AASHTO specifications. However, the specifications require that bridges complement their surroundings, be graceful in form, and present an appearance of adequate strength. While building design often has architects covering their aesthetics, bridge design usually does not involve other professionals to cover the aesthetic aspect. There may be exceptions, for example, for bridges with a very long span or spans, although not all long-span bridges have been designed using professional aesthetic assistance. To that end, bridge engineers are charged to consider the bridge's aesthetics in the design. This fact makes the subject of aesthetics even more critical to discuss.

The AASHTO specifications also note that significant improvements in appearance can often be made with small changes in shape or position of structural members at negligible cost. For prominent bridges, additional cost to achieve improved appearance is often justifiable, considering that the bridge will likely be a feature of the landscape for its expected life span or even more years to come. For more background information, further knowledge, and design guidance, the reader is referred to the Transportation Research Board (TRB) *Bridge Aesthetics Around the World* (1991).

Nevertheless, the AASHTO specifications do offer some conceptual guidelines, which are not meant to be rigid. A more pleasant appearance of the bridge is suggested by improving the shapes and relationships of the structural components themselves. Extraordinary and nonstructural embellishments should be avoided. The following more specific guidelines should be considered:

❏ Alternative bridge designs without piers or with few piers should be studied during the site and location selection stage and refined during the preliminary design stage.

❏ Pier form should be consistent in shape and detail with the super-structure.

❏ Abrupt changes in the form of components and structural type should be avoided. Where the interface of different structural types cannot be avoided, a smooth transition in appearance from one type to another should be attained.

❏ Attention to details, such as deck drain downspouts, should not be overlooked.

❏ If the use of a through structure is dictated by performance and/or economic considerations, the structural system should be selected to provide an open and uncluttered appearance.

❏ The use of the bridge as a support for message or directional signing or lighting should be avoided wherever possible.

❏ Transverse web stiffeners, other than those located at bearing points, should not be visible in elevation.

❏ For spanning deep ravines, arch-type structures should be preferred.

The AASHTO specifications also include the following further comments on highway bridge aesthetics.

The most admired modern structures are those that rely for their good appearance on the forms of the structural components:

❏ Components are shaped to respond to the structural function. They are thick where the stresses are greatest and thin where the stresses are smaller.

❏ The function of each part and how the function is performed is visible.

❏ Components are slender and widely spaced, preserving views through the structure.

❏ The bridge is seen as a single whole, with all members consistent and contributing to that whole; for example, all elements should come from the same family of shapes, such as shapes with rounded edges.

❏ The bridge fulfills its function with a minimum of material and minimum number of elements.

❑ The size of each member compared with the others is clearly related to the overall structural concept and the job the component does.

❑ The bridge as a whole has a clear and logical relationship to its surroundings.

Several procedures have been proposed to integrate aesthetic thinking into the design process (TRB, 1991). Because the major structural components are the largest parts of a bridge and are easily seen first, they determine the appearance of a bridge. Consequently, engineers should seek excellent appearance in bridge parts in the following order of importance:

❑ Horizontal and vertical alignment and position of the bridge in the environment

❑ Superstructure type: arch, girder, and so on.

❑ Pier placement

❑ Abutment placement and height

❑ Superstructure shape, that is, haunched, tapered, depth

❑ Pier shape

❑ Abutment shape

❑ Parapet and railing details

❑ Surface colors and textures

The engineer should determine the likely position of the majority of viewers of the bridge, then use that information as a guide in judging the importance of various elements in the appearance of the structure.

Perspective drawings or photographs taken from the important viewpoints can be used to analyze the appearance of proposed structures. Models are also useful. The appearance of standard details should be reviewed to make sure that they fit the bridge's design concept.

2.9 Summary

This chapter has focused on the overall requirements for highway bridges. They are applicable to design and evaluation of these bridges, although more of these requirements appear to target the design practice. In design, most of the topics or sections in this chapter are addressed in the first stage of preliminary design, while selecting site, location, structural type, major materials, and so on. This stage requires a significant amount of experience with a wide variety of these choices. Some of such experience is documented in the form of statistics, research reports on design practice, specific bridge design cases, and so on. A portion of the experience is not documented, especially that associated with the local condition. For example, the low fabrication cost for a specific type of bridge component in a local area often is not easily available to engineers in another remote state.

Further coverage of the preliminary design of highway bridges is considered beyond the scope of this book for a first course of bridge design and evaluation. Instead, the following chapters will focus on the second stage of highway bridge design, namely the detailed design stage, given that the overall decisions have been made regarding the location, structural type, major materials and so on. Actually one needs to thoroughly understand the second stage in order to perform well in the first stage of bridge design. In fact, most, if not all, engineers involved in preliminary design have a number of years of practice in detailed design.

References

American Association of State and Highway Transportation Officials, (AASHTO) (2012), *LRFD Bridge Design Specifications*, 6th ed., AASHTO, Washington, DC.

American Association of State and Highway Transportation Officials (AASHTO) (2011a), *Manual for Bridge Evaluation*, 2nd ed., AASHTO, Washington, DC.

American Association of State and Highway Transportation Officials (AASHTO) (2011b), *A Policy on Geometric Design of Highways and Streets*, 6th ed., AASHTO, Washington, DC.

American Association of State and Highway Transportation Officials (AASHTO) (2002), *Standard Specifications for Highway Bridges*, 17th ed. AASHTO, Washington, DC.

American Association of State and Highway Transportation Officials (AASHTO) (1983), *Manual for Bridge Evaluation*, 2nd ed., AASHTO, Washington, DC.

Kulicki, J. M., Prucz, Z., Clancy, C. M., Mertz, D. R., and Nowak, A. S. (2007), "Updating the Calibration Report for AASHTO LRFD Code," NCHRP 20-07/186, Transportation Research Board, Washington, DC.

Moses, F. (2001), "Calibration of Load Factors for LRFR Bridge Evaluation," NCHRP Report 454, Transportation Research Board, Washington, DC.

Nowak, A. (1999), "Calibration of LRFD Bridge Design Code," NCHRP Report 368, Transportation Research Board, Washington, DC.

Rackwitz, R. and Fiessler, B. (1978), "Structural Reliability under Combined Random Load Sequences," *Computers and Structures*, Vol. 9, pp. 489–494.

Shinozuka, M. (1983), "Basic Analysis of Structural Safety," *Journal of Structural Engineering*, Vol. 109, No. 3, pp. 721–739.

Transportation Research Board (TRB) (1991), *Bridge Aesthetics Around the World*, TRB, Washington, DC.

Problems

2.1 Assume that you are a representative of a bridge owner. Develop a set of specific requirements for a highway bridge to replace an existing one. Examples of your requirements may be length, width, traffic volume to be carried, and required life span. Your grade for this problem depends on how comprehensive your list of requirements is.

2.2   Assume that you are a project engineer working for a firm that performs highway bridge design. To prepare a bid for a highway bridge design project, make a list of information that you would like to have to start your preparation. Examples of information you may want to have may be type of material for the superstructure, foundation type, site condition, and project starting date and ending date. Your grade for this problem depends are how complete your list is.

2.3   Use available information sources including the Internet to find bridge structure types with their economically appropriate span length range. For example, suspension bridge spans are economically advantageous or viable for long spans.

2.4   Consider a highway bridge beam bending strength design. Assume that a safety margin can be written as $g = R - L - D$, where R is the moment capacity of the beam and L and D are the live-load and dead-load effects in that beam. These random variables are assumed to be normally distributed and independent of one another. They have mean values and standard deviations as follows: $\mu_R = 3520$ kft, $\sigma_R = 300$ kft; $\mu_L = 1200$ kft, $\sigma_L = 245$ kft; $\mu_D = 800$ kft, $\sigma_D = 65$ kft. Find the reliability index β for the beam's bending failure.

2.5   Consider a highway bridge beam for shear strength design. Assume that a safety margin can be written as $Z = R - L - D$, where R is the shear capacity of the beam and L and D are the shear live-load and dead-load effects in the beam. Find the reliability index β for this beam's shear failure using the following mean values and standard deviation variations for R, D, and L: $\mu_R = 475$ k, $\sigma_R = 45$ k; $\mu_L = 200$ k, $\sigma_L = 50$ k; $\mu_D = 125$ k, $\sigma_D = 10$ k.

3 Loads, Load Effects, and Load Combinations

3.1 Introduction

Chapters 3 through 6 cover detailed design for highway bridges. This chapter focuses on the loads, their effects, and their combinations in the detailed design of highway bridges. Chapters 4 through respectively deal with the design of superstructure, bearings, and substructure of typical highway bridges.

Bridge structural design needs to cover the effect of a variety of loads or actions, such as gravitational force of the components of the structure, centrifugal force due to the vehicles on the bridge if the bridge is on a horizontally curved roadway, force applied by flooding or earthquake, and so on. It is important for the bridge engineer to exhaustively identify all the loads that can possibly be applied to the structure during its expected life span including its construction stage, to use appropriate models in load effect analysis, and to reasonably combine different load effects for modeling and covering possibly simultaneous application of these loads or actions. The AASHTO LRFD specifications cover a variety of these loads and their effects. This chapter focuses on a number of these loads commonly encountered in the design and evaluation of highway bridges according to the AASHTO specifications.

It should be noted that these loads and load effects are often referred to in the specifications as "loads" without mentioning the word "effects," although the context unambiguously indicates one or the other. The reader should keep this in mind when reading the specifications.

The code-specified load effects can be divided into two categories: permanent and transient. Examples of permanent load effects are the self-weight of bridge members (the deck, wearing surface, parapets, and so on) and earth pressure. Transient load effects are more variable, including but not limited to those due to vehicular loads, friction forces, and ice loads. These two categories of load effects are discussed separately in Sections 3.2 and 3.3.

3.2 Permanent Loads

Permanent loads can be further divided into bridge component self-weight loads and earth loads, as treated below.

3.2.1 Dead Loads DC, DW, and DD

In the AASHTO LRFD specifications, the dead-load effect is treated in three subgroups according to their severity of random variation: dead load due to structural and nonstructural components (abbreviated as DC for component dead load), dead load due to wearing surface and utilities (DW for wearing surface dead load), and the down drag force (DD for down drag). Please note that these abbreviated forms do not follow the convention of combined first letters of the words. The specification provisions have been calibrated to consider their different variations for the associated strength limit state in a conceptual way. Accordingly, different dead-load factors have been prescribed to cover corresponding variations.

In the LRFD specifications DC refers to the self-weight of structural and nonstructural components, such as a truss member, a beam, a concrete deck, and a fence. Unit weights of these materials have been given in the specifications so that they can be conveniently and consistently used if no further detailed information is available. Table 3.2-1 displays such data for a number of commonly used materials in bridge construction. Examples 3.1 and 3.2 include estimation of DC applied to respectively two different bridge components, the reinforced concrete deck slab and steel beams in a beam bridge superstructure.

Table 3.2-1
Unit weights of commonly used materials

Material	Unit Weight (k/ft³)
Aluminum alloys	0.175
Bituminous wearing surfaces	0.14
Cast Iron	0.45
Compacted sand, silt, or clay	0.12
Concrete	
Lightweight	0.11
Normal weight	
$f'_c \leq 5\,\text{k/in.}^2$	0.145
Normal weight	
$5\,\text{k/in.}^2 < f'_c \leq 15\,\text{k/in.}^2$	$0.14 + 0.001 f'_c$
Loose sand, silt, or gravel	0.1
Soft clay	0.1
Rolled gravel, macadam, or ballast	0.14
Steel	0.49
Stone masonry	0.17
Water	
Fresh water	0.0624
Salt water	0.064

Example 3.1 Reinforced Concrete Deck Design (Dead-Loads DC and DW)

❑ **Design Requirement**

To design a cast-in-place reinforced concrete deck for a steel beam bridge with five girders spaced at 8 ft 10 in. and an overhang 4 ft 5 in. wide, as shown in Figure Ex3.1-1, find the dead loads DC and DW on a typical strip of the deck that is 1 ft wide.

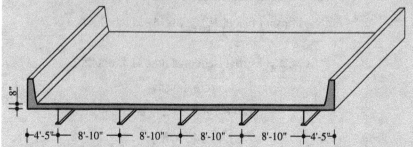

8"

4'-5"　8'-10"　8'-10"　8'-10"　8'-10"　4'-5"

Figure Ex3.1-1
Superstructure dimensions.

For the concrete deck design, a typical strip of the deck is considered as shown in Figure Ex3.1-2

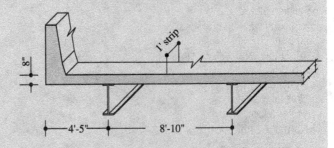

Figure Ex3.1-2
Typical design strip.

☐ Deck Parameters

Density of concrete: $W_{concrete} = 0.145 \, \text{k/ft}^3$
(normal weight $f'_c \leq 5 \, \text{k/in.}^2$)   *3.5.1*

Density of future wearing surface (FWS): $W_{FWS} = 0.14 \, \text{k/ft}^3$
(asphalt concrete)   *3.5.1*

Thickness of future wearing surface: $t_{FWS} = 2.5 \, \text{in.}$

Thickness of top integral wearing surface: $t_{IWS} = 0.5 \, \text{in.}$ *Counted as dead load, not strength*

Thickness of slab: $t_{slab} = 8 \, \text{in.}$ (Minimum 7 in.)   *9.7.1.1*

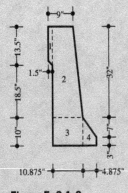

Figure Ex3.1-3
Bridge parapet
dimensions.

☐ Parapet DC

A crash-tested parapet is selected for the bridge as shown in Figure Ex3.1-3:

Base width $B_{base} = 1 \, \text{ft} \, 3\frac{3}{4} \, \text{in.}$

Parapet height $H_{parapet} = 42 \, \text{in.} = 3.5 \, \text{ft}$

$DC_{parapet}$ = cross-sectional area (1 ft depth)

(concrete density)

= (area 1 + area 2 + area 3 + area 4)

1 ft (0.145 k/ft^3)

$$\left(1.5 \text{ in.} \frac{13.5 \text{ in.} + 15 \text{ in.}}{2} + 32 \text{ in.} \frac{7.5 \text{ in.} + 10.875 \text{ in.}}{2} \right.$$
$$\left. + 10.875 \text{ in.} (10 \text{ in.}) + 4.875 \text{ in.} \frac{10 \text{ in.} + 3 \text{ in.}}{2} \right)$$
$$= \frac{}{144 \text{ in.}^2/\text{ft}^2} 1 \text{ ft} (0.145 \text{ k/ft}^3)$$

$$= 0.457 \text{ k}$$

A $DC_{parapet}$ is applied at each deck overhang as shown in Figure Ex3.1-4.

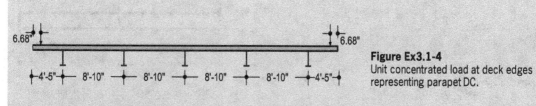

6.68"   6.68"

|←4'-5"→|← 8'-10" →|← 8'-10" →|← 8'-10" →|← 8'-10" →|←4'-5"→|

Figure Ex3.1-4
Unit concentrated load at deck edges representing parapet DC.

Deck DC in Figure Ex3.1-5 as Uniformly Distributed Load

A typical 1-ft-wide deck is considered for analysis and design, as shown in Figure Ex3.1-2:

$$DC = \text{dead load of slab} = t_{slab} \, (1 \text{ ft}) \, W_{concrete}$$

$$= \left(\frac{8 \text{ in.}}{12 \text{ in./ft}} \right) 1 \text{ ft} (0.145 \text{ k/ft}^3) = 0.097 \text{ k/ft}$$

Wearing Surface DW in Figure Ex3.1-5 as Uniformly Distributed Load

$$DW = \text{dead load of FWS} = t_{FWS} \, (1 \text{ ft}) \, W_{FWS}$$

$$= \left(\frac{2.5 \text{ in.}}{12 \text{ in./ft}} \right) 1 \text{ ft} (0.140 \text{ k/ft}^3) = 0.029 \text{ k/ft}$$

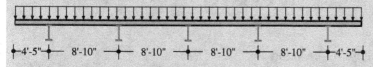

|←4'-5"→|← 8'-10" →|← 8'-10" →|← 8'-10" →|← 8'-10" →|←4'-5"→|

Figure Ex3.1-5
Uniformly distributed load representing deck DC or wearing surface DW.

Example 3.2 Steel-Rolled Beam Bridge Design (Load Effects DC and DW)

◻ Design Requirement

To design a composite rolled steel beam bridge superstructure, determine the dead loads DC and DW on an interior beam. The span length is 40 ft and its five-beam cross section is shown in Figure Ex3.2-1. Include a deck overhang of 3 ft 11 in.

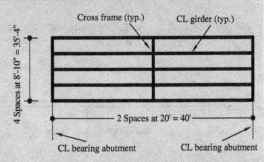

Figure Ex3.2-1
Framing plan.

◻ Design Parameters

Total deck thickness $t_{slab} = 9.0$ in.
Deck design thickness $t_{design\ slab} = 8.5$ in.
Steel density $W_{steel} = 0.490$ k/ft^3 *3.5.1*
Concrete density $W_{concrete} = 0.145$ k/ft^3 *3.5.1*
Miscellaneous structural steel dead load (per beam)
   $W_{miscellaneous} = 0.02$ k/ft
Stay-in-place deck form weight $W_{deck\ forms} = 14.98$ lb/ft $= 0.015$ k/ft
Parapet weight (each) $W_{parapet} = 0.53$ k/ft
Future wearing surface density $W_{FWS} = 0.14$ k/ft^3 *3.5.1*
Future wearing surface thickness $t_{FWS} = 2.5$ in.
Deck width $w_{deck} = 43$ ft 2 in. $= 43.20$ ft
Roadway width $w_{roadway} = 40$ ft 4 in. $= 40.33$ ft
Haunch depth $d_{haunch} = 1$ in.

◻ Dead Loads for Interior Beam

1. Concrete deck
$$DL_{deck} = W_{concrete}St_{slab} = 0.145 \text{ k/ft}^3 (8 \text{ ft } 10 \text{ in.})9 \text{ in.}/(12 \text{ in./ft})$$

$$= 0.96 \text{ k/ft}$$

2. Stay-in-place forms
$$W_{deck\ forms} = 0.015 \text{ k/ft}^2 \quad S = 8 \text{ ft } 10 \text{ in.} \quad w_{top\ flange} = 14 \text{ in.}$$

The top flange width is assumed:

$$DL_{deck\ forms} = W_{deck\ forms}(S - W_{top\ flange})$$

$$= 0.015\ k/ft^2(8\ ft\ 10\ in. - 14\ in.) = 0.12\ k/ft$$

3. Miscellaneous weights:

$$DL_{miscellaneous} = 0.02\ k/ft$$

The miscellaneous weight is for the weight of traffic signs, illumination, and so on.

4. Concrete parapet:

$$\text{Assumed } W_{parapet} = 0.457\ k/ft \quad N_{girders} = 5$$

$$DL_{parapet} = 2W_{parapet}/N_{girders} = 0.183\ k/ft$$

For the concrete parapets, the dead load per unit length is computed assuming that the superimposed dead load of the two parapets is distributed uniformly among all the girders.

5. Future wearing surface:

$$W_{road\ way} = 40\ ft\ 4\ in. \quad N_{girders} = 5$$

$$DL_{FWS} = \frac{W_{FWS}(t_{FWS}/12\ in./ft)w_{roadway}}{N_{girders}}$$

$$= \frac{0.140\ k/f^3\ (2.5\ in./12\ in./ft)\ 40.33\ ft}{5} = 0.24\ k/ft$$

6. Steel:

$$DL_{steel} = 0.194\ k/ft$$

Assume a W 27 × 194 section.

7. Total dead load from beam (for noncomposite cross section):

$$DL_{beam} = DL_{steel} + DL_{deck} + DL_{deck\_forms} + DL_{miscellaneous}$$

$$= 0.194\ k/ft + 0.96\ k/ft + 0.12\ k/ft + 0.02\ k/ft$$

$$= 1.29\ k/ft$$

❑ **Dead-Load Moments and Shears for Design of Interior Beams:**

1. Total-beam DC:

$$M_{beam} = 1.29 \, k/ft \, (40 \, ft)^2/8 = 258 \, kft$$

$$V_{beam} = 1.29 \, k/ft \, (40 \, ft)/2 = 25.8 \, k$$

2. Parapet DC:

$$M_{parapet} = 0.183 \, k/ft \, (40 \, ft)^2/8 = 36.6 \, kft$$

$$V_{parapet} = 0.183 \, k/ft \, (40 \, ft)/2 = 3.66 \, k$$

3. Future wearing surface DW:

$$M_{FWS} = 0.24 \, k/ft \, (40 \, ft)^2/8 = 48 \, kft$$

$$V_{FWS} = 0.24 \, k/ft \, (40 \, ft)/2 = 4.8 \, k$$

In the specifications DW refers to the self-weight of the wearing surface, typically made of asphalt concrete. Separation of this self-weight from DC is treateds differently than DC because such a wearing surface is constructed at the site and the variation in its as-built state is usually more significant. Therefore DW is given a higher load factor than DC, since DC is usually fabricated in the shop and its quality control and quantity control are more rigorous. For example, steel beams, prestressed concrete beams, guard rails, and so on, are normally manufactured using standard plates, shapes, or forms and then shipped to the site to be installed. As a result, their dimensions are much more precisely controlled and thus can be more accurately estimated in design. The different treatments of the dead loads DC and DW also represent a noticeable change from the previous generation of the AASHTO bridge design and evaluation specifications. Examples 3.1 and 3.2 also include DW for a reinforced concrete deck and an interior steel beam in a beam bridge superstructure.

In addition, the weight of utilities carried by the bridge is also included in DW. For example, they may include pipes for sewage, oil, gas, communication wire cables, and so on. Figure 3.2-1 shows pipes placed between two steel beams (with shear studs) before the concrete deck is constructed. Figure 3.2-2 demonstrates utility pipes carried by a highway bridge in service.

Figure 3.2-1
Utility pipes placed between two steel beams before concrete deck is placed.

Figure 3.2-2
Utility pipes supported by a bridge in service.

Downdrag DD (also known as negative skin friction) to piles and shafts in bridge foundation is also addressed in the AASHTO specifications. It can be caused by soil settlement due to loads applied after the piles or shafts are installed. This force may be triggered by new fill, groundwater lowering, liquefaction, and so on. Figure 3.2-3 illustrates the situation. The specifications

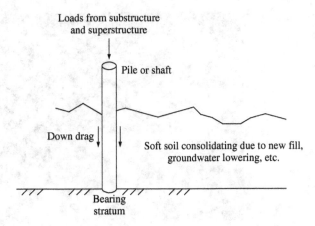

Figure 3.2-3
Downdrag forces on piles or shafts.

identify the following situations as examples where possible downdrag shall be evaluated:

- ❏ Sites are underlain by compressible material such as clays, silts, or organic soils.
- ❏ Fill will be or has recently been placed adjacent to the piles or shafts, such as is frequently the case for bridge approach fills.
- ❏ The groundwater can be substantially lowered.
- ❏ Liquefaction of loose sandy soil can occur.

It should be noted that there are other miscellaneous items carried on the bridge that need to be included as dead load acting on bridge structural components. Examples are traffic sign structures, illumination facilities, bracing members for the primary members, and so on. It is important not to miss any in design. Namely the loads discussed in this chapter should not be treated as the content of a checklist but they illustrate a concept on how these loads need to be identified and estimated. This concept is applicable to all possible loads to be applied to the bridge structure at various stages of its life. These loads may be different from bridge to bridge. Figure 3.2-4 gives more examples of loads on highway bridges.

3.2.2 Permanent Earth Loads EH, EV, and ES

Earth forces belong to another group of permanent forces expected to be applied on bridge components. The AASHTO specifications identify EH, EV, and ES in this group. EH refers to the horizontal earth pressure normally relevant to substructure components, such as an abutment. Figure 3.2-5 shows soil held by an abutment backwall parallel to the traffic below the bridge whose superstructure is yet to be erected. The soil is also retained by the abutment's wingwall perpendicular to the traffic direction. Figure 3.2-6 displays the model for earth forces on an abutment along the longitudinal axis of the bridge (or the traffic direction if there is no skew).

Figure 3.2-4
Miscellaneous weights carried by
bridge such as traffic signs.

Figure 3.2-5
Soil retained by abutment seen in construction before
superstructure erection.

EV represents the vertical earth pressure applied to substructure components by refill after the components are completed and buried as designed.
A typical bridge component subjected to EV is the footing of an abutment
or pier. Figure 3.2-6 includes a footing under load EV.

To estimate these earth loads and determine their effects in bridge
design, soil pressure p is needed. It is assumed to be linearly proportional
to the depth of the earth:

$$p = k\gamma_s z \qquad (3.2\text{-}1)$$

where k is a coefficient depending on the wall stiffness and soil type. It is
defined as k_o for walls that do not deflect or move, k_a for walls that deflect

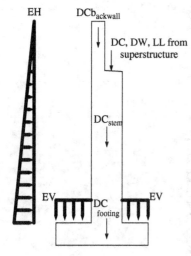

Figure 3.2-6
Models of earth load EH and EV on a bridge abutment in elevation.

or move sufficiently to reach minimum active conditions, and k_p for walls that deflect or move sufficiently to reach a passive condition. The term γ_s is the unit weight of soil and the values in Table 3.2-1 may be used if no site-specific data are available, and z is the depth below the surface.

More details for the coefficient k are given next. For normally consolidated soils, vertical walls, and level ground, the coefficient of at-rest lateral earth pressure k_o may be taken as a function of φ_f', the effective friction angle of the soil:

3.11.5.2 $$k_o = 1 - \sin \varphi_f' \qquad (3.2\text{-}2)$$

For overconsolidated soils, the coefficient of at-rest lateral earth pressure may be assumed to vary as a function of the overconsolidation ratio OCR:

3.11.5.2 $$k_o = (1 - \sin \varphi_f')\text{OCR}^{\sin\varphi_f'} \qquad (3.2\text{-}3)$$

In many instances, OCR may not be known with enough accuracy to calculate k_o. Based on information on this issue, in general, for lightly overconsolidated sands (OCR 1 to 2), k_o is in the range of 0.4 to 0.6. For highly overconsolidated sand, k_o can be on the order of 1.0.

Values for the coefficient of active lateral earth pressure k_a may be taken as

3.11.5.3 $$k_a = \frac{\sin^2(\theta + \varphi_f')}{\Gamma \sin^2 \theta \sin(\theta\text{-}\delta)} \qquad (3.2\text{-}4)$$

3.11.5.3 $$\Gamma = \left[1+\sqrt{\frac{\sin(\varphi_f' + \delta)\sin(\varphi_f' - \beta)}{\sin(\theta - \delta)\ \sin(\theta + \beta)}}\right]^2 \qquad (3.2\text{-}5)$$

where δ = friction angle between fill and wall taken as specified in Table 3.2-2 in degrees

         β = angle of fill to the horizontal as seen in Figure 3.2-7 in degrees

         θ = angle of back face of wall to horizontal shown in Figure 3.2-7 in degrees

For noncohesive soils, values of the coefficient of passive lateral earth pressure k_p may be taken from the charts provided in AASHTO Article 3.11 for the case of a sloping or vertical wall with a horizontal backfill or a vertical wall and sloping backfill.

Table 3.2-2
Friction angle δ

Interface Materials	Friction Angle δ, deg
Mass concrete or masonry on the following foundation materials:	
Clean sound rock	35
Clean gravel, gravel–sand mixtures, coarse sand	29–31
Clean fine to medium sand, silty medium to coarse sand, silty or clayey gravel	24–29
Clean fine sand, silty or clayey tine to medium sand	19–24
Fine sandy silt, nonplastic silt	17–19
Very stiff and hard residual or preconsolidated clay	22–26
Medium stiff and stiff clay and silty clay	17–19
Steel sheet piles against the following soils:	
Clean gravel, gravel–sand mixtures, well-graded rock fill with spalls	22
Clean sand, silty sand–gravel mixture, single-size hard rock fill	17
Silty sand, gravel, or sand mixed with silt or clay	14
Fine sandy silt, nonplastic silt	11

In the AASHTO design code ES, or surcharge through earth, refers to those loads and their effects on a wall buried in soil due to forces applied on the surface of the backfill soil behind the wall, as shown in Figure 3.2-8. As an example, ES may be due to stockpiles of material, often applied during bridge construction. As seen in Figure 3.2-8, such surcharge load is estimated as a horizontally and uniformly applied load to the wall. However, similar surcharge due to vehicular load is separated in the AASHTO specifications as LS (surge due to live load) to be discussed below, because the variation associated with vehicular load is considerably different. Therefore, its load factor is the same as that for live load and not as that for earth load.

Where a uniform surcharge is present, a constant horizontal earth pressure shall be added to the basic earth pressure. This constant earth pressure may be taken as

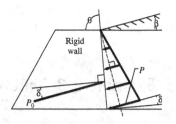

Figure 3.2-7
Notation for Coulomb active earth pressure.

3.11.6.1
$$\Delta p = k_s q_s \qquad (3.2\text{-}6)$$

where $\Delta p =$ constant horizontal earth pressure due to uniform surcharge (k/ft^2)

      $k_s =$ coefficient of earth pressure due to surcharge

      $q_s =$ uniform surcharge applied to upper surface of active earth wedge (k/ft^2)

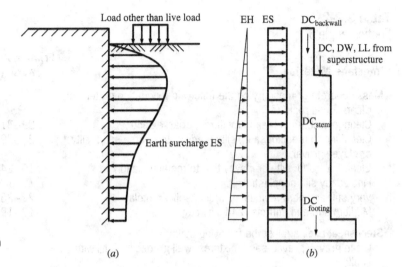

Figure 3.2-8
Earth surcharge load ES to
abutment: (a) distribution of earth
surcharge (b) modeled distribution
of earth surcharge ES.

For active earth pressure conditions, k_s shall be taken as k_a. For at-rest conditions, k_s shall be taken as k_o. Otherwise, intermediate values appropriate for the type of backfill and amount of wall movement may be used.

Example 3.3 shows the application and estimation of EH and EV for a bridge abutment. DC and a few other loads are applicable to that abutment, while quantitative estimation is not shown to allow focusing on the earth loads.

Example 3.3 Abutment Design (Earth Loads EH and EV)

❑ **Problem Statement** To design the abutment for a steel beam bridge superstructure shown in Figure Ex3.3-1 according to the AASHTO specifications, estimate the earth loads EH and EV on its backwall, stem, and footing. Assume that the abutment is supported on a pile foundation.

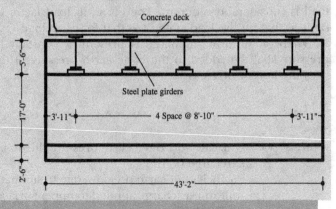

Figure Ex3.3-1
Front view of reinforced concrete cantilever
abutment.

☐ **Abutment Dimensions**   The preliminary dimensions of the abutment are shown in Figures Ex3.3-2, Ex3.3-3, and Ex3.3-4.

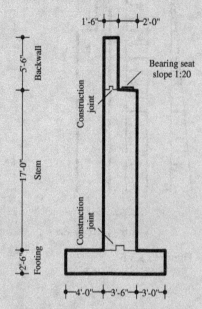

Figure Ex3.3-2
Side view of reinforced concrete cantilever abutment.

Figure Ex3.3-3
Details of abutment and superstructure elevation.

The dimensions are preliminarily determined based on previous designs and experience.

☐ **Earth Loads on Backwall**

Lateral earth pressure EH: *3.11.5*

At bottom of backwall as shown in Figure Ex3.3-4:

3.11.5.1; 3.5.1 $p = k\gamma_s z = 0.3(0.11 \text{ k/ft}^3)\, 5.5 \text{ ft} = 0.18 \text{ k/ft}^2$

$$EH_{backwall} = \frac{1}{2}\left(0.18 \text{ k/ft}^2\right) 5.5 \text{ ft} = 0.50 \text{ k/ft}$$

at height 1.83 ft from bottom of backwall

☐ **Earth Loads on Stem**

Lateral earth pressure EH: *3.11.5*

At bottom of stem as seen in Figure Ex3.3-5,

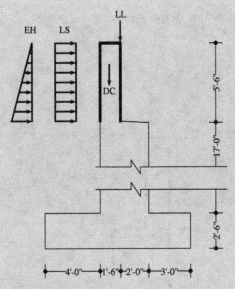

Figure Ex3.3-4
Loads on abutment backwall.

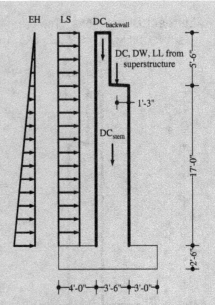

Figure Ex3.3-5
Loads on abutment stem.

$$3.11.5.1; 3.5.1 \quad p = k\gamma_s z = 0.3(0.11 \text{ k/ft}^3)\,22.5 \text{ ft} = 0.74 \text{ k/ft}^2$$

$$EH_{backwall} = \frac{1}{2}\left(0.74 \text{ k/ft}^2\right)22.5 \text{ ft} = 8.3 \text{ k/ft}$$

at height 7.5 ft from bottom of stem

□ **Earth Loads on Footing**
Vertical earth load EV: Using $\gamma_S = 0.11 \text{ k/ft}^3$,
Table 3.2-1. Average of loose gravel and compacted sand.   *3.5.1*

$$DC_{earthback} = 22.5 \text{ ft}\,(4 \text{ ft})\,0.11 \text{ k/ft}^3 = 9.90 \text{ k/ft}$$

$$DC_{earthfront} = 1 \text{ ft}\,(3 \text{ ft})\,0.11 \text{ k/ft}^3 = 0.33 \text{ k/ft}$$

Lateral earth pressure EH: *3.11.5*
At bottom of footing as shown in Figure Ex3.3-6:

$$3.11.5.1; 3.5.1\, p = k\gamma_s z = 0.3(0.11 \text{ k/ft}^3)\,25 \text{ ft} = 0.83 \text{ k/ft}^2$$

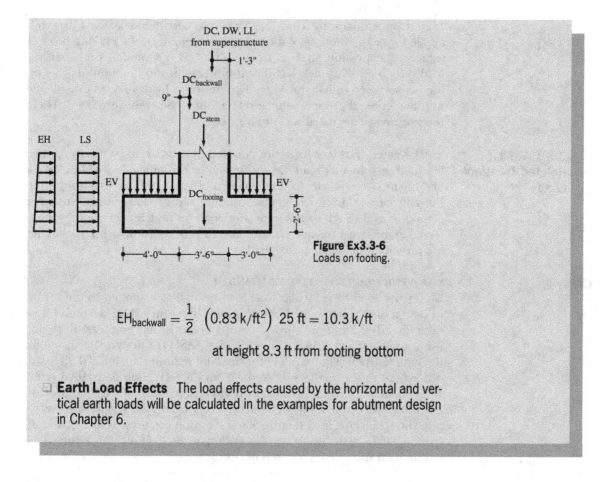

Figure Ex3.3-6
Loads on footing.

$$EH_{backwall} = \frac{1}{2}\ (0.83\,k/ft^2)\ 25\,ft = 10.3\,k/ft$$

at height 8.3 ft from footing bottom

☐ **Earth Load Effects** The load effects caused by the horizontal and vertical earth loads will be calculated in the examples for abutment design in Chapter 6.

DIFFERENTIAL SHRINKAGE SH AND CREEP CR

Differential shrinkage strains between concrete of different ages and compositions and between concrete and steel or wood may induce load effects causing distressing and cracking. Such behavior also needs to be covered in the design to prevent adverse consequences. Creep strains for concrete and wood may also need to be addressed in the design. These issues are unique for concrete and will be discussed in detail when, design of concrete members is presented later in the book in Chapter 4.

3.2.3 Other Permanent Loads

3.3 Transient Loads

Transient loads to highway bridges are those applied over a short period of time, typically ranging from a few seconds (e.g., vehicular loads) to possibly several months (e.g., seasonal thermal expansion or contraction forces caused by temperature change). They may also be repeatedly or rarely applied over the lifetime of the bridge. For example, the braking force of vehicles may possibly apply as frequently as every several seconds. Seismic load due to an extremely significant but rare ground motion may act on the bridge once in its entire life. The transient loads may be categorized into two general groups: vehicle related and non–vehicle related. These two groups are discussed separately next.

3.3.1 Vehicle-Related Transient Loads

Truck-related transient loads are denoted here as LL for statically applied live load or vehicular load, IM for impact or dynamic effect of LL, BR for braking force of vehicle, CE for vehicular centrifugal force to a bridge horizontally curved, LS for vehicular-load-induced surcharge to substructure components through soil, WL for wind load on vehicles and thereby transferred to the structure, and CT for load effect of collision with a truck. They are presented below in more detail.

LIVE (VEHICULAR) LOAD EFFECTS LL AND IM

LL for live load or vehicular load is to be statically applied to the bridge in design or evaluation, with an additional allowance IM to account for its dynamic effect while moving across the span. LL is computed in design using the standard load specified in the AASHTO design specifications: the fatigue truck (Figure 3.3-1) for the fatigue limit state and HL93 for all other limit states (Figures 3.3-2, 3.3-3, and 3.3-4). Note that HL93 refers to a collection of several vehicle load models, namely the HL93 truck load (Figure 3.3-2) or the HL93 tandem load (Figure 3.3-3), whichever governs, plus the HL93 lane load (Figure 3.3-4). In addition, these vehicle loads are specified to have the transverse dimensions given in Figure 3.3-5 when analysis in the transverse direction is performed.

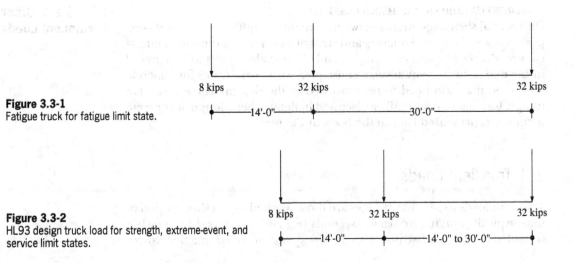

Figure 3.3-1
Fatigue truck for fatigue limit state.

Figure 3.3-2
HL93 design truck load for strength, extreme-event, and service limit states.

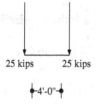

25 kips   25 kips

|◄—4'-0"—►|

Figure 3.3-3
HL93 design tandem load for
strength, extreme-event, and
service limit states.

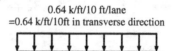

0.64 k/ft/10 ft/lane
=0.64 k/ft/10ft in transverse direction

Figure 3.3-4
HL93 design lane load for strength,
extreme-event, and service limit
states.

Vehicle load
(design truck and design tandem)

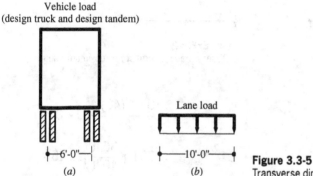

|◄—6'-0"—►|
(a)

Lane load

|◄——10'-0"——►|
(b)

Figure 3.3-5
Transverse dimensions of vehicular load.

Example 3.4 illustrates application of the HL93 load along with IM to a
simple span bridge. Note that the design load effect for a specific beam will
require a realistic distribution of the HL93 load to the beam (presented in
Chapter 4 for various bridge components).

Example 3.4 Steel Rolled Beam Bridge Design (Live-Load Effects)

❏ **Design Requirement** For the simply supported bridge superstructure
in Examples 3.1 and 3.2, find the live-load effect LL for both moment
and shear in the beams under one lane of HL93 load. The span length
is 40 ft.

❏ **Live-Load Effects for Design** The following two cases need to be
calculated to find the maximum of the two to be used in design:

1. Truck load (with IM) + lane load
2. Tandem load (with IM) + lane load

❏ **Maximum Live-Load Shear (see Figures Ex3.4-1–Ex3.4-3)**

$$V_{truck} = 32 + \frac{32\,(26)}{40} + \frac{8\,(12)}{40} = 55.2\,k$$

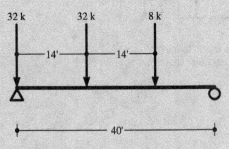

Figure Ex3.4-1
HL-93 truck load placement for maximum shear.

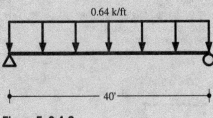

Figure Ex3.4-2
HL-93 lane load placement for maximum shear.

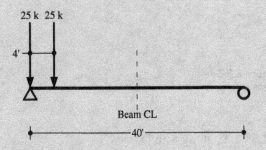

Figure Ex3.4-3
Tandem loading placement for maximum shear.

$$V_{lane} = \frac{0.64\,(40)}{2} = 12.8\,k$$

$$V_{tandem} = 25\,k\left(\frac{36\,ft}{40\,ft}\right) + 25\,k$$

$$= 22.5\,k + 25\,k = 47.5\,k$$

This maximum shear is less than the truck load maximum shear of 55.2 k; thus the truck load controls. Thus, the design shear under one lane load is

$$V_{LL\ one\ lane} = (1 + IM)\ 55.2\,k + 12.8\,k$$

$$= 1.33\,(55.2\,k) + 12.8\,k = 86.2\,k$$

❏ **Maximum Live-Load Moment**

Truck Load Find the center of gravity (CG) of the HL-93 truck in Figure Ex3.4-4 as detailed in Table Ex3.4-1 and use below. Take moments about the 32-k axle to the right:

$$Y' = \frac{sum\ of\ moments}{sum\ of\ forces} = \frac{672\ kft}{72\ kft} = 9\ ft\ 4\ in.$$

$$Y = 14 - Y' = 4.67\ ft = 4\ ft\ 8\ in.$$

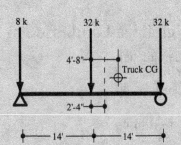

Figure Ex3.4-4
Center of gravity of HL-93 truck.

Table Ex3.4-1
Calculation for center of gravity of HL-93 truck

Force	Arm	Moment
8 k	28 ft	224 kft
32 k	14 ft	448 kft
32 k	0 ft	0 kft
Total		672 kft

To find the maximum moment of the HL–93 truck load, the midspan point needs to bisect the distance between the center of gravity and the nearest 32-k load, as shown in Figure Ex3.4-5.

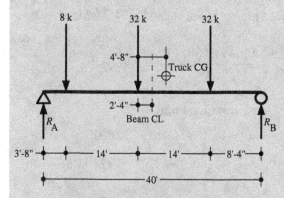

Figure Ex3.4-5
HL-93 truck placement for maximum moment.

Reactions

$$R_B = \frac{1}{40}\left[8\,k\,(3.67\,ft) + 32\,k\,(17.67\,ft) + 32\,k\,(31.67\,ft)\right] = 40.2\,k$$

$$R_A = (8\,k + 32\,k + 32\,k) - 40.2\,k = 31.8\,k$$

Truck load moment

$$M_{truck} = R_A(17.67\,ft) - 8\,k\,(14\,ft) = 449.7\,kft$$

Tandem Load To determine the maximum moment of the tandem load, the two 25-k axles need to be bisected by the center of the span, as shown in Figure Ex3.4-6:

Maximum tandem load moment

$$M_{tandem} = 25\,k\,(18\,ft) = 450.0\,kft$$

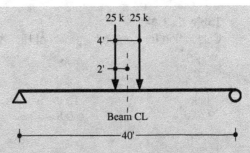

Figure Ex3.4-6
Tandem loading placement for maximum moment.

This is approximately the same moment as the truck load.

Lane Load Lane load moment (see Figure Ex3.4-2 for load positioning):

$$M_{lane} = \frac{0.64\ k/ft\ (40\ ft)^2}{8} = 128.0\ kft$$

Total Moment for One-Lane Live Load

$$M_{LL\ one\ lane} = (1 + IM)\ 450\ kft + 128\ kft$$

$$= 1.33\ (450) + 128 = 726\ kft$$

Compared with trucks in traffic on U.S. roadways, the AASHTO fatigue truck in Figure 3.3-1 is viewed as a model for real five-axle semitrailer trucks as shown in Figure 3.3-6. Nevertheless, the fatigue truck has only three axles, with the heavier two (32 kips each) modeling the two tandems each with two axles, as seen in Figure 3.3-6. Note also that the federal legal axle load is 20 kips and legal tandem load (with two axles 4 ft apart) is 34 kips. Overall statistics show that such five-axle (18-wheel) trucks represent about 70% of the truck traffic on rural highways in the United States.

The HL93 design truck appears to intend to cover more widely variable truck configurations and use the worst case for bridge design and evaluation. Accordingly, the distance between the two heavier axles is prescribed as a variable from 14 to 30 ft, instead of the constant 30 ft for the fatigue truck in Figure 3.3-1. With respect to the federal weight limit mentioned above, the HL93 design truck is actually not legal according to the federal bridge formula widely used to enforce compliance. The truck violates the federal axle load limit (20 kips) as well as the axle distance requirement (being too short at 14 ft between the two heavier axle weights). So the HL93

Figure 3.3-6
Popular five-axle trucks on U.S. highways.

design truck actually would be required to have a special permit to operate depending on roadway jurisdiction. However, designing for a truck load that exceeds the legal limit is conservative, particularly with consideration to possible overloading. In addition, this violation is not very significant.

The impact factor IM covers the additional load effect due to dynamically applying the truck load to bridge components. Field dynamic test measurement results show that this additional load effect is a function of the smoothness of the roadway surface. It is accordingly estimated at a level of a few to about 15% for primary members such as main girders in short- and medium-span bridges. The AASHTO design specifications prescribe 15% for the fatigue limit state and 33% for the other limit states. These limit states will be discussed in more detail later in this chapter. The fatigue limit state is to covers a sort of weighted average of the routine truck load effect, and thus the 15% IM value is close to the daily maximum level. On the other hand, the higher value of 33% apparently ensures that the real impact effect will never exceed this level. In particular, this higher IM value in the code covers the strength limit states for the maximum load effects over the entire expected life span of 75 years. It is obviously more conservative.

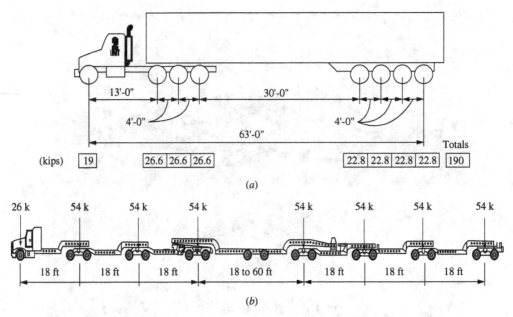

Figure 3.3-7
Example permit truck loads for design and evaluation: (a) Wisconsin; (b) California.

When special trucks are concerned (for design under the strength II limit state or evaluation under permit checking, to be further discussed below) with drastically different dimensions from those in Figures 3.2-2 to 3.2-5, different requirements of the jurisdiction will need to be satisfied as to their transverse placement and even load factor. Figure 3.3-7 shows two examples of overweight trucks used in design for Wisconsin and California bridges, respectively.

To guide consideration in design and evaluation to possible simultaneous occupation of trucks on the bridge span, a multiple presence factor (MPF) is prescribed in the AASHTO LRFD specifications. The values of this factor are displayed in Table 3.3-1 as a function of number of loaded lanes. It indicates that the two-lane occupation is used as a reference case with MPF equal to 1. In other words, two lanes loaded with the HL93 load in each lane along with

Table 3.3-1
AASHTO multiple presence factor

Number of Loaded Lanes	Multiple Presence Factors, m
1	1.2
2	1
3	0.85
>3	0.65

a live-load factor ($\gamma_L = 1.75$) to be seen in the next section are considered the reference case for others with more or less loaded lanes. For example, for the case of one lane, one lane of HL93 along with $\gamma_L = 1.75$ is considered to be inadequate to cover the live load. Thus an additional factor of 1.2 is required to fully cover the possible maximum load. This is indicated in Table 3.3-1. Similarly, a reduction of load reflected in MPF lower than 1 is prescribed in the AASHTO LRFD specifications for the cases of three and more lanes loaded as shown in Table 3.3-1. This is because more lanes simultaneously and fully loaded by the HL93 load are unlikely. Therefore, 0.85 and 0.65 are given in the specifications to quantify this lower likelihood. For instance, for a three-lane bridge span, Table 3.3-1 requires the design load to be 0.85(3 lanes) = 2.55 HL93 loads. Namely a three-lane bridge span is required to carry only 2.55 lanes of load; each lane is the standard HL93 load.

In addition, a dynamic impact factor of 33% for superstructure components is required in the design specifications to cover impact induced by vehicular loads. While it is uniformly applied to all span lengths within the applicability range in the specifications, IM is allowed to vary in bridge evaluations depending on the smoothness of the riding surface found to be the controlling factor of dynamic impact.

This contrast is another example showing different treatments for new bridges and existing bridges, as discussed earlier in Chapter 2. New bridges (yet to be constructed) are required to have a higher capacity and existing bridges are subjected to a lower requirement. The higher requirement may appear to be very conservative or much higher than that for daily operation, partially because increasing the bridge capacity in new bridges yet to be constructed is not nearly as expensive as strengthening existing structures. The latter can be as expensive as replacing the entire bridge if no reliable methods/technologies are available for the required strengthening. For example, steel members can be more readily strengthened by bolting additional steel plates or shapes to the existing members. Reinforced concrete members, on the other hand, are much more difficult to strengthen because positive bonding between new and old concrete cannot be always realized and anchorage of the new reinforcement is equally uncertain.

BRAKING FORCE BR

Load BR denotes the force applied by vehicles on a bridge when using the brakes to slow down or to stop. The braking force is taken as the greater of (1) 25% of the axle weights of the design truck or design tandem or (2) 5% of the design truck plus the lane load or 5% of the design tandem plus the lane load. This braking force shall be placed in all design lanes carrying traffic headed in the same direction, as illustrated in Figure 3.3-8 for a two-lane configuration. These forces shall be assumed to act horizontally at a distance of 6 ft above the roadway surface, also shown in Figure 3.3-8,

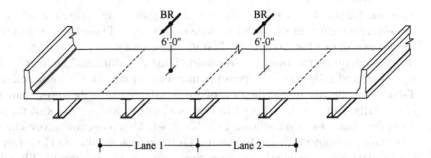

Figure 3.3-8
Break force BR applied on
bridge.

in either longitudinal direction (forward or backward) to cause the worst
force effect in the component being analyzed and designed. All design
lanes shall be simultaneously loaded for bridges to be used in one direction
and likely to become one directional in the future. Note also that the
(MPF) in Table Ex3.4-1 applies according to the number of loaded lanes.

CENTRIFUGAL FORCE CE
Load CE refers to the centrifugal force applied by a vehicle on a bridge with
a horizontally curved profile. A bird's-eye view of the road profile is shown
in Figures 3.3-9 and 3.3-10, along with the force generated from the vehicle
to the road or bridge. This force is estimated according to the curvature of
the road profile, as the product of the axle weights of the design truck or
tandem and a factor C. According to the AASHTO specifications,

3.6.3-1
$$C = f\frac{v^2}{gR}$$
(3.3-1)

where v = highway design speed (ft/s)
        f = 1.0 for fatigue
        $= \frac{4}{3}$ otherwise
        g = gravitational acceleration: 32.2 (ft/s^2)
        R = radius of curvature of traffic lane (ft)

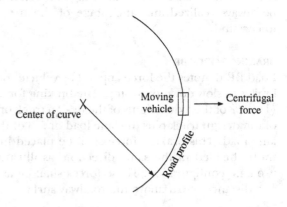

Figure 3.3-9
Generation of centrifugal force.

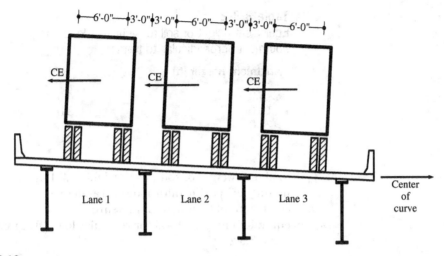

Figure 3.3-10
Centrifugal force by vehicles on a horizontally curved bridge.

EARTH PRESSURE DUE TO LIVE-LOAD SURCHARGE LS

This earth pressure is applied to substructure members, but generated by vehicular load, as shown in Figure 3.3-11. According to the AASHTO specifications, the increase in earth pressure due to live load can be estimated as

3.11.6.4
$$\Delta p = k\gamma_s h_{eq} \qquad (3.3\text{-}2)$$

where Δp = constant horizontal earth pressure due to live-load
surcharge (k/ft^2)

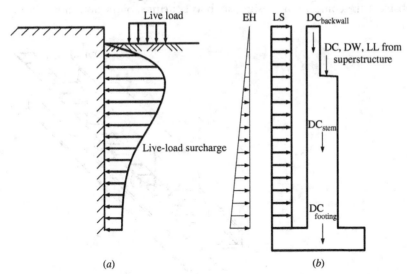

Figure 3.3-11
Live-load surcharge to abutment: (*a*) distribution of live-load surcharge; (*b*) modeled distribution of live-load surcharge LS.

Table 3.3-2
Equivalent height of soil for vehicular load on
abutment perpendicular to traffic

Abutment Height (ft)	h_{eq} (ft)
5.0	4.0
10.0	3.0
$\geq$20.0	2.0

γ_s = total unit weight of soil (k/ft^3); Table 3.2-1 may be used if
no further specific information is available

k = coefficient of lateral earth pressure

h_{eq} = equivalent height of soil for vehicular load (ft) given in
Table 3.3-2

The wall height shall be taken as the distance between the surface of
the backfill and the bottom of the footing along the pressure surface being
considered.

WIND LOAD ON VEHICLE WL

Load WL is another load effect related to vehicle load (or live load), but due
to wind, as shown in Figure 3.3-12. This load acts on the vehicles present
on the bridge span and is transferred to the vehicle wheels, the deck, the
deck supporting system, and then the substructure and foundation. Since
the superstructure components (the deck and deck-supporting system) are
quite strong and rigid in the plane of WL, induced stresses in these components will not govern their design. As a result, this load is more relevant
to substructure component design, such as bearings, piers, abutments, and

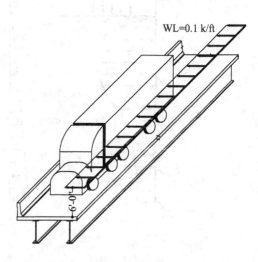

Figure 3.3-12
Wind load on vehicles.

piles. As seen in Figure 3.3-12, this load acts horizontally on the vehicles that happen to be on the bridge, and it is simplified as a linear load along the length of the vehicles and thus the length of the bridge span. Its magnitude is estimated at 0.1 k/ft and is simplified as acting at a height of 6 ft above the road surface, like the braking force BR. Therefore, WL can cause moments, shear forces, and axle forces in substructure members.

TRUCK COLLISION LOAD CT

Load CT in the AASHTO LRFD specifications refers to truck collision forces to bridge components. For example, the design for deck overhang often needs to consider this load, transferred through the traffic railing consisting of reinforced concrete parapets, metal pipes, and so on. Collision is considered a rare load in the AASHTO specifications; thus it is categorized as one of the extreme-event limit states. They deal with loads that appear seldomly and are elaborated further in Section 3.4.3 on load combination in this chapter. Accordingly, the magnitude of this load prescribed in the design code is not uniform for all bridges as for many other loads discussed above, such as the live load, wind load, and so on. Instead, the magnitude of CT is given in the design specifications dependent on the likely size and mass of the truck colliding on the bridge railing. Therefore, for different bridges on different roads, the likely size and mass are different and are discussed next.

Figure 3.3-13 shows the forces prescribed in the design specifications modeling truck collision to the traffic railing. Five parameters are given in the specifications to describe the collision force CT: three magnitudes of transverse, longitudinal, and vertical collision forces (F_t, F_L, and F_v) and

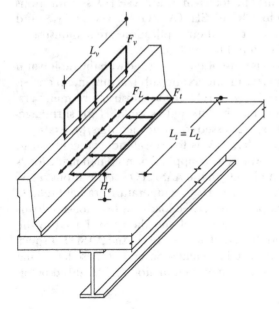

Figure 3.3-13
Truck collision load to bridge.

Table 3.3-3
CT load magnitude depending on crash requirement for railing

Design Forces and Designations	Railing Test Levels					
	TL-1	TL-2	TL-3	TL-4	TL-5	TL-6
F_t transverse (k)	13.5	27	54	54	124	175
F_L longitudinal (k)	4.5	9	18	18	41	58
F_v vertical (k) down	4.5	4.5	4.5	18	80	80
L_t and L_L (ft)	4	4	4	3.5	8	8
L_v (ft)	18	18	18	18	40	40
H_e (min) (in)	18	20	24	32	42	56
Minimum H, height of rail (in)	27	27	27	32	42	90

their two lengths of distribution ($L_t = L_L$ and L_v). Table 3.3-3, taken from the AASHTO specifications, gives the values of these parameters for design as functions of the railing crash test levels designated as TL-1 through TL-6. As seen, TL-5 represents the highest and TL-1 the lowest test levels. To be seen in Chapter 4 and related design examples, a typical design approach for deck overhang subjected to such a load level is to (1) ensure that the railing system has adequate capacity and then (2) design the deck overhang for the maximum of the railing capacity and the load requirement in Table 3.3-3. This approach will produce a deck overhang system that will not fail before the railing system fails if such a truck collision ever occurs.

3.3.2 Non-Vehicle-Related Transient Loads

The non-vehicle related transient loads in the AASHTO specifications include, but are not limited to, CR, SE, SH, CV, EQ, IC, TG, TU, WS, and PL. They are briefly discussed next and their applications are demonstrated in the design examples in Chapters 4 through 6.

CR for creep-induced forces is most often concerned with the design and evaluation of prestressed concrete members and substructure members. SE for settlement-induced forces is more relevant to substructure members, while superstructure members may be affected as well. SH for shrinkage-related forces is also relevant to prestressed concrete members, for example, for shrinkage- related prestress loss. CV is for vessel-induced collision to members of bridges over waterway. Nature-applied or non-man-made loads are mostly transient. For example, EQ in the AASHTO specifications refers to seismic loads, IC to ice-related loads, TG to temperature gradient-related loads, and TU to uniform temperature-caused loads, and WS to wind loads on the structure (not those on the live load and then transferred to the structure, denoted as WL and discussed above). PL in the AASHTO specifications denotes pedestrian-induced live loads. Several of these loads that are more often encountered in design and evaluation are discussed below in more detail.

TEMPERATURE LOADS TU AND TG   *3.12.2, 3.12.3*

The uniform temperature change TU refers to the deformation-induced effects that result when the entire bridge is subjected to temperature change between seasons. Figure 3.3-14 shows a beam thermal expansion causing a force to the abutment as an example of uniform temperature load effect TU. To facilitate load effect estimation, the AASHTO specifications provide two optional procedures, A and B.

Procedure A is simpler than B and uses the uniform temperature changes shown in Table 3.3-4 to estimate the deformation and then the induced forces and stresses if the deformation is constrained. As seen, this temperature change depends on climate and the material subjected to temperature change and deformation.

Procedure B is applicable to only two structure types: (1) steel girder bridges with concrete deck and (2) concrete girder bridges with concrete deck. The AASHTO specifications provide the extreme temperatures $T_{\text{MaxDesign}}$ and $T_{\text{MinDesign}}$ in contour maps for the entire country. Thus the temperature range

$$\Delta_T = T_{\text{MaxDesign}} - T_{\text{MinDesign}}$$

shall be used in place of Table 3.3-4 to estimate the induced deformations and forces.

Procedure B has a higher resolution compared with procedure A with respect to the bridge location. The extreme temperature maps in

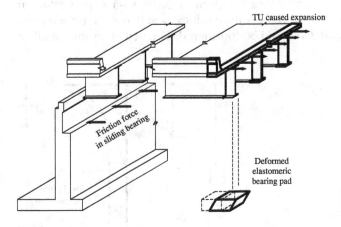

TU caused expansion

Friction force in sliding bearing

Deformed elastomeric bearing pad

Figure 3.3-14
TU-caused deformation and forces.

Table 3.3-4
Temperature ranges for load TU for procedure A (°F)

Climate	Steel or Aluminum	Concrete	Wood
Moderate	0–120	10–80	10–75
Cold	−30–120	0–80	0–75

the specifications allow determination of Δ_T for a specific location by reading off $T_{\text{MaxDesign}}$ and $T_{\text{MinDesign}}$ from the maps. Interpolation between contour lines may be needed if the location happens to not be on a contour line. In contrast, Table 3.3-4 allows only two climates: moderate or cold. Nevertheless, Procedure A is applicable to more material types, although with a low resolution with respect to bridge site location.

Temperature gradient TG is also temperature driven but is different than TU. Similar to the adverse effect resulting from uneven settlement, structures can suffer from stresses resulting from uneven temperature distribution in the cross section, modeled here using TG. For example, if the top side of a structure is continually exposed to a heat source, such as the sun, and the bottom side is not, the temperature differential in the cross section can cause high stresses since the colder side constrains the hotter side from elongation. The AASHTO LRFD specifications specify the vertical temperature gradient in concrete and steel superstructures with concrete decks as shown in Figure 3.3-15.

According to the AASHTO specifications, length A in Figure 3.3-15 is:

- ❑ 12 in. for concrete superstructures with a depth of 16 in. or deeper
- ❑ 4 in. less than the actual depth for concrete sections shallower than 16 in.
- ❑ 12 in. with $t = $ depth of concrete deck for steel superstructures

Temperatures T_1 and T_2 in Figure 3.3-15 are decided from a solar radiation zone map and their values are given in Table 3.3-5 as prescribed in the specifications. Positive temperature values as well as negative temperature values in T_1 and T_2 should be considered. The negative values shall be

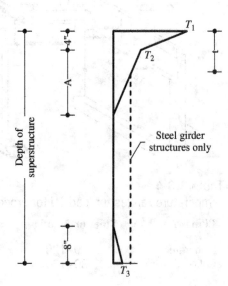

Figure 3.3-15
Positive vertical temperature gradient in concrete and steel superstructures.

Table 3.3-5
Basis for temperature gradient TG

Zone	T_1 (°F)	T_2 (°F)
1	54	14
2	46	12
3	41	11
4	38	9

obtained by multiplying the values in Table 3.3-5 by −0.20 for decks with an asphalt concrete overlay and −0.30 for concrete decks without asphalt concrete overlay. The negative values can cause the positive bending moment in the midspan area of the bridge span possibly to be superimposed with other stresses and therefore should not be neglected.

WIND LOAD WS   *3.8.1; 3.8.2*
Wind load WS refers to the load effect due to wind that directly acts on the bridge structure and is transferred to the component being designed. Table 3.3-6 displays the base pressure of wind acting on bridge structures specified in the AASHTO design speciications.

These wind pressure values are based on a base wind speed of 100 mph. If site specific data are available and different from 100 mph, the code allows modification of the wind pressure as a function of the squared ratio of the site's wind speed to the reference 100-mph speed. This is because wind pressure is understood in physics as approximately proportional to wind speed squared.

Nevertheless, the total wind load shall not be less than 0.30 k/ft in the plane of a windward chord and 0.15 k/ft in the plane of a leeward chord on truss and arch components and not less than 0.30 k/ft on beam or girder spans. However, for most short- and medium-span bridges focused in this book such site-specific data are often not available. Thus Table 3.3-6 is more often used to estimate the wind load effect WS along with these minimum values for the scope of short- to medium-span highway bridges.

WS can be divided into two categories: horizontal (*3.8.1*) and vertical (*3.8.2*) wind loads. They are separately discussed next. The vertical wind load is relatively simpler to deal with since it is rarely encountered for the

Table 3.3-6
Base wind pressure on structure members

Superstructure Component	Windward (k/ft²)	Leeward (k/ft²)
Trusses, columns, and arches	0.05	0.025
Beams	0.05	NA
Large flat surfaces	0.04	NA

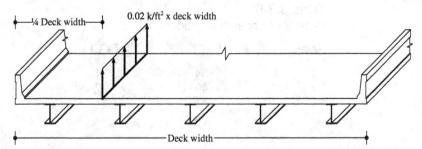

Figure 3.3-16
Vertical wind load application.

short- and medium-span bridges focused on in this book. Figure 3.3-16 illustrates how to apply vertical wind load to such bridges as specified in the AASHTO design code. As seen, the wind is acting upward along a longitudinal line marking one quarter of the deck width away from the outer face of the railing system. This force apparently will not govern the deck's strength due to its small magnitude compared with vehicle load LL. It may only contribute to one of the critical load combinations for substructure components.

The horizontal wind load may be applied to super- and substructures of a bridge in different magnitudes due to their respective distances to the ground. Note that the wind velocity at the ground is zero and thus no wind pressure or force is expected.

As an example of superstructure in Figure 3.3-17, horizontal WS is shown applied to the parapet's wall and a fascia beam side. Note that WS can be applied in each of the two directions. Thus the orientation causing the worst combined load effect needs to be designed for. In some cases the worst orientation can be easily identified and the other can be eliminated without further analysis. One example is shown in Figure 3.3-17 where the

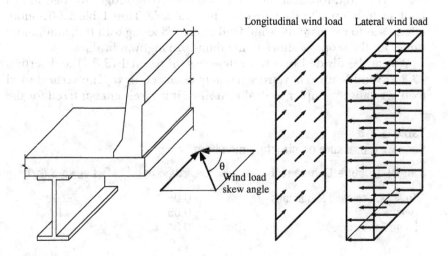

Figure 3.3-17
Application example of
horizontal WS to
superstructure.

Table 3.3-7
Base wind pressure depending on attack skew angle defined in Figure 3.3-17

| Skew Angle of Wind (deg) | Trusses, Columns, and Arches | | Girders | |
	Lateral (k/ft²)	Longitudinal (k/ft²)	Lateral (k/ft²)	Longitudinal (k/ft²)
0	0.075	0	0.05	0
15	0.07	0.012	0.044	0.006
30	0.065	0.028	0.041	0.012
45	0.047	0.041	0.033	0.016
60	0.024	0.05	0.017	0.019

opposite orientation is symmetric to the one shown, and thus there is no need to analyze and design for the other direction if the components are designed symmetric about the longitudinal center line of the bridge. The wind forces in Figure 3.3-17 will also never govern the design of the parapet and the beam, but they may govern the substructure component design when combined with other forces. They also may control the design for bracing members to the beams. Note that wind force may attack the structure randomly from any direction. Thus, the AASHTO design code offers the base wind pressure values shown in Table 3.3-7 depending on the attack angle, defined as that between the perpendicular to the bridge longitudinal axis and the wind direction, indicated as angle θ in Figure 3.3-17. Both directions of force need to be applied to find the maximum load effect for the component being designed. The pressures in Table 3.3-7 may also be updated if site-specific wind speed data are used. Then the wind pressures in Table 3.3-7 can be updated by multiplying them by the squared ratio of the site wind speed to the reference 100-mph wind speed.

For regular girder-supporting-deck bridges having span length shorter than 125 ft and height lower than 30 ft above the ground or water level, the WS determination can be simplified as 0.05 k/ft² in the transverse direction and 0.012 k/ft² in the longitudinal direction. Both directions of force need to be applied simultaneously in the design.

For substructure application of wind load, Figures 3.3-18 and 3.3-19 illustrate two typical examples of WS respectively applied to an abutment and a pier. In Figure 3.3-19 for a pier, two directions of force are included. The resultant of the two components shall be 0.04 k/ft² according to the AASHTO specifications (3.8.1.2.3). In Figure 3.3-18, for an abutment with a wingwall, the transverse direction of

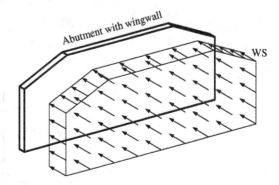

Figure 3.3-18
Wind load on an abutment with wingwall.

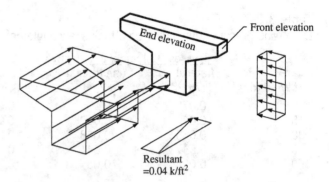

Figure 3.3-19
Wind load on a pier.

wind pressure is deemed to be negligible, compared with the load-carrying capacity in that direction. Therefore, that load is ignored.

FRICTION FORCE FR   3.13

FR should be estimated based on extreme values of the friction coefficient between the sliding surfaces. A typical friction force in bridge components is due to planar or spherical sliding of surfaces in bearings and is then transferred to substructure components. Therefore FR is most relevant to substructure component design, as shown in Figure 3.3-20. The effect of moisture and possible degradation or contamination of sliding or rotating surfaces upon the friction coefficient needs to be considered in the process of load effect estimation for design. If warranted, physical testing may need to be considered in determining the maximum and minimum friction coefficients, especially if the surfaces are expected to be roughened along with an increase of bridge age.

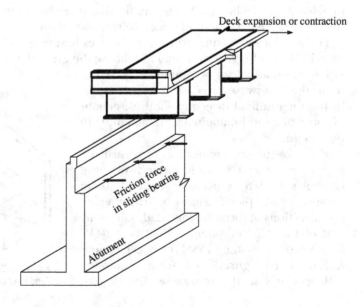

Figure 3.3-20
Example of friction forces on abutment.

EARTHQUAKE LOAD EQ   *3.10*

Earthquakes represent one of the most devastating forces to highway bridges. However, tremendously large seismic forces causing bridge collapse and associated life loss rarely occur. It is thus important to design bridges with a balance between the tolerable risk and the extra cost just for the seismic load. The AASHTO specifications prescribe peak ground accelerations as the source of load to accomplish this goal. They are given in maps to indicate the design acceleration for the United States. These peak ground accelerations are given with a 7% of probability to be exceeded for 75 years. They and the local soil condition control the shape of the design response spectrum, which is used to determine the magnitude of the forces in the bridge components. In general, the superstructure of a typical short- or medium-span highway bridge is assumed to move in a vertical plane or a number of vertical planes during a seismic event with horizontal accelerations and thereby induced forces. The support structure, mainly the substructure including bearings if present, has to take on the seismic forces. These forces are to be addressed in design under the so-called extreme-event limit state, to be discussed below in the section of load combinations.

VESSEL COLLISION FORCE CV   *3.14*

CV in the AASHTO design specifications refers to the force induced by accidental collision between a vessel and a bridge. All bridge components in a navigable waterway crossing located in design water depths not less than 2.0 ft are required to be designed for vessel impact.

   The minimum design impact load for substructure design shall be determined using an empty hopper barge drifting at a velocity equal to the yearly mean current for the waterway location. The design barge shall be a single 35 × 195-ft barge, with an empty displacement of 200 tons, unless approved otherwise by the bridge owner. More details for collision loads are given in Article 3.14 in the AASHTO design specifications.

3.4 Load Combinations   *3.4.1*

The AASHTO specifications prescribe a general format of load combination as follows:

1.3.2.1
$$Q_n = \sum \eta_i \gamma_i Q_i \tag{3.4-1}$$

where Q_n = total design load effect (nominal value)
   Q_i = load effects including those discussed in Section 3.3
   γ_i = load factors to be defined below
   η_i = load modifier associated with η_D related to ductility, η_R related to redundancy, and η_I related to operational classification

3.4.1 General Formulation of Load Combination

Depending on the load, the AASHTO specifications allow a range of load factor γ_i values from which the design engineer can choose. For loads for which a maximum value of γ_i is appropriate,

1.3.2.1 $$\eta_i = \eta_D \eta_R \eta_I \geq 0.95 \tag{3.4-2}$$

and for loads for which a minimum value of γ_i is appropriate,

1.3.2.1 $$\eta_i = \frac{1}{\eta_D \eta_R \eta_I} \leq 1.0 \tag{3.4-3}$$

The values of η_D, η_R, and η_I are given in the AASHTO specifications, as shown in Tables 3.4-1 to 3.4-3.

Table 3.4-1
Load modifier relating to ductility, η_D

For strength limit state	
≥1.05	For nonductile components and connections
1.00	For conventional designs and details complying with these specifications
≥0.95	For components and connections for which additional ductility-enhancing measures have been specified beyond those required by specifications
For all other limit states: 1.00	

Table 3.4-2
Load modifier relating to redundancy, η_R

For strength limit state	
≥1.05	For nonredundant members
1.00	For conventional levels of redundancy, foundation elements where φ already accounts for redundancy as specified
≥0.95	For exceptional levels of redundancy beyond girder continuity and a torsionally closed cross section
For all other limit states: 1.00	

Table 3.4-3
Load modifier relating to operational classification, η_I

For strength limit state	
≥1.05	For critical or essential bridges
1.00	For typical bridges
≥0.95	For relatively less important bridges
For all other limit states: 1.00	

A total of 13 limit states that are included in the AASHTO design specifications are required to be checked in typical bridge design to control the risk of failures with respect to various aspects and requirements as discussed in Chapter 2. These limit states are determined based on previous experience and observation of the behavior and failure modes of highway bridges. The prescribed load factors γ_i are said to be calibrated, meaning that they have been determined considering statistical data of the quantities involved and the risk to the bridge components, instead of based on experience only. These limit states are discussed in more detail below.

The AASHTO-specified strength limit states refer to the ultimate limit states related to strength and stability during the expected design life. The load factors γ_i for the five strength limit states are given in Table 3.4-4. These limit states are discussed below.

3.4.2 Strength Limit States and Load Factors

For permanent load effects, Table 3.4-4 gives the load factor γ_p without a value because a range of the load factor can be used according to the AASHTO specifications. Table 3.4-5 provides theses ranges from the specifications. These ranges are offered for selection based on whether the load increases or decreases stability of the member being designed. The correct choice should lead to the critical loading for design. For example, dead load may act as a resistance in a failure mode for stability. In such a load combination the dead-load factor γ_p should be taken as the minimum value in Table 3.4-5.

In addition, Table 3.4-4 also gives two values for the load effect TU or uniform temperature load: 0.5 or 1.2. The larger of the two values shall be used for deformations and the smaller for all other effects. For simplified analysis

Table 3.4-4
Load factors γ_i for strength limit states

Strength Limit State	DC DD DW EH EV ES EL PS CR SH	LL IM CE BR PL LS	WA	WS	WL	FR	TU	TG	SE	EQ	BL	IC	CT	CV
I	γ_p	1.75	1	—	—	1	0.5/1.2	γ_{TG}	γ_{SE}	—	—	—	—	—
II	γ_p	1.35	1	—	—	1	0.5/1.2	γ_{TG}	γ_{SE}	—	—	—	—	—
III	γ_p	—	1	1.4	—	1	0.5/1.2	γ_{TG}	γ_{SE}	—	—	—	—	—
IV	γ_p	—	1	—	—	1	0.5/1.2	—	—	—	—	—	—	—
V	γ_p	1.35	1	0.4	1	1	0.5/1.2	γ_{TG}	γ_{SE}	—	—	—	—	—

Table 3.4-5
Ranges of γ_p permanent load effects

Load Type, Foundation Type	Load Factor	
	Maximum	Minimum
DC: Component and attachments	1.25	0.90
DC: Strength IV only	1.50	0.90
DW: Wearing surfaces and utilities	1.50	0.65
EH: Horizontal earth pressure		
☐ Active	1.50	0.90
☐ At rest	1.35	0.90
EL: Locked-in construction stresses	1.00	1.00
EV: Vertical earth pressure		
☐ Overall stability	1.00	N/A
☐ Retaining walls and abutments	1.35	1.00
☐ Rigid buried structure	1.30	0.90
☐ Rigid frames	1.35	0.90
☐ Flexible buried structures other than metal box culverts	1.95	0.90
☐ Flexible metal box culverts	1.50	0.90
ES: Earth surcharge	1.50	0.75

of concrete substructures in the strength limit state, a value of 0.50 for γ_{TU} may be used when calculating force effects but shall be taken in conjunction with the gross moment of inertia in the columns or piers. When a refined analysis is performed for concrete substructures in the strength limit state, a value of 1.0 for the load factor for TU shall be used in conjunction with a partially cracked moment of inertia determined by analysis.

Furthermore, the load factor for temperature gradient γ_{TG} is given in the AASHTO specifications to be selected on a project-specific basis. The reliability of available information can be covered by selecting an appropriate value. If required information for selection does not exist, the following values may be used: 0.0 at the strength and extreme-event limit states, 1.0 at the service limit state when live load is not considered, and 0.50 at the service limit state when live load is considered.

The load factor for settlement SE, γ_{SE}, should also be selected on a project-specific basis. In lieu of such project-specific information to the contrary, it may be taken as 1.0. Load combinations which include settlement shall also be applied without settlement.

Strength I This is the basic load combination for ordinary use of the bridge under vehicular load without wind for the expected life span of 75 years. Therefore, the live load factor 1.75 is meant to cover the maximum load effect over the 75-year period. In addition, the magnitude of load factor usually reflects the level of uncertainty – the more uncertain the model and the involved quantities, the higher the load factor needed. This live-load factor of 1.75, being the largest among all the load factors, is required to cover the highest level of uncertainty associated with these quantities. Determination of this load factor value is a result of calibration based on structural reliability as discussed in Chapter 2.

Strength II This load combination refers to the use of the bridge by the owner-specified special vehicles, such as the owner's special design vehicles, special evaluation vehicles, and permit vehicles, also without wind load combined. Two examples of such jurisdiction-dependent vehicles are shown in Figure 3.3-7. Note also that there may be other vehicles on the bridge when an owner-specific vehicle (design, evaluation, or permit vehicle) is placed on the bridge for design. Therefore, an additional vehicle or additional vehicles may be needed alongside the special vehicle in design. These jurisdiction-dependent vehicles are relatively less frequently observed on the road, although they are usually more severe, inducing higher load effects than the standard HL93 load. The live-load factor for these loads is reduced to 1.35 from 1.75 for the HL93 load, justifiable when permit loads are better known or controlled and of small quantity.

Strength III This load combination is different from the previous two that have a focus on vehicular loads without wind. It considers the bridge use subjected to an extreme wind load with a velocity exceeding 55 mph. This wind speed is considered to be high enough so that simultaneous occurrence with a full live load such as HL93 on the bridge is unlikely. Accordingly, the live-load factor is set at zero and the load factor for wind load WS is prescribed at 1.40.

Strength IV This load combination is for relatively longer bridge spans where the dead-load effect is much higher than the vehicular load effect. This load combination is required because the calibration process discussed in Chapter 2 and performed for the AASHTO design specifications did not cover the span range with higher dead-to-live load ratios. It is thus required in the specifications to have this additional check. Accordingly, the live-load factor is set at zero to ignore live load.

Strength V This load combination covers the bridge's use subjected to normal vehicular load with wind of 55 mph velocity. With such a wind speed, not all full truck loads are expected to be on the span, and thus the live-load factor is set at 1.35, decreased from 1.75 in the Strength I limit state. The wind load factor is prescribed at 0.4, in contrasting with 1.4 for the Strength III limit state, where the wind speed exceeds 55 mph.

The extreme-event limit state shall be taken to ensure the structural survival of a bridge during a major earthquake or flood or when collided by a vessel, vehicle, or ice flow. Table 3.4-6 displays the load factors γ_i for the

3.4.3 Extreme-Event Limit States and Load Factors

Table 3.4-6

Load factors for extreme-event limit states

Extreme-Event Limit State	DC DD DW EH EV ES EL PS CR SH	LL IM CE BR PL LS	WA	WS	WL	FR	TU	TG	SE	EQ[a]	BL	IC[a]	CT[a]	CV[a]
I	γ_p	γ_{EQ}	1	—	—	1	—	—	—	1	—	—	—	—
II	γ_p	0.5	1	—	—	1	—	—	—	—	1	1	1	1

[a]Use one at a time.

extreme-event limit states in the AASHTO specifications. They are for loads that rarely occur and are mainly due to natural disasters or man-made accidents, including those due to earthquakes (EQ), ice (IC), truck collision (CT), and vessel collision (CV). BL for blast load is a new addition in the 2012 edition of the specification. These load cases are thought to have return periods in excess of the design life of the bridge. Compared with those loads covered in the strength limit states above, the frequencies of these loads are much lower. For many structures, some of these loads may never occur during the lifetime of the structure, such as earthquake and vessel collision load.

In Table 3.4-6, the load factor for dead load γ_p has been defined earlier in Table 3.4-5; γ_{EQ} is the live-load factor for live loads simultaneously present with earthquake loads. Past editions of the AASHTO Standard Specifications (2002) used $\gamma_{EQ} = 0.0$. This issue of what may be an appropriate γ_{EQ} value has not been resolved. The possibility of partial live load, that is, $0 < \gamma_{EQ} < 1.0$, should be considered.

Extreme Event I This limit state is designed to cover seismic loads. It is thus expected that for bridges located in seismically active regions this load combination may control the design of certain seismic-load-sensitive components, such as the abutments, piers, bearings, and connections. Note also that the seismic forces are factored at 1.0 without amplification, reflecting the low frequency of occurrence compared with those loads combined in the strength limit states. Also, it is interesting to note that the load factor γ_{EQ} in the specifications is meant to be applied to the transient live (vehicle) loads, including LL, IM, CE, BR, PL, and LS.

Extreme Event II This limit state is another load combination for extreme events of load during use of the bridge, including, applied one at a time, blast (BL), ice load (IC), collision by vessels (CV) and trucks (CT), and check floods and certain hydraulic events, along with live loads factored at 0.5. This reduced vehicular load factor accounts for the low likelihood that both a credible highest extreme load and the full truck load occur simultaneously.

The strength limit states consider the bridge use under the worst loading situation over the planned life span of 75 years. The service limit state combinations focus on daily or routine loads to prevent failures that may adversely affect service and/or reduce bridge life. Such failures are related to stress, deformation, and cracking under regular operating conditions. Note also that the risk for these failure modes to occur is allowed to be higher than those global and/or ultimate failure modes causing collapse and/or interruption to service. Table 3.4-7 shows the load factors in the AASHTO specifications for these service limit states. Note, however, that the service failure modes included in the current AASHTO specifications have not been explicitly calibrated as done for the strength failure modes discussed in Chapter 2.

3.4.4 Service Limit States and Load Factors

Service I Service I load combination is to be used to check stress, deformation, cracking occurrence, and/or cracking width in normal operation. Thus the loads included are taken at their nominal values along with a 55-mph wind load. The wind load factor is 0.30, to be consistent with other combinations involving wind load. This load combination

Table 3.4-7
Load factors for service limit states

Service Limit State	DC DD DW EH EV ES EL PS CR SH	LL IM CE BR PL LS	WA	WS	WL	FR	TU	TG	SE	EQ	BL	IC	CT	CV
I	1	1	1	0.3	1	1	1/1.2	γ_{TG}	γ_{SE}	—	—	—	—	—
II	1	1.3	1	—	—	1	1/1.2	—	—	—	—	—	—	—
III	1	0.8	1	—	—	1	1/1.2	γ_{TG}	γ_{SE}	—	—	—	—	—
IV	1	—	1	0.7	—	1	1/1.2	—	1	—	—	—	—	—

is also intended to be used to control deflection in buried metal structures, tunnel liner plates, and thermoplastic pipes and crack width in reinforced concrete structures. It is for transverse analysis relating to tension in concrete segmental girders as well. This load combination is also used for the investigation of slope stability.

Service II The Service II load combination corresponds to the overload provisions in the previous AASHTO specifications and hence the higher-than-1.0 live load factor. It focuses on two particularly important types of failure mode: (1) yielding of steel structures and (2) slip of slip-critical connections under a load level approximately halfway between Service I and Strength I limit states. The occurrence and repeated occurrence of these failures may cause a serious consequence. Therefore, control of these failure modes is exercised here and at the level of load given.

Service III This limit state is to cover the risk of tension-caused cracking in prestressed concrete superstructures. Prestressing in these members is critical to their load-carrying capacity. Tensile cracking may detrimentally reduce this capacity. Therefore, controlling such cracking is considered important to be exercised in design according to the AASHTO specifications.

Specially permitted vehicles by jurisdiction are covered in this limit state, with a live load factor of 0.8, reduced from 1.0 in the Service I limit state, to account for their relatively lower frequencies of appearing on a bridge span.

Service IV This limit state has a unique focus on the tension in prestressed concrete columns also for cracking control. The purpose is identical to that of Service III, but for substructure components. The 0.70 factor on wind represents an 84-mph wind speed for zero tension in prestressed concrete columns for 10-year mean reoccurrence winds. The prestressed concrete columns must still meet strength requirements as set forth in load combination Strength III.

3.4.5 Fatigue Limit States

Fatigue is a failure mode causing fatigue cracking as a result. When such a crack occurs at a location in a bridge, the consequence may or may not be serious, ranging from bridge system collapse to minor stress redistribution without noticeable change in behavior or safety. The fatigue limit states are to prevent fatigue cracking. However, no quantitative analysis is required in the AASHTO specifications as to how such cracking may affect the system behavior of the bridge structure.

There are now two fatigue limit states in the AASHTO specifications, one (Fatigue I) for infinite load-induced fatigue life and the other (Fatigue II) for finite load-induced fatigue life design. Table 3.4-8 shows the corresponding load factors 1.50 and 0.75 for live load, respectively, for only truck live load LL, impact or dynamic factor IM, and centrifugal force CE. Only

Table 3.4-8
Load factors for fatigue limit states

Fatigue Limit State	DC DD DW EH EV ES EL PS CR SH	LL IM CE BR PL LS	WA	WS	WL	FR	TU	TG	SE	EQ	BL	IC	CT	CV
I–LL, IM and CE only	—	1.5	—	—	—	—	—	—	—	—	—	—	—	—
II–LL, IM and CE only	—	0.75	—	—	—	—	—	—	—	—	—	—	—	—

one lane of truck loading is checked in these two limit states, ignoring the load effect from other lanes. As a result, the multiple presence factor in Table 3.3-1 does not apply. This is partially because the life maximum is not of particular concern but routine load is.

Fatigue cracking has been categorized as load induced and non–load (or distortion) induced. Non-load-induced fatigue cracking is considered to be caused by excessive out-of-plane deformation or distortion, as illustrated in Figure 3.4-1 for an example of a steel beam. Steel beam design

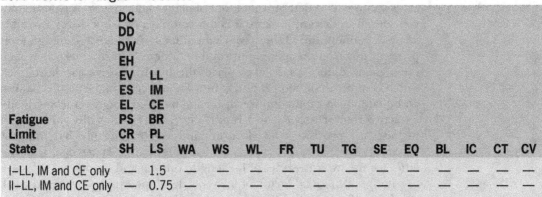

(a)

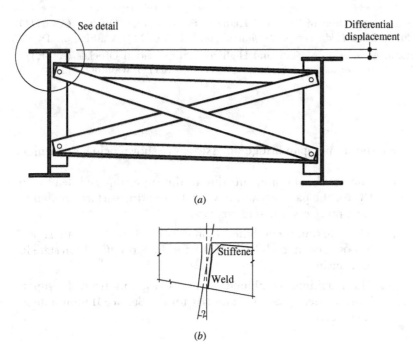

(b)

Figure 3.4-1
An example of distortion (out-of-beam-plane deformation) of a steel beam: (a) differential displacement between beams; (b) out-of-plane deformation in beam.

for this case usually only considers stress perpendicular to the cross section as well as deformation in the plane of the vertical symmetry axis of the cross section. The demonstrated out-of-plane deformation is typically not considered in design, because it is too complicated to perform a reliable analysis. Namely, the higher requirement for knowledge and design tools prevents this from happening in routine design. Nevertheless, when such out-of-plane deformation takes place, the involved stress can be higher, which is never designed for and causes fatigue cracking. The mechanism and quantitative prediction for distortion-induced fatigue cracking is still an active research topic. As a result, the AASHTO specifications do not include any provisions for distortion-induced fatigue life but only for load-induced fatigue life. The specified fatigue limit states deal with the latter for the two categories of life length: infinite and finite lives.

The finite life load factor of 0.75 was developed using measured truck weights available at the time and an empirical relation of variable stress range due to these truck loads and their equivalent constant stress range. Then the infinite life load factor is derived to be 1.50 (doubling 0.75) based on an assumption that the maximum stress range in the random variable spectrum is twice the effective stress range caused by Fatigue II load combination.

References

American Association of State and Highway Transportation Officials (AASHTO) (2012), *LRFD Bridge Design Specifications*, 6th ed., AASHTO, Washington, DC.

American Association of State and Highway Transportation Officials (AASHTO) (2011), *Manual for Bridge Evaluation*, 2nd ed., AASHTO, Washington, DC.

Problems

3.1   For the simple span bridge superstructure shown below, determine:

(a) The maximum moments due to the superimposed dead load DC for the parapets and DW for the wearing surface applied to each prestress concrete box beam.

(b) The maximum combined moments resulting from the super-imposed dead loads DC and DW for the Strength I limit state in each beam.

(c) The maximum combined moments resulting from the super-imposed dead loads DC and DW for the Service II limit state in each beam.

Assume that the deck consists of a 9-in.-think structural Portland cement concrete including a 0.5-in. sacrificial surface and a 2.5-in.-thick bituminous wearing surface. Use the following data for material weight: $W_{box\_beam} = 0.784\,\text{k/ft}$, $W_{concrete} = 0.145\,\text{k/ft}^3$, $W_{asphalt} = 0.14\,\text{k/ft}^3$. The parapet weight may be evenly distributed among the five beams. Note also that you are not required to include live load.

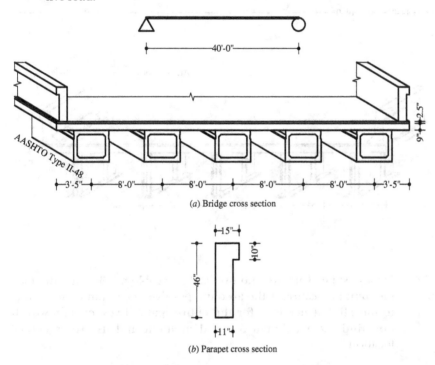

(a) Bridge cross section

(b) Parapet cross section

3.2   For the bridge superstructure in Problem 3.1, determine:

(a) The maximum unfactored shear V_{LL} resulting from one lane of HL93 live load LL.

(b) The maximum unfactored moment M_{LL} resulting from one lane of HL93 live load LL.

3.3   For the continuous steel plate girder bridge superstructure shown below and the given locations in the girder, determine the maximum unfactored moment M_{LL} at location A due to one lane of HL93 truck live load. Find the maximum unfactored negative moment M_{LL} at the middle support between the two spans under the HL93 design truck load (whose axle distance between the two heavier axles is variable). Use reasonable assumptions if needed.

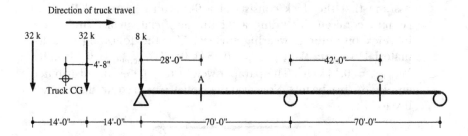

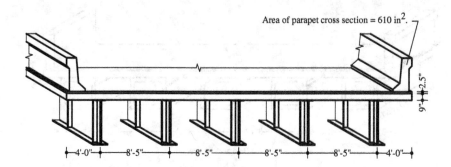

3.4   For the superstructure and live-load in Problem 3.3, is the live-load moment at location A the absolute positive maximum moment due to the HL93 truck load for the entire span? If yes, explain why. If not, find the maximum live-load moment and its cross-sectional location.

3.5   For the simply supported bridge span of 110 ft shown below, find:

(a) The dead-load moments DC and DW to a typical composite prestressed concrete box beam in the cross section.

(b) The maximum combined moments resulting from the superimposed dead loads DC and DW for the Strength I limit state.

(c) The maximum combined moments resulting from the superimposed dead loads DC and DW for the Service II limit state.

The parapet weight may be evenly distributed among the nine beams. Assume that the deck consists of 3-in.-thick asphalt concrete wearing surface and 5-in. Portland cement concrete deck on top of the AASHTO Type I-48 box beams. Also use $W_{\text{box beam}} = 0.721\,\text{k/ft}$, $W_{\text{concrete}} = 0.145\,\text{k/ft}^3$, and $W_{\text{grout}} = 0.125\,\text{k/ft}^3$. The live load is not addressed in the exercise.

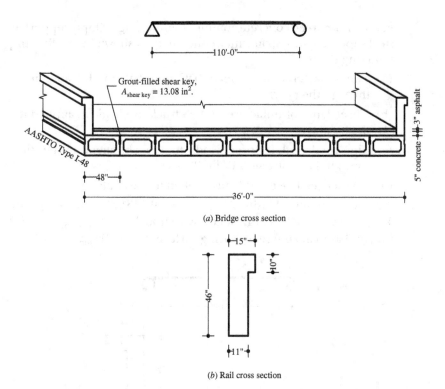

(a) Bridge cross section

(b) Rail cross section

3.6   For the simply supported bridge superstructure shown below, determine:

(a) The unfactored braking force BR for each lane.

(b) Which lane(s) the obtained BR should be applied to.

(c) The maximum reaction forces on the bearings supporting the
beams resulting from the braking force BR.

Assume that the roadway consists of three 12-ft lanes. Lanes 1 and 3
are for vehicles traffic in opposite directions, and lane 2 is used as a
turning lane.

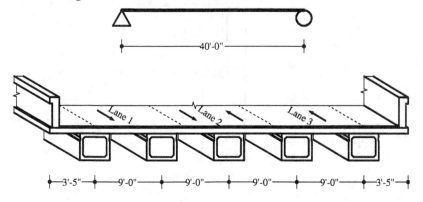

3.7 For the reinforced concrete abutment supporting a simply supported
     steel superstructure span with a concrete deck shown below, find the
     following items.

   (a) The unfactored dead loads DC and DW to the abutment
       through the bearings.

   (b) Three lanes of unfactored HL93 truck live load LL applied to
       the abutment.

   (c) The combined factored loads to the abutment under the
       strength I and strength IV limit states.

   (d) The service I factored loads applied to the abutment.

Assume concrete density $W_{concrete} = 0.145\,k/ft^3$, steel density
$W_{steel} = 0.490\,k/ft^3$, steel girder cross-sectional area $A_{steel\_girder} = 74.2$
in.^{2}, and asphalt concrete wearing surface density $W_{asphalt} = 0.14\,k/ft^3$.

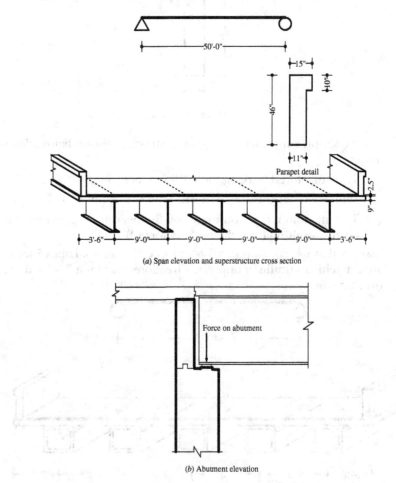

(a) Span elevation and superstructure cross section

(b) Abutment elevation

3.8 For fatigue design of shear studs in the steel girder bridge span shown below, determine the fatigue live-load shear range V_f at the following locations from the left support: (1) $x = 0$, (2) $x = 5$, and (3) $x = 15$ ft. Shear range is the positive maximum shear minus the negative maximum shear.

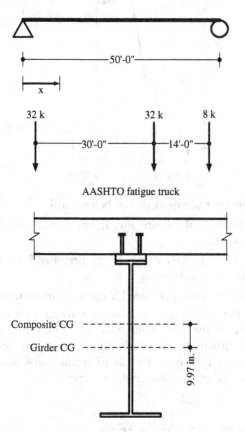

AASHTO fatigue truck

3.9 For the bridge abutment shown below, find:

(a) The active horizontal soil pressure EH applied along the depth of the abutment.

(b) The active horizontal earth surcharge pressure ES applied along the depth of the abutment.

(c) The active vertical soil pressure EV applied to the abutment footing.

Assume that the soil density remains approximately constant through the depth of the abutment $\gamma_{soil} = 0.090\,\text{k/ft}^3$, the fill material consists of a well-graded gravel–sand mix, and preliminary soil analysis reveals a soil friction angle $\varphi_f' = 30°$.

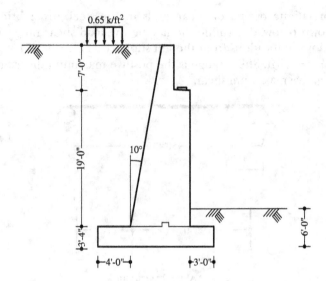

3.10 For the bridge abutment shown below, find

(a) The horizontal soil pressure EH applied along the depth of the abutment.

(b) The horizontal earth surcharge pressure ES applied along the depth of the abutment.

(c) The vertical soil pressure EV on the abutment footing.

Assume that soil density remains approximately constant through the depth of the abutment and the abutment does not deflect or move sufficiently to reach an active condition. Use soil density $\gamma_{soil} = 0.11 \, \text{k/ft}^3$. The fill material consists of a well-graded gravel–sand mix. The soil friction angle $\varphi_f' = 30°$.

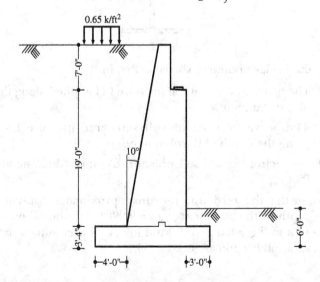

3.11 Find the total maximum moment due to two lanes of the HL93 load in a simply supported beam modeling a bridge span with span length of 65 ft.

3.12 Find the maximum shear force due to three lanes of the HL93 load in a simply supported beam modeling a bridge span with span length of 65 ft.

3.13 Find the maximum positive moment at a cross section at a distance equal to 40% of the span length from the first support due to one lane of the HL93 load in a two-span continuous beam modeling a continuous-span bridge with each span length equal to 175 ft. Assume that the two spans have the same constant stiffness along the length. The bridge span arrangement is the same as sketched in Problem 3.3, except the span length.

3.14 Find the maximum negative moment due to one lane of the HL93 load in a two-span continuous beam modeling the same bridge as Problem 3.13, with span length of 175 ft. Assume that the two spans have the same constant stiffness along the length. The bridge span arrangement is the same as in Problem 3.3, except the span length.

3.15 For a two-span continuous beam modeling the same bridge in Problem 3.13, with span length of 175 ft, determine the maximum shear force at the left support due to three lanes of the HL93 load. Assume that the two spans have the same constant stiffness along the length. The bridge's span arrangement is the same as in Problem 3.3, except the span length.

3.16 For a two-span continuous beam modeling the same bridge as Problem 3.13, with span length of 175 ft, determine the maximum shear force at the left section of the middle support due to three lanes of the HL93 load. Assume that the two spans have the same constant stiffness along the length. The bridge's span arrangement is the same as in Problem 3.3, except the span length.

4 Superstructure Design

4.1 Introduction

Typical field construction of a highway bridge progresses through excavation, foundation building, substructure construction, and then superstructure erection and finishing. Many components may be prefabricated elsewhere, such as prestressed concrete members and steel arches. Typical bridge design starts from the top of the structure. It is more convenient or sure if the components on the top are determined (designed) prior to those supporting them, because the loads on supporting members are more certain to estimate. This sequence also follows the load transfer path from the top component to the ground. Therefore, for short- and medium-span highway bridges focused on in this book, the design process often follows the sequence of (1) the deck, (2) the deck-supporting system (beams, trusses, arches, etc.), (3) bearings if any, (4) piers and/or abutments, and (5) foundations. Chapters 4 through 6 follow this sequence in presentation and discussion. Chapter 4 deals with the design of the bridge superstructure above the bearings. Chapter 5 presents bearing design. Chapter 6 discusses the design of the bridge superstructure below the bearings. Note that some bridge spans and systems may not have bearings. Then special arrangements need to be made to allow the structure system to respond to required movements, often by allowing additional stress and strain that need to be designed for.

This general design sequence from the top (supported) members to the bottom (supporting) members can avoid otherwise needed iterations because a supporting member can be designed with the supported members already designed and their weights are known as loads to the supporting member as well as what forces applied to the supported members. Nevertheless, iterations may still be required in the design of a component, while more experienced engineers can reduce the number of iterations by selecting the preliminary member size without detailed calculation, which does meet the proportioning requirements.

It should also be noted that in some bridges certain components appearing at a higher location may not be the ones supported by those below them. For example, cable structures such as suspension bridges and cable-stayed bridges have cables as the supporting members for the deck system and they are higher in elevation than the deck being supported by the cables.

Highway bridge superstructures need to be designed to function safely in carrying possible loads to be applied on them. Some of these loads are illustrated in Figure 4.1-1, with the arrows indicating directions of the loads. They all have a main structural system to accomplish that. It is important to understand the particular system before designing it. To that end, bridges are often classified according to that main load-carrying system or superstructure system. For example, beam bridges refer to a system of parallel (or almost parallel) beams to transfer the vehicular loads to the foundation. Truss bridges use a system of two or more trusses to accomplish that, arch bridges use a system of two or more arches, and so on. Depending on the requirement for span length due to geometric and economic considerations, different systems are selected for particular span ranges to be economical.

This chapter is organized as follows. Section 4.2 briefly discusses a variety of main superstructure systems. Section 4.3 presents a discussion on typical components in these systems. Section 4.4 focuses on deck systems

Figure 4.1-1
Typical loads on highway bridge superstructure.

in superstructure and Section 4.5 on deck-supporting systems. Section 4.6 presents details of reinforced concrete deck slab design as the overwhelmingly popular deck system for highway bridges in the United States. Design examples with calculation details are included in Section 4.6. Sections 4.7 and 4.8 include details of beam design respectively for steel and prestressed concrete beams, being the most popular beam types in the United States. Design examples with detailed calculations are included in these sections according to the AASHTO specifications. Bearing and substructure design are covered in Chapters 5 and 6, respectively.

4.2 Highway Bridge Superstructure Systems

Several commonly used highway bridge superstructure systems are introduced next to provide general background information. More detailed component design concepts and procedures are presented later.

Beam bridges are overwhelmingly the majority of highway bridges in the United States and the world. Thus beam bridge systems receive intensive attention in this book. A typical beam bridge superstructure under construction is shown in Figure 4.2-1. As seen, the primary members of a beam bridge superstructure are beams to support a deck or deck slab, which is yet to be constructed in Figure 4.2-1. The deck in turn provides a driving surface for vehicle traffic and often also pedestrian traffic if needed. The beams are braced using diaphragms or cross bracings to make the beams form a frame system as shown, so that they work together as an integrated system, not as

4.2.1 Beam Bridges

Skewed diaphragm

Plate girder

Diaphragm

Figure 4.2-1
Beam bridge superstructure frame.

individual beams. The beam ends are then supported on piers and/or abutments (or bearings on piers and/or abutments). The piers or abutments are then supported by the foundation consisting of footings and/or piles. Thus vehicle loads carried by the bridge transfer from the deck to the beams, bearings, piers/abutments, and then to the ground. For short and medium span lengths, beam bridges are economical since their self weight is not significant, compared with the vehicle load.

4.2.2 Truss Bridges

When the span length needs to be longer over a wide waterway (e.g., a significant creek or river), roadway (e.g., between two hills), and so on, other span types may become more economical, such as trusses and arches. Typically, two trusses or arches tied together are used to form a stable system. In such a span, the two trusses or two arches can be viewed as two very long and deep beams with significant reduction of self-weight. Namely, many parts of the "long beam's web" are taken away, leaving only a few vertical and diagonal members to carry the shear. The top and bottom chords or members of the trusses or arches have the role of resisting the moment for this "long beam." A simple truss bridge system is shown in Figure 4.2-2 to illustrate the concept. Note also that truss systems may compete with beam systems in cost-effectiveness within a certain span length range.

4.2.3 Arch Bridges

Arch bridge systems have a very pleasing appearance of curved upper chords, also making them compression members. Curved members are preferred in the AASHTO specifications when there are significant aesthetic requirements. The compressive stress condition of the upper chord allows use of several materials particularly strong in compression, such as masonry and Portland cement concrete, which are typically less expensive. Nevertheless, steel arches are also popular.

Figure 4.2-3 demonstrates an arch bridge system of steel boxes. Steel and reinforced concrete arches are commonly seen, and composite arches using

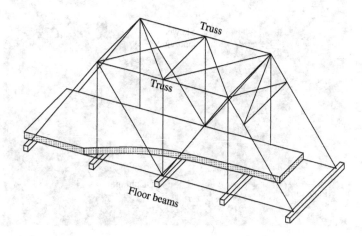

Figure 4.2-2
Truss bridge system.

Figure 4.2-3
Arch bridge system.

concrete-filled steel tubes as primary arch members have also been used in many parts of the world.

Truss and/or arch systems can only economically span a range of length. When longer spans are required a more revolutionary configuration is needed, such as a cable-stayed and suspension span. These systems take advantage of cables being extremely strong in tension and subject them to tension all the time. This approach eliminates the need for preventing buckling due to compression in primary members. Therefore the cost increase rate with span length is reduced, and thus it is more cost effective with respect to the span requirement. Figure 4.2-4 exhibits a typical cable-stayed bridge system. Note that there is also a deck system in cable-stayed bridges that provides a driving surface for vehicle traffic on the roadway. Bridges with a long span such as cable-stayed bridges often are required to carry significant traffic, and thus the deck system needs to be wider to accommodate the vehicle traffic. Therefore the deck system has to possess much higher stiffness and relatively lighter weight with respect to the spanned space. As a result, the so-called orthotropic deck is often selected in long-span bridges due its relative lighter weight and higher stiffness, which is to be discussed further in Section 4.3 when more details of deck systems are presented.

4.2.4 Cable-Stayed Bridges

Figure 4.2-4
Cable-stayed bridge system.

To span further longer than cable stayed bridges, suspension bridge systems are the most capable ones at this time. As seen in Figure 4.2-5, suspension bridge systems consist of main cables that are typically made of several

4.2.5 Suspension Bridge Systems

Figure 4.2-5
Suspension bridge system.

thousand or more high-strength steel wires. This system can provide support to longer decks than that in cable-stayed bridges. In contrast, the cables in cable-stayed bridges have less and smaller wires.

4.3 Primary Components of Highway Bridge Superstructure

In the superstructure of highway bridges, there are mainly two essential structural systems: the deck system and the deck-supporting system. The deck system provides the driving surface for vehicle traffic and sometimes also pedestrian traffic. The deck-supporting system transfers the vehicle load and other loads from the deck system to the substructure and then to the ground. Therefore, the deck-supporting system is also often referred to as the primary superstructure system.

Figure 4.3-1 is a photograph taken during construction of a simply supported bridge span with spread prestressed concrete box beams to support a reinforced concrete deck yet to be placed. A number of the structural components indicated in Figure 4.3-1 are typical in beam bridges. For example, the reinforcements on the top of the beams and anchored in the beams will work as shear connectors between the deck and the beam. Note also that epoxy coating of steel reinforcement has become very popular in many areas in the United States. Figure 4.3-2 shows epoxy-coated steel reinforcement (sometimes referred to as rebars) in a bridge construction site, including straight bars of different lengths and hoops of different shapes. They are used in concrete components cast at the site, such as footings, pile caps, abutments, piers, decks, and traffic barriers.

Many deck-supporting systems are available that are cost effective for fabrication. Therefore their total cost, including fabrication, transportation to the site, and erection, is largely dependent on the transportation cost to the

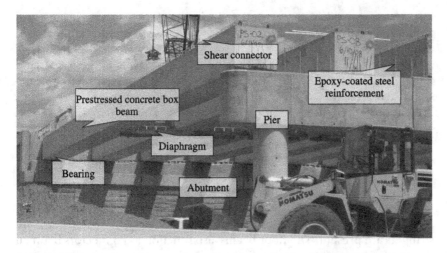

Figure 4.3-1
Superstructure beams to support concrete deck yet to be constructed.

Figure 4.3-2
Epoxy-coated steel reinforcement at bridge construction site.

site. Namely, the fabrication costs of these different systems are competitive but the transportation costs of the components to the bridge site may vary depending on the distance from the fabrication facility to the site. Examples of beam bridge deck-supporting systems are steel beams, prestressed concrete I beams, prestressed concrete box beams adjacent to each other (as opposed to spread from each other in Figure 4.3-1), timber beams, and aluminum beams. Some of the popular options will be discussed later in this chapter in detail.

For truss bridges, the deck-supporting system consists of the trusses and the floor beams connecting the trusses, possibly along with floor-beam-supported stringers (i.e., small longitudinal beams). The floor beams are supported by two or more main trusses. This system is illustrated in Figure 4.3-3 for a case with stringers to reduce the deck slab span. Sometimes they are not used when the floor beams are spaced closer to each other and the deck slab can economically span the floor beam spacing. The deck slab main span is different depending on whether there are stringers and needs to be designed accordingly. For example, if a reinforced concrete slab is used as the deck, the main reinforcement will need to be across the stringers if present or across the floor beams if no stringers are present. The floor beams also function as connectors between the two main trusses to form a stable system of two (or more) trusses, as seen in Figure 4.3-3.

The stringers in the longitudinal direction, if used, are usually significantly smaller than the main beams or girders shown in Figure 4.3-1, also with smaller spacings between them. They are needed to reduce the span of the deck (in the transverse direction perpendicular to the traffic direction). When the floor beams are spaced far away from each other for larger trusses, the deck would have to span between floors beams, which would make the deck too thick with high dead load and higher cost. The stringers can reduce the span requirement for the deck and thus the cost requirement.

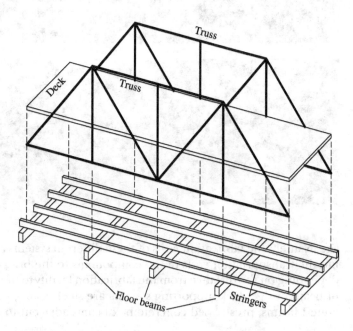

Figure 4.3-3
Main structural components in truss bridge.

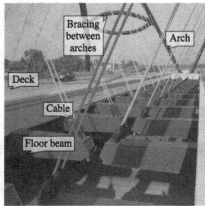

Figure 4.3-4
Main structural components in arch bridges: (top) elevation, (bottom) details.

Arch bridge superstructure systems function similarly with truss bridges, except that the arches take the place of trusses. An example is shown in Figure 4.3-4, with the main components marked. Arch systems are often more esthetically pleasing but with a higher requirement for fabrication technology and quality control. The main arches need a connection system between them, which also functions as the direct support for the deck

Figure 4.3-5
Main structural components in cable-stayed bridges.

Figure 4.3-6
Main structural components in suspension bridges.

system. Therefore this subsystem is often viewed as part of the deck system carried by the arches.

To expand span length further, the cable-stayed bridge deck-supporting system is the cable-tower system. The suspension bridge deck-supporting system consists of the main cables and thereon supported suspensions. These two systems are respectively shown in Figures 4.3-5 and 4.3-6. Their deck systems are usually required to be stiffer than those used on beam, truss, and arch bridges. While their deck system span lengths are longer or much longer than those for short- and medium-span bridges (from 20 ft to approximately 250 ft), their design is considered to be out of the scope of this introductory book to highway bridge design and evaluation.

The two structural subsystems in the superstructure, the deck system and the deck-supporting system, are further discussed separately next.

4.4 Deck Systems

A number of deck systems are used in highway bridges. They are briefly discussed next to set the stage for detailed design steps and techniques for some commonly used components in later sections of this chapter. It is noted that understanding the construction process should be part of the design process for highway bridges and their subsystems and components. Then design can be approached with adequate consideration to the construction process, especially when constructability is of concern.

The deck supports or directly provides a driving surface to vehicle traffic on a highway bridge. The surface is required to be aligned and consistent with the roadway surface connecting to the bridge for a smooth and comfortable ride. The reinforced concrete deck slab is certainly the most popular choice for highway bridges in the United States, particularly on bridges on the interstate highway system carrying significant vehicle traffic. The slab is also often referred to as a reinforced concrete deck or concrete deck, while the word *deck* sometimes includes part of the deck system or deck-supporting system. For low-traffic-volume and/or low-speed-limit roads, a timber deck may also be used for lower cost. However, the timber deck is not considered composite to the deck-supporting system in this case. The AASHTO specifications prefer composite deck systems. Continuous deck systems that minimize joints are also favored. Such systems can reduce the requirement for maintenance cost because the risk of leakage accelerating substructure deterioration is reduced when joints are eliminated.

4.4.1 Reinforced Concrete Slab System

The most popular deck system in modern highway bridge construction in the world is the cast-in-place reinforced concrete slab composite or noncomposite to the supporting members. Design examples for this type of deck slab according to AAHSTO specifications are included in this chapter. This group also includes the system of precast concrete panel systems with an additional cast-in-place layer for a monolithic deck structure. This system has been used in the U.S. interstate highway system as well as relatively less intensively traveled roads such as county and local roads. Figure 4.4-1 displays a typical cross section of reinforced concrete deck, and Figure 4.4-2 shows a typical procedure of concrete placement and curing for cast-in-place reinforced concrete slab decks.

The cast-in-place reinforced concrete deck system has the following advantages:

❏ Relatively easy to construct for any size (dimensions) and shape

❏ Cost effective for a practical range of beam spacings

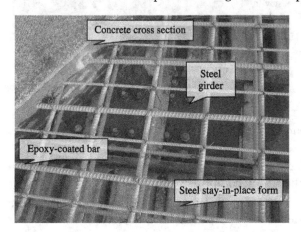

Figure 4.4-1
Cross section of reinforced concrete deck on steel girder.

(a)

Figure 4.4-2
Concrete deck
construction procedure:
(a) cleaning using
vacuum after deck
forming (steel studs on
steel beam shown); (b)
placing chairs for
transverse reinforcement
to ensure cover depth;
(c) placing epoxy-coated
main (transverse)
reinforcement on chairs;
(d) placing epoxy-coated
longitudinal
reinforcement in bottom
and then top layers; (e)
placement of concrete
using pump and vibrator;
(f) concrete finishing
using an
across-bridge-width
paver; (g) applying curing
compound after concrete
finishing.

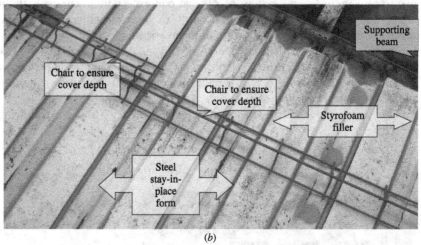

(b)

❑ Seamless as a monolithic structure able to "roof" the entire bridge
  system

❑ Able to provide a composite cross section with the supporting beams
  taking advantage the strong compressive strength of concrete

❑ Stiff in providing a framing function for the supporting beams,
  although unintended in noncomposite decks

(c)

(d)

Figure 4.4-2
(Continued).

The first two advantages above have traditionally been the main reasons for using this deck system. The monolithic nature of cast-in-place concrete deck is not obvious until it is compared with other deck systems, such as a timber deck. A "water-tight" deck made of concrete can protect other bridge components from the accelerated deterioration caused by water leakage.

(e)

Figure 4.4-2
(*Continued*).

(f)

In fact, making the concrete deck slab more water tight has been one of the important efforts in improving concrete decks. The composite action increases the load-carrying capacity in supporting beams and thus reduces cost. The system analysis of highway bridges has shown that the concrete deck also provides significant stiffness so that it can transfer load to other beams when one of the beams fails to carry more load. This function also

(g)

Figure 4.4-2
(*Continued*).

increases the redundancy of the superstructure system and its structural system reliability against failure.

The reinforced concrete deck system also has the following disadvantages compared with some other available systems:

❑ Relatively heavy, because of its solid concrete cross section

❑ Susceptible to cracking and, if cracked, vulnerable to accelerated deterioration for the deck and entire bridge system

Research on reducing concrete deck cracking potential, controlling crack width, and protecting steel reinforcement in decks has been active in recent decades. The AASHTO specifications also include specific provisions to control these factors in design.

Figure 4.4-2 illustrates a typical reinforced concrete deck construction process prior to covering the concrete for curing. For concrete deck construction, curing is also an important step for ensuring a high-quality deck with no or little cracking. Figure 4.4-3 shows a deck covered with burlap

Figure 4.4-3
Burlap-covered concrete deck for curing at site.

Figure 4.4-4
Sealed concrete cylinders cast at site for lab testing.

for curing, which is a critical step to minimize cracking for deck durability. Figure 4.4-4 displays several 6 × 12 cylinders made at the site of the concrete slab deck, to be tested for concrete quality control.

Depending on the bridge owner or jurisdiction, an asphalt concrete surface on the concrete deck may or may not be required to open the bridge to traffic. In addition, the asphalt concrete overlay remains an option to be placed after years of service and the deck surface has deteriorated and is compromising ridabiliy. The new rider surface provided by the overlay will improve rideability but will not structurally strengthen the deck.

As mentioned earlier, one of the major advantages of this full-depth and cast-in-place concrete deck system over the other deck system is that it is very easy to produce a composite deck. The composite action takes advantage of the high compressive strength of concrete compared with its low tensile strength and consequently can save material for the cross section. Figure 4.3-1 shows shear connectors between prestressed concrete beams and their concrete deck made of epoxy-coated steel reinforcement embedded in the beam. Figures 4.4-2a and c show steel studs welded to steel beams as shear connectors between the beams and their concrete deck. The composite action that results significantly increases the cross-sectional moment capacity and thus can reduce material consumption.

Corrugated steel stay-in-place forms have been widely used in constructing cast-in-place full-depth concrete decks. Figure 4.4-5 shows such forms in place with Styrofoam filling the corrugates. The corrugation increases the form's stiffness in the transverse direction spanning on the supporting beams. However, filling the entire cross section for a flat surface would require much more structural concrete than needed. To avoid such waste, Styrofoam is often used to fill the space below the highest surface of the form as seen (also the bottom surface of the concrete deck after concrete placement).

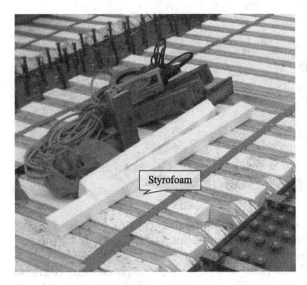

Styrofoam

Figure 4.4-5
Corrugated steel stay-in-place form filled with Styrofoam.

Timber deck systems have the following advantages:

❑ More cost effective, particularly when the site is close to the supply source

❑ Lower weight

The lighter self-weight can be critical in selecting this deck system, especially when the deck-supporting system has inadequate load-carrying capacity. This may occur when an old bridge is inadequately load rated. A concrete slab deck, for example, would be much heavier and reduces the load-carrying capacity margin for the live (truck) load. Therefore, a lighter timber deck will provide more capacity for the live load and thus mitigate or eliminate the inadequate capacity. More details of such situations are discussed in Chapter 7 on highway bridge evaluation. Figure 4.4-6 shows an old truss bridge with a timber deck, providing relatively higher live-load carrying capacity.

As mentioned above, when the floor beams supporting the timber deck system have a large spacing, small beams are used on top of the floor beams, referred to as stringers. They are placed longitudinally in the traffic direction, so that the timber pieces are placed transversely and are supported by the stringers. When the floor beams have a relatively small spacing, stringers are not needed, and the timber pieces can be placed directly on the floor beams. Thus the timber pieces are placed in the longitudinal direction parallel to the traffic. These two ways of spanning will require the timber deck to be designed differently. With stringers, the timber deck is designed for axle or wheel load since the timber pieces are not continuous in the

Figure 4.4-6
Timber deck on old truss bridge.

traffic direction. Without stringers, the timber pieces will need to be designed considering vehicle load with several axles, depending on the length of the timber pieces.

Of course, timber decks have disadvantages compared with concrete ones. For example, since timber has a lower strength than reinforced concrete per unit cross section, the support spacing for a timber deck needs to be smaller than for a concrete deck. Often this spacing becomes too small to be practical and cost effective and thus a concrete deck is required. In addition, timber has a relatively shorter life span than reinforced concrete, especially if maintenance is provided at the same time intervals. As a result of these requirements, timber decks are much more often used on local bridges carrying much lower truck traffic volumes. The total area of timber decks on U.S. highway bridges is much smaller than that of Portland cement concrete decks.

4.4.3 Metal Grid Deck System

Figure 4.4-7 shows a typical metal grid deck system used on highway bridges. It is made of a steel grid supporting a concrete top. This top is thinner than

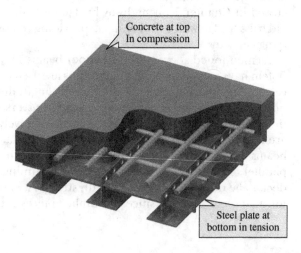

Figure 4.4-7
Metal grid deck system (courtesy of The D. S. Brown Company, North Baltimore, OH).

the typical full-depth concrete bridge decks discussed above. There are a number of variations to this system for different applications, while they all have the following advantages for certain site conditions:

❏ Lighter than the full-depth reinforced concrete deck system due to shallower concrete depth compared with full-depth concrete deck (but heavier than timber deck systems)

❏ Capable of carrying higher load than timber deck systems

With these advantages, metal grid deck systems have the niche for bridges that require the deck to be lighter but not compromise load-carrying capacity. A number of existing long-span truss and beam bridges have used such a metal deck system or a similar one to maximize the capacity for live load. This increases the bridge's load rating as discussed in Chapter 7. Certainly, it is much less popular than the full-depth concrete deck system.

Although more expensive, the so-called orthotropic steel deck system has been almost exclusively used in long-span bridges due to its relatively lighter weight and higher load-carrying capacity. Figure 4.4-8 illustrates this system.

4.4.4 Orthotropic Steel Deck System

As seen, the deck plate (driving surface) is supported by a number of small trapezoidal ribs or small "box beams" in the longitudinal direction only – hence the name of "orthotropic deck." In turn, these ribs are supported on floor beams carried by the main girders. Using these small hollow trapezoidal box beams significantly reduces the deck weight compared with a reinforced concrete deck slab. A challenge to this deck system is to ensure that the pavement material be well adhered to the steel surface plate and be able to sustain a high volume of truck traffic typically experienced in long-span bridges. Another challenge to this deck system is possible fatigue failure (cracking) at welds joining these steel plates and beams. Both types of

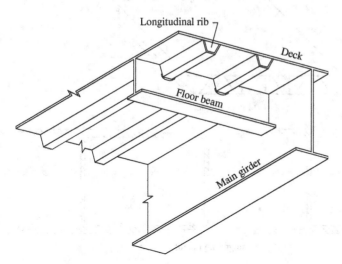

Figure 4.4-8
Steel orthotropic deck.

failure, premature pavement cracking and steel fatigue cracking, have been observed. In addition, this full-depth steel cross section of orthotropic deck is more expensive than the concrete deck while it can span much wider spacing between and along the main girders.

While the deck is made of steel, a skid-resisting surface should be adhered to the steel plate for traffic safety consideration. It is usually made of asphalt concrete with special treatment to the steel for reliable bonding and durability. However, there have been reports of premature failure of the surface on long-span bridges with high traffic volumes.

4.4.5 Fiber-Reinforced Polymer Deck System

Fiber-reinforced polymer (FRP) has been proposed to be a new material for bridge construction for some years. The complete systems of FRP material that have been tried on field highway bridges are deck systems, although there are no provisions in the AASHTO specifications for its design. This is different from other deck systems discussed above. Figure 4.4-9 exhibits a modular deck system that consists of modules in the longitudinal (traffic) direction of the bridge. As many modules as needed can be added for the required width. The modules are connected using adhesive. A surface with adequate skid resistance is also required, usually bonded to the FRP surface using adhesive. The longevity of this bond is still a subject for research.

Such an FRP system may be used on a variety of deck-supporting systems, similar to the timber deck system and steel grid deck system discussed above. For example, it can be used on steel beams and concrete beams. However, the fastening system for securing the deck system to the supporting system (e.g., a beam) may vary depending on the material to which it is fastened. A common way is bolting, either directly to steel beams or indirectly to concrete beams through steel plates anchored in the concrete. Reports indicate

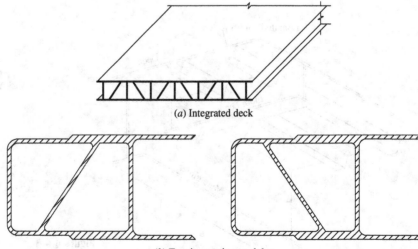

(a) Integrated deck

(b) Two integrating modules

Figure 4.4-9
An FRP bridge deck system.

that FRP decks have been used only on bridges carrying noncritical roads with low traffic volume.

4.5 Deck-Supporting Systems

There are many ways to support the deck system. As mentioned earlier, beam systems are most popular by number of bridges and more so by number of spans. In beam bridges, the beams are the primary superstructure members supporting the deck and transferring the load to the bearings and/or the substructure. There is a large variety of beams available for beam bridges. Crrently the most popular in the United States for new construction are steel beams and prestressed concrete beams due to their cost effectiveness and convenience for field construction based on their ready-to-be-placed nature. In comparison, reinforced concrete beams, which were once widely used, are no longer preferred, mainly because of the significant field work required if fabricated in place and higher self-weight compared with steel and prestressed concrete options. Accordingly, in this section, the main focus will be on the steel and prestressed concrete beams.

For load carrying, beam bridges usually have a number of parallel beams that form a frame using some connections between them, such as diaphragms. Figures 4.2-1 and 4.3-1 show respectively connected steel beams and prestressed box concrete beams in place. Such multibeam systems are advantageous in providing a redundant structural frame. Namely, if one of the beams is damaged or even completely fails (e.g., due to an over height truck collision or fatigue crack development), the system will still be stable and will not collapse. Due to this high redundancy intentionally built in, the structural analysis of the superstructure system required in the design becomes complex because the system is statically indeterminant. In other words, statics is inadequate to reasonably estimate the worst load effect in a beam. This load effect is needed to proportion the cross section or to determine the required reinforcement.

To avoid this need for complex structural analysis, the AASHTO specifications have adopted a simplified approach to facilitate design. This approach uses a concept of load effect distribution, approximating 3D analysis of the superstructure system. So-called distribution factors are given in the AASHTO specifications for routine design for various superstructure arrangements (combinations of different beams and deck systems), span lengths, beam spacings, deck thicknesses, and so on. Using these distribution factors, the total load effect (moment, shear, or other effects) of the bridge approximated as a "beam line" is distributed among the parallel beams being designed. This approach has effectively avoided sophisticated analysis that currently would have been done using more advanced tools like finite-element analysis software programs requiring advanced knowledge and more time to complete.

Note also that the distribution factors given in the specifications for this purpose are meant to give the worst load effects, which are convenient to use in design. Therefore it should be emphasized that the real load effect distribution to a particular beam in a given bridge likely is different from what the code-specified load distribution factors indicate. The load effect distribution in a beam superstructure is a function of many factors, including, but not limited to, span length, spacing between beams, and stiffness of the deck. For example, a very stiff deck will distribute the total load effect more evenly to all the beams. An extremely flexible deck will distribute the total load effect much less evenly among the parallel beams, causing the beam close to the truck load to carry much more than those further away from the loaded location. In addition, longer span lengths tend to distribute the load effect more completely compared with shorter spans, since shorter spans provide shorter paths to transfer down to the ground in the longitudinal direction. These load distribution factors are presented below when dealing with specific superstructure systems, such as concrete deck on steel beam and concrete deck on prestressed concrete I beam.

4.5.1 Prestressed Concrete Beams

Figure 4.5-1 exhibits two typical cross sections of prestressed concrete I beams used in the United States. For standardization, the AASHTO has specified six such cross sections as shown in Figures 4.5-1 to 4.5-3. These

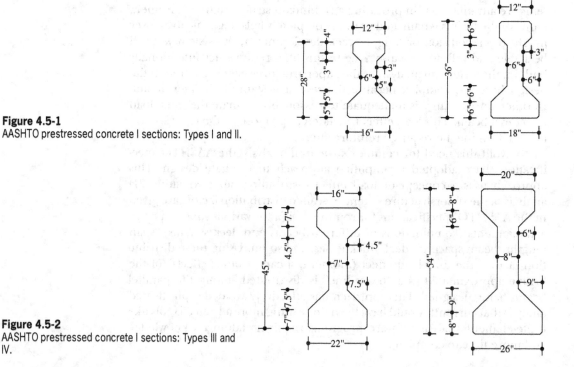

Figure 4.5-1
AASHTO prestressed concrete I sections: Types I and II.

Figure 4.5-2
AASHTO prestressed concrete I sections: Types III and IV.

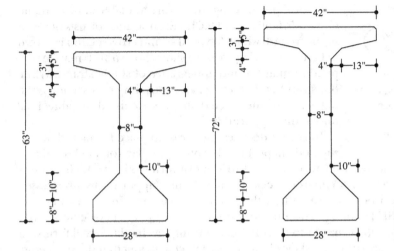

Figure 4.5-3
AASHTO prestressed concrete I
sections: Types V and VI.

cross sections are for different span lengths to be economical, with larger cross sections for longer spans. This standardization reduces the significant cost for form fabrication because the form cost is distributed to many bridges using these standardized cross sections. Using these sections, the designer is required to determine the number and location of prestressing steel strands to meet all the requirements in the AASHTO specifications. For routine application, Table 4.5-1 provides commonly needed properties of these sections as well as their recommended maximum span lengths.

For composite action between the beams and a concrete deck, shear connectors are required and can be provided quite conveniently in pre-stressed concrete beams. Steel reinforcement has been seen to be used for this function. It is employed as bent loops mechanically connected with rein-forcement in the prestressed concrete beam and cast into the concrete as part of the beam. Figure 4.3-1 indicates similar shear connectors where the beams are placed on an abutment and a pier for a simply supported span.

Table 4.5-1
Properties of AASHTO prestressed concrete I-beam cross sections

Type	Area (in.^{2})	Y_{bottom} (in.)	Inertia (in.^{4})	Weight (k/ft)	Maximum Span (ft)
I	276	12.59	22,750	0.287	48
II	369	15.83	50,980	0.384	70
III	560	20.27	125,390	0.583	100
IV	789	24.73	260,730	0.822	120
V	1,013	31.96	521,180	1.055	145
VI	1,085	36.38	733,320	1.13	167

Figure 4.5-4
Cross section
of steel
prestressing
strand.

Typical prestressing strands are available at three diameters: 0.5, 0.5625, and 0.6 in. Each strand consists of seven high-strength wires. A typical strand cross section is shown in Figure 4.5-4. Figure 4.5-5 shows how strands may be supplied in quantity and appearance of single strand. While the strand cross sections are not solid cross sections, the net areas for the three diameters are listed in Table 4.5-2 for routine application.

In recent decades, steel cast-in-place forms with a corrugated shape have become popular for deck concrete due to the reduced work needed to remove forms. Figure 4.5-6 shows a bottom view of a concrete deck on such forms supported by prestressed concrete I beams. Notice that this is an internal bay between two beams. The AASHTO design specifications do not allow the fascia or external bay of the deck (overhang) to be constructed using stay-in-place steel forms. In addition, also note the skew of the bridge by recognizing that the deck edge is not perpendicular to the beams' longitudinal axes (also the traffic direction). Such skew is formed where the beam-supporting system (abutments

Figure 4.5-5
(Left) Steel prestressing strands in roll; (right) single strand.

Table 4.5-2
Cross sections of commercially available prestressing strands

Strand Diameter (in.)	Cross Section Area (in.^{2})
0.5	0.153
0.5625	0.192
0.6	0.217

Figure 4.5-6
Corrugated steel stay-in-place form on prestressed concrete I beams.

and/or piers) has to follow the direction of the roadway or waterway underneath the bridge, not perpendicular to the roadway carried by the bridge. This minimizes the span length requirement and/or acquisition of right of way, as illustrated in Figure 4.5-7. Figure 4.5-8 also shows the skewed deck's reinforcement in a bridge.

The AASHTO specifications require the main (transverse) reinforcement to follow the skew angle when the skew angle is smaller than 25°

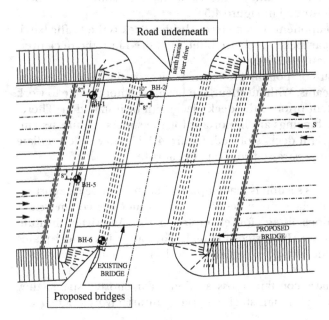

Figure 4.5-7
Two intersecting roads requiring skewed bridges.

Figure 4.5-8
Reinforcement layout in
skewed bridge.

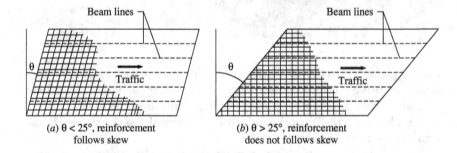

Figure 4.5-9
Reinforcement
requirements for skew
angle <25° (left) and
>25° (right).

and be perpendicular to traffic if the skew angle is above 25°. These
requirements are illustrated in Figure 4.5-9.

An important component on a concrete bridge deck is the traffic barriers, guard rails, or parapets, which are often anchored into the deck overhang, as shown in Figure 4.5-10.

Another essential component for beam bridges is the diaphragms or
bracings between beams. Without them, the beams will not be able to work
as a monolithic system prior to the deck placement or installation. Therefore diaphragms or bracings are important for the superstructure system,
especially before a rigid deck is placed. Figure 4.5-11 shows intermediate
diaphragms between prestressed concrete I beams.

4.5.2 Steel Beams

Based on how the cross section is made up, steel beams can be categorized
as rolled shapes and plate girder sections. The former use commercially
available steel sections listed, for example, in the American Institute
of Steel Construction (AISC) steel manual. The latter have steel plates
as determined by the bridge designer. Typical cross sections are shown
in Figure 4.5-12.

Rolled shapes have constant cross sections. For shorter spans, they
are more economical although steel material is invariably used along the

Figure 4.5-10
Steel reinforcement for parapet anchored in deck overhang.

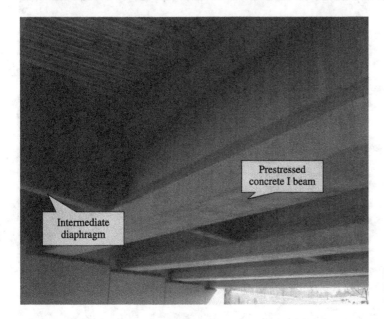

Figure 4.5-11
Diaphragms between prestressed concrete I beams.

beam length. Apparently some material is not needed, for example in the end regions of a simply supported beam where the moment requirement diminishes.

Plate girder sections can save material cost by cutting material in those areas. However, their higher fabrication cost may offset the cost saved on material. For longer spans, the amount of saved material will increase and

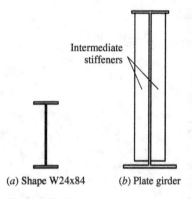

Intermediate stiffeners

(a) Shape W24x84   (b) Plate girder

Figure 4.5-12
Comparison of steel rolled shape and plate girder cross sections.

thus its cost saving may exceed the higher cost of fabrication to realize a total cost saving for the bridge. Thus, plate girders are used in longer spans for cost effectiveness.

Figures 4.5-12 compares two examples typical of the rolled shape and plate girder. Both are at the cross section where the bending moment is maximum along the length of the beam. As seen, the rolled shape is more bulky and thus more stable and the plate girder section is skinnier and less stable but saves material while providing higher moment capacity since the two flanges are placed further apart than the rolled shape. As a result, plate girders require stiffeners to stabilize the cross section as shown. Note also that the plate girder option can change its cross section (e.g. by changing the thickness of the flanges and/or the height of the web) along the beam length to meet different requirements of load effects (usually moment and shear). In contrast, the rolled shape does not.

Figures 4.5-13 and 4.5-14 show photographs of a rolled shape steel beam bridge and a plate girder steel beam bridge, respectively. Design examples

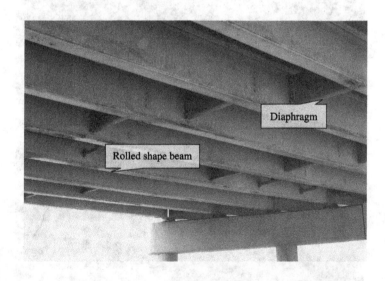

Diaphragm

Rolled shape beam

Figure 4.5-13
Steel rolled beam highway bridge.

Concrete parapet

Plate girder

Intermediate stiffener

Figure 4.5-14
Steel plate girder highway bridge.

Figure 4.5-15
Deck construction with underneath roadway open to traffic.

Figure 4.5-16
Corrugated steel stay-in-place form for concrete deck on steel plate girders.

for each case are included later in this book to illustrate respective typical design processes. Figure 4.5-15 illustrate a situation of deck construction for highway bridge.

It is also very typical that steel beams support a reinforced concrete deck to provide the driving surface. Figure 4.5-16 exhibits the bottom view of such a deck after completion of construction. Also note the bridge skew by recognizing the deck edge not perpendicular to the supporting beams. Figure 4.5-17 shows this skew from the deck top before concrete was placed. Note again that the reinforcement did not follow the skew since the skew angle is larger than 25°, per AASHTO specifications.

For composite sections that are very popular now in the United States, shear studs are overwhelmingly used in steel bridges as shear connectors between the concrete deck and the steel beams. Figure 4.5-18 shows a steel stud being welded to a steel beam's top flange, and Figure 4.5-19 shows a welded stud. Shear studs need to be designed to carry the shear force between the concrete deck and the steel beam. They also need to be checked to prevent possible fatigue failure at the weld as well as other possibly concerned failure modes. They are treated as an important structural

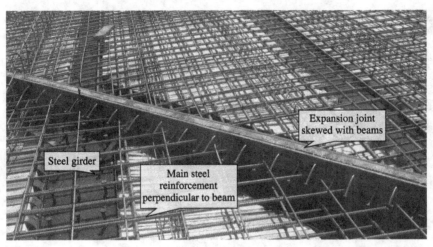

Figure 4.5-17
Skewed span with straight reinforcement.

Figure 4.5-18
Welding steel shear stud to steel beam top flange.

Figure 4.5-19
Shear stud welded on steel beam top flange.

component in the design of steel beam bridges. The specific steps of shear stud design are included in the design examples with calculation details in this chapter.

For both rolled shapes and plate girders, end stiffeners are usually needed. They are welded or bolted to the steel beam to strengthen the shear capacity. An example is shown in Figure 4.5-20 for a plate girder. Similar stiffeners placed away from the bean ends may or may not be needed

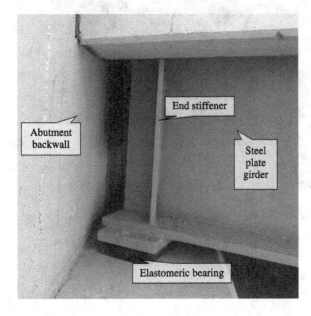

Figure 4.5-20
End of steel plate girder.

for strengthening the web, referred to as intermediate stiffeners. They also need to be included in plate girder design. Another important group of steel components is the bracing members, as seen in Figure 4.5-21. Both of these two types of steel components need to be designed as structural members.

Figure 4.5-22 exhibits a special steel bridge component for continuous spans, referred to as pin and hanger as an expansion joint to accommodate thermal expansion and contraction. This type of connection is still being used, especially for rehabilitation of old steel bridges without replacing or

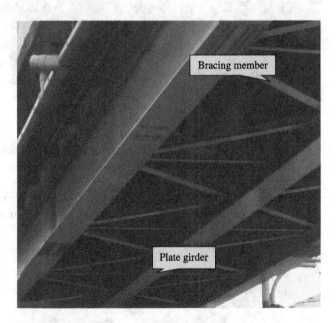

Figure 4.5-21
Bracing system for steel plate girders.

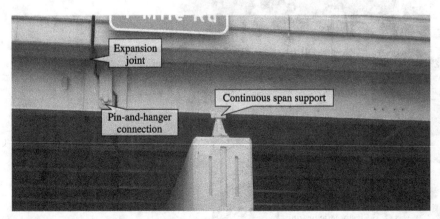

Figure 4.5-22
Pin-and-hanger system for continuous steel plate girder spans.

changing the superstructure system configuration. Pin-and-hanger connections need to be designed as structural components.

Slab bridges here refer to those bridges with a reinforced concrete slab as the superstructure without beams. Namely the slab actually functions as both the beams and the deck in deck-on-beam bridges. Apparently, the slab would have to be thicker than the deck in beam bridges for the span length, since there are no beams supporting the slab now. This would increase the self-weight and thus the cost for the bridge.

4.5.3 Slab Superstructure

As a result of considering cost effectiveness, slab bridges are only used to span shorter distances than beam bridges. Figure 4.5-23 shows a typical example of such short spans of highway bridge. For this type of application, a slab bridge saves the beams, compared with beam bridges.

Reinforced concrete slabs possess high shear capacity. Therefore, slab bridges may represent a viable option when high axle load is of concern. To that end, some slab bridges in airports have been constructed to support runways and roads for airplane traffic.

Steel trusses were popular before the 1950s in the United States. Trusses serve as large beams. Their top and bottom chords act as a beam's top and bottom flanges resisting moment. Truss web members (vertical or diagonal) perform the same function as the web in beams to carry shear force. The main advantages of truss bridges are:

4.5.4 Steel Trusses

❏ Significantly reduced self-weight compared with beams of the same span length.

❏ Able to span longer than beams due to relatively lighter weight.

❏ Relatively easier construction compared with arches

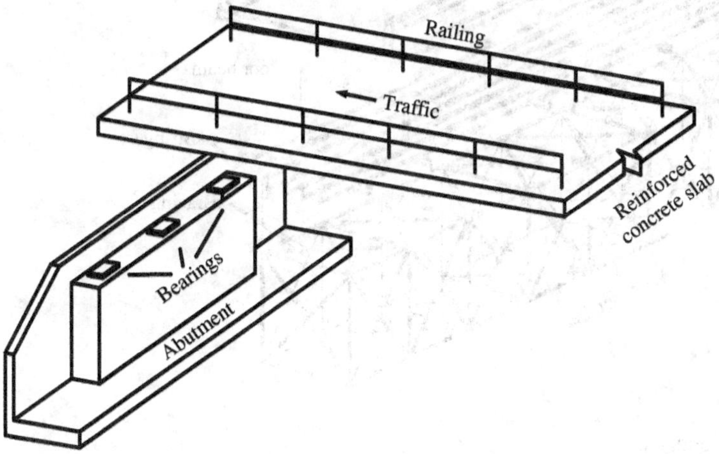

Figure 4.5-23
Slab highway bridge.

The weight reduction is realized by taking away a large amount of material from the web.

Besides the top and bottom cords and web members, bracing members between the trusses are also major structural members that need to be carefully designed. Their main function is to connect the two trusses to form a stable spatial truss frame to support the deck system, as shown in Figure 4.5-24. Similar to the example in Figure 4.3-3, floor beams and stringers are commonly used to support the deck system, as seen in Figure 4.5-24 but arranged a little differently.

4.5.5 Concrete Arches

Reinforced concrete arches represent another type of deck-supporting system that has been used in highway bridge superstructure. They may be prestressed as well to save material and cost. On the other hand, they are less popular now in the United States perhaps because steel has become much less expensive.

4.5.6 Steel Arches

Steel arches are also a viable solution for highway bridge superstructure to support a deck system. The span length of steel arches varies significantly. Simple steel arches may be made of steel tubes, with concrete filling as an option. More complex steel arches include truss members as a member of the arch for much longer spans.

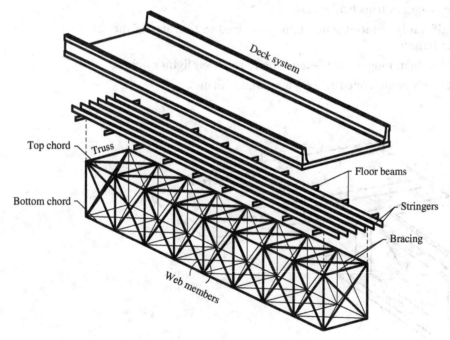

Figure 4.5-24
Typical superstructure components in truss bridges.

4.6 Design of Reinforced Concrete Deck Slabs

For monolithic concrete bridge deck slabs satisfying certain conditions, an empirical design, requiring no analysis, is permitted in the AASHTO specifications. Continuity between the deck and its supporting components is desired and encouraged. Furthermore, composite action between the deck and its supporting components is required where technically feasible.

It should be noted that the deck is preferred to be jointless or continuous to improve the weather and corrosion-resisting effects of the whole bridge. This will reduce inspection efforts and maintenance cost and will increase structural effectiveness and redundancy.

Figure 4.6-1 gives a flowchart for the procedure of reinforced bridge deck slab design, according to the AASHTO specifications. For simplicity, the reinforced concrete deck slab is referred to as an RC deck or a deck hereafter. Note also that such a design is required not only for brand new construction but also for new decks on existing structures for deck replacement. Such a deck replacement often takes place along with repair and/or rehabilitation of other bridge components, such as beams, bearings, piers,

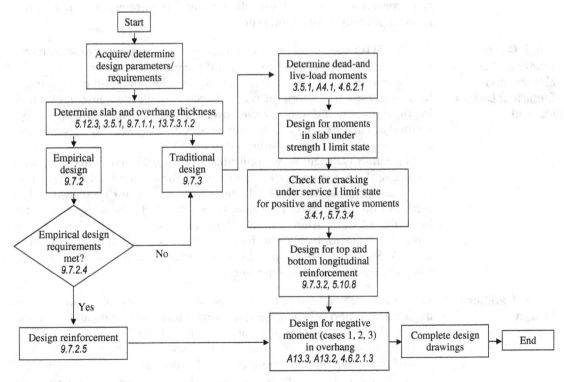

Figure 4.6-1
Typical concrete deck design procedure.

and abutments. While typical RC deck life is about a half or less of the bridge design life of 75 years, deck replacement is almost always required more frequently than bridge replacement.

4.6.1 Design Requirement and Parameters

The flowchart in Figure 4.6-1 indicates that the design process starts with identification of design requirements and corresponding parameters. These requirements include the bridge geometry, type of beams, and beam spacing. The bridge geometry will determine the deck's geometry, such as length, width, and skew if any. The beam type will dictate whether and how composite action can be realized and designed. The beam spacing is actually the supporting span length for the deck, which obviously is a major controlling factor for the deck's thickness and reinforcement amount. With these requirements made clear, a number of other design parameters can be accordingly determined, such as concrete strength, reinforcement steel strength, deck thickness, and so on. It should be noted that some of these parameters may or may not necessarily be decided by the deck designer. They can be constraints given by the bridge owner and/or the site condition as well. For example, a number of state bridge owners specify an RC deck thickness or thickness range based on the requirement for the deck life span. With these parameters determined, detailed calculations can proceed to satisfy code-specified general requirements such as those for strength and service limit states.

4.6.2 General Traditional Design Method and Empirical Design Method

The AASHTO specifications provide two general approaches for concrete deck design, the traditional method and the empirical method.

The traditional method is based on the bending moment design. It has been used for deck design for many years, although the approach to finding the design moment has evolved over these years. The empirical design method was initiated in Canada upon recognition of excessive capacity provided using the traditional design method.

The empirical method is applicable when certain conditions are met regarding the bridge and deck geometry to be detailed below. No structural analysis for finding the load effects is required in the empirical design method. Only the steel reinforcement amounts need to be designed according to the AASHTO specifications. However, the empirical method is applicable to the interior bays of the deck, not the deck overhang, also according to the AASHTO specifications. The deck overhang still needs to be designed using the traditional design method.

4.6.3 Traditional Design   9.7.3

The traditional design has a focus on the deck's flexural capacity since bending is the assumed failure mode. In contrast with the empirical design, the traditional design is mechanistic using structural analysis.

The analysis in the traditional design is based on a strip of typical deck carried by the supporting system (usually beams), as seen in Figure 4.6-2.

As indicated in the concrete deck design procedure flowchart in Figure 4.6-1, the process can be divided into two parts: (1) interior bay

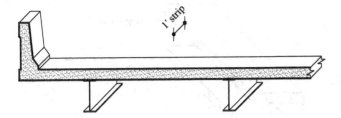

Figure 4.6-2
Strip of concrete deck on steel beams considered for design.

design and (2) overhang design. The interior bay refers to the part of the deck between two supports (i.e., two beams in Figure 4.6-2 as an example). The overhang is the cantilever part of the deck beyond a support (i.e., a beam in Figure 4.6-2).

The interior bay design considers the dead-load and vehicular load effects in normal condition. The overhang design needs also to consider truck collision with the railing, and the overhang is designed to be stronger than the railing in such a situation. Namely, should unfortunately such a collision take place, the deck overhang will not fail before the railing. As seen in the flowchart in Figure 4.6-1, the interior bay design for a deck using the traditional method may be substituted by the empirical method if the specified conditions are met. However, the deck overhang design has to be done using the traditional method. There are no other approaches given in the AASHTO specifications for deck overhang design. These two parts of deck design are discussed separately in detail next. Example 4.1 demonstrates the traditional design procedure with detailed calculations for interior bays.

The major loads on the deck are (1) dead load of railing, (2) wearing surface to be constructed either now or possibly in the future, (3) the self-weight of the deck, including a possible layer of sacrificial concrete depending on jurisdiction, (4) truck wheel load, and (5) possible future concrete overlay if applicable. This list attempts to cover those loads possibly to be applied over the life span of the deck, and it may not be exhaustive since not every bridge is identical.

4.6.4 Dead- and Live-Load Effects for Interior Bays

The four kinds of dead load listed above are conventionally modeled as uniformly distributed and concentrated loads on the strip of deck, as shown in Figure 4.6-3. While a 2D analysis is commonly used in design to find the dead-load moment, 3D analysis can be more precise but also more costly in the analysis effort.

On the other hand, the wheel live load is not as simple as the dead load to analyze, because a width wider than the 1-ft strip of the deck is participating in carrying the load of a typical wheel with a tire print of 20 in. (in the transverse direction) × 8 in. (in the longitudinal direction). Assume the entire 16-k wheel acting on the 1-ft strip would overestimate the load effect and result in an overconservative design. A detailed 3D analysis is required to

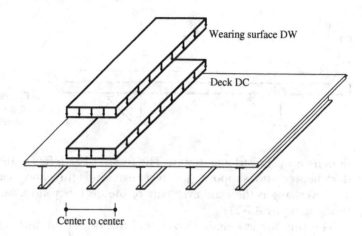

Figure 4.6-3
Tributary area for dead-load effects.

Center to center

Table 4.6-1
Live-load moment (kft) in 1-ft strip of deck   *A4.1*

S	Positive Moment	Negative Moment by Distance from CL of Girder to Design Section for Negative Moment						
		0.0 in.	3 in.	6 in.	9 in.	12 in.	18 in.	24 in.
4'-0"	4.68	2.68	2.07	1.74	1.60	1.50	1.34	1.25
4'-6"	4.63	3.00	2.58	2.19	1.90	1.65	1.32	1.18
5'-0"	4.65	3.74	3.20	2.66	2.24	1.83	1.26	1.12
5'-6"	4.71	4.36	3.73	3.11	2.58	2.07	1.30	0.99
6'-0"	4.83	4.88	4.19	3.50	2.88	2.31	1.39	1.07
6'-6"	5.00	5.31	4.57	3.84	3.15	2.53	1.50	1.20
7'-0"	5.21	5.98	5.17	4.36	3.56	2.84	1.63	1.37
7'-6"	5.44	6.26	5.43	4.61	3.78	3.15	1.88	1.72
8'-0"	5.69	6.48	5.65	4.81	3.98	3.43	2.49	2.16
8'-6"	5.99	6.66	5.82	4.98	4.14	3.61	2.96	2.58
9'-0"	6.29	6.81	5.97	5.13	4.28	3.71	3.31	3.00
9'-6"	6.59	7.15	6.31	5.46	4.66	4.04	3.68	3.39
10'-0"	6.89	7.85	6.99	6.13	5.26	4.41	4.09	3.77
10'-6"	7.17	8.52	7.64	6.77	5.89	5.02	4.48	4.15
11'-0"	7.46	9.14	8.26	7.38	6.50	5.62	4.86	4.52
11'-6"	7.74	9.72	8.84	7.96	7.07	6.19	5.22	4.87
12'-0"	8.01	10.28	9.40	8.51	7.63	6.74	5.56	5.21
12'-6"	8.28	10.81	9.93	9.04	8.16	7.28	5.97	5.54
13'-0"	8.54	11.31	10.43	9.55	8.67	7.79	6.38	5.86
13'-6"	8.78	11.79	10.91	10.03	9.16	8.28	6.79	6.16
14'-0"	9.02	12.24	11.37	10.50	9.63	8.76	7.18	6.45
14'-6"	9.25	12.67	11.81	10.94	10.08	9.21	7.57	6.72
15'-0"	9.47	13.09	12.23	11.37	10.51	9.65	7.94	7.02

Source: AASHTO LRFD Bridge Design Specifications, 2012. Used by permission.

more precisely estimate the live-load effect LL. The AASHTO specifications offer a table of wheel load effect for a typical 1-ft strip as a function of beam spacing. This table is based on 3D analysis results to simplify design and is shown in Table 4.6-1. Previous AASHTO specifications also offer a simplified approach to estimating the live-load effect in the 1-ft deck strip, which is also known to conservatively overestimate.

Example 4.1 demonstrates the traditional design procedure with detailed calculations for interior bays, and Example 4.2 illustrates an alternative design method for the same part of concrete deck.

This traditional design procedure is demonstrated in Example 4.1 with detailed calculations for interior bays of an RC deck.

4.6.5 Strength I and Service I Limit State Design for Interior Bays

Example 4.1 Reinforced Concrete Deck Design (Interior Bay)

□ **Design Requirement**

Design a cast-in-place reinforced concrete deck for a steel beam bridge with five girders spaced at 8 ft 10 in. and an overhang 4 ft 5 in. wide as shown in Figure Ex4.1-1. Assume $f_c' = 4$ k/in.^{2} and $f_y = 60$ k/in.^{2} (This example addresses the deck portions between internal bays between beams. Examples 4.3 and 4.4 deal with the overhang cantilevers.)

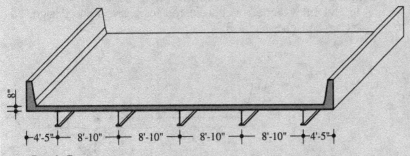

8"

+4'-5"+ 8'-10" + 8'-10" + 8'-10" + 8'-10" +4'-5"+

Figure Ex4.1-1
Superstructure dimensions.

□ **Deck Parameters**

5.12.3   Top cover $C_{top} = 2.5$ in.   (exposure to deicing salts)

5.12.3   Bottom cover $C_{bottom} = 1$ in.   (up to No. 11 bars)

3.5.1   Concrete density $W_{concrete} = 0.145$ k/ft^3 $\left(\text{normal weight } f_c' \leq 5 \text{ k/in.}^2\right)$

3.5.1   Future wearing surface (FWS) density $W_{FWS} = 0.14\,k/ft^3$ (bituminous)

Future wearing surface thickness $t_{FWS} = 2.5$ in.

Top integral wearing surface thickness $t_{IWS} = 0.5$ in

Counted as dead load not strength.

9.7.1.1   Slab thickness $t_{slab} = 8$ in.   (minimum 7 in.)

13.7.3.1.2   Deck overhang thickness $t_{overhang} = 9$ in.   (minimum 8 in.)

Girder spacing $S = 8\,ft\,10\,in. = 8.83\,ft$

Load modifier $\eta = 1.0$

❑ **Parapet Parameters**

Weight per foot of parapet: $W_{parapet} = 0.457\,k/ft$

See Example 3.1 and Figure Ex4.1-2

Base width $B_{base} = 1\,ft\,3\tfrac{3}{4}\,in.$

Parapet height $H_{parapet} = 42\,in. = 3.5\,ft$

❑ **Design Procedure**

(a) Design for internal bays, including the following limit states:

1. Strength I limit state for both positive and negative moment
2. Service I limit state for flexural cracking

(b) Design for overhang to be covered in Examples 4.3 and 4.4.

❑ **Strength I—Positive Dead-Load Effect**   *3.4.1*

A typical 1-ft-wide deck is considered for analysis and design, as shown in Figure Ex4.1-3.

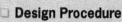

Figure Ex4.1-2
Bridge parapet dimensions.

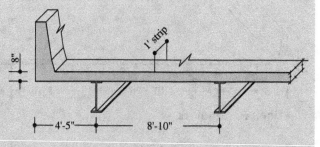

Figure Ex4.1-3
Typical design strip.

Load factor for slab and parapet weight (DC) for strength I limit state:

Maximum $\gamma_{p,\,DC} = 1.25$

Load factor for future wearing surface weight (DW) for strength I limit state:

$$\text{Maximum } \gamma_{p, DW} = 1.5$$

Dead Load

$$\text{Slab dead load} = t_{slab}(1 \text{ ft}) W_{concrete} = \left(\frac{8 \text{ in.}}{12 \text{ in./ft}}\right) 1 \text{ ft} (0.145 \text{ k/ft}^3) = 0.097 \text{ k/ft}$$

$$\text{FWS dead load} = t_{FWS}(1 \text{ ft}) W_{FWS} = \left(\frac{2.5 \text{ in.}}{12 \text{ in./ft}}\right) 1 \text{ ft} (0.140 \text{ k/ft}^3) = 0.029 \text{ k/ft}$$

Strength I—Positive Live Load effect

For girder spacing $S = 8$ ft 10 in. the maximum unfactored positive live-load moment is $M_{LL} = 6.19$ kft for the 1 ft strip, *A4.1*
This 6.19-kft moment includes dynamic load allowance IM = 0.33.

Strength I Total Positive Factored Design Moment

$$M_{POS} = \gamma_{p,DC} M_{DL \text{ slab}} + \gamma_{p,DC} M_{DL \text{ parapet}} + \gamma_{p,DW} M_{DL \text{ FWS}} + \gamma_{LL} M_{LL}$$

$$= 1.25(0.097 \text{ k/ft})3.48 \text{ ft}^2 + 1.25(0.46 \text{ k/ft})0.28 \text{ ft} (1 \text{ ft})$$

$$+ 1.5(0.029 \text{ k/ft})3.48 \text{ ft}^2 + 1.75(6.19 \text{ kft}) = 11.57 \text{ kft/ft}$$

The maximum positive-moment values at midbay of bay 2 from Figures Ex4.1-4 and Ex4.1-5 are used here.

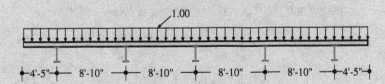

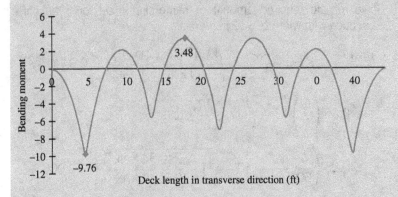

Figure Ex4.1-4
Moment diagram for unit uniformly distributed load.

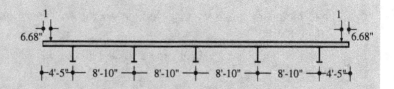

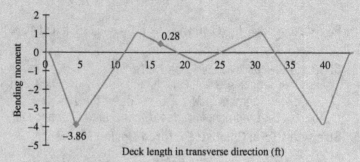

Figure Ex4.1-5
Moment diagram for unit concentrated
load at deck edges representing
parapet.

☐ **Design for Strength I Positive Flexure for Interior Bays**
   Assume No. 6 bars

$$\text{Bar diameter} = 0.75 \text{ in.}$$

$$\text{Bar area} = 0.44 \text{ in.}^2$$

*Using this assumed bar size, the required area of steel, A_s, is cal-
culated and the required bar spacing is found as follows:*

$$\text{Effective depth } d_e = t_{slab} - C_{bottom} - \left(\frac{\text{bar diameter}}{2} \right) - t_{IWS}$$

$$= 8 \text{ in.} - 1 \text{ in.} - \left(\frac{0.75 \text{ in.}}{2} \right) - 0.5 \text{ in.} = 6.13 \text{ in.}$$

Solve for the required amount of reinforcing steel for a typical
section with width $b = 12$ in.:

5.5.4.2.1 $R_n = \dfrac{M_{POS}}{\varphi_f b d_e^2} = \dfrac{11.57 \text{ kft} (12 \text{ in./ft})}{(0.9)\, 12 \text{ in.} \,(6.13 \text{ in.})^2} = 0.34 \text{ k/in.}^2$

$$\rho = 0.85 \left(\frac{f_c'}{f_y} \right) \left[1 - \sqrt{1 - \frac{2R_n}{0.85 f_c'}} \right]$$

$$= 0.85 \left(\frac{4 \text{ k/in.}^2}{60 \text{ k/in.}^2} \right) \left[1 - \sqrt{1 - \frac{2 \,(0.34 \text{ k/in.}^2)}{0.85 \,(4 \text{ k/in.}^2)}} \right] = 0.006$$

The above two equations are the general reinforced concrete design equations.

$$A_s = \rho b d_e = 0.006(12 \text{ in.})6.13 \text{ in.} = 0.44 \text{ in.}^2/\text{ft}$$

$$\text{Required bar spacing} = \frac{\text{bar area}}{A_s} = \frac{0.44 \text{ in.}^2}{0.44 \text{ in.}^2/\text{ft}} = 12 \text{ in.}$$

Use No. 6 bars at 12 in. as shown in Figure Ex4.1-6.
OK for Strength I positive moment in interior bays.

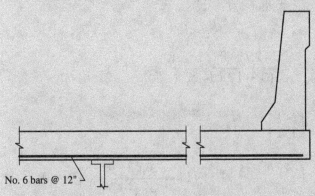

No. 6 bars @ 12"

Figure Ex4.1-6
Positive flexure reinforcement for interior bays.

☐ **Strength I—Negative Live-Load Effect**
For girder spacing $S = 8$ ft 10 in. the unfactored negative live-load moment at a check section 6 in. from the beam centerline is $M_{LL} = -5.11$ kft, assuming a 12-in. top flange of the beam. A4.1.
This −5.11-kft moment includes dynamic load allowance IM = 0.33.

☐ **Strength I Total Negative Factored Design Moment**

$$M_{NEG} = \gamma_{p,DC} M_{DL\,slab} + \gamma_{p,DC} M_{DL\,parapet} + \gamma_{p,DW} M_{DL\,FWS} + \gamma_{LL} M_{LL}$$

$$= 1.25(0.097 \text{ k/ft})9.76 \text{ ft}(1 \text{ ft}) + 1.25(0.46 \text{ k/ft})3.86 \text{ ft}^2$$

$$+ 1.5(0.029 \text{ k/ft})9.76 \text{ ft}(1 \text{ ft}) + 1.75(5.11 \text{ kft}) = 12.76 \text{ kft/ft}$$

The maximum negative-moment values at the fascia beam from Figures Ex4.1-4 and Ex4.1-5 are used here for the controlling section.

☐ **Design for Strength I Negative Flexure for Interior Bays**
Assume No. 7 bars:

Bar diameter = 0.875 in.   Bar area = 0.60 in.^{2}

From this assumed bar size, the required area of steel, A_s, is calculated as follows. The required bar spacing is then found.

$$d_e = t_{slab} - C_{top} - \left(\frac{\text{bar diameter}}{2}\right) - t_{IWS}$$

$$= 8 \text{ in.} - 2.5 \text{ in.} - \left(\frac{0.875 \text{ in.}}{2}\right) - 0.5 \text{ in.} = 4.56 \text{ in.}$$

Solve for the required amount of reinforcing steel for $M_{NEG} = 12.76$ kft/ft and the strip of 1 ft, that is, $b = 12$ in.:

5.5.4.2.1 $\qquad\qquad\qquad\qquad \varphi_f = 0.9$

$$R_n = \frac{M_{NEG}}{\varphi_f bd_e^2} = \frac{12.76 \text{ kft} (12 \text{ in.}/\text{ft})}{(0.9) \, 12 \text{ in.} (4.56 \text{ in.})^2} = 0.68 \text{ k/in.}^2$$

$$\rho = 0.85 \left(\frac{f_c'}{f_y}\right) \left[1 - \sqrt{1 - \frac{2R_n}{0.85 f_c'}}\right]$$

$$= 0.85 \left(\frac{4 \text{ k/in.}^2}{60 \text{ k/in.}^2}\right) \left[1 - \sqrt{1 - \frac{2 \, (0.68 \text{ k/in.}^2)}{0.85 \, (4 \text{ k/in.}^2)}}\right] = 0.013$$

The above two equations are the reinforced concrete design equations.

$$A_s = \rho b d_e = 0.013(12 \text{ in.})4.56 \text{ in.} = 0.70 \text{ in.}^2$$

$$\text{Required bar spacing} = \frac{\text{bar area}}{A_s} = \frac{0.60 \text{ in.}^2}{0.70 \text{ in.}^2/\text{ft}} = 10.29 \text{ in.}$$

Use No. 7 bars at 8 in., as shown in Figure Ex4.1-7.

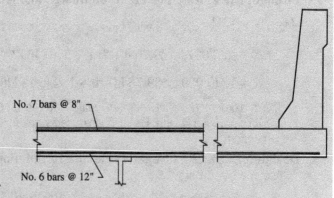

No. 7 bars @ 8"

No. 6 bars @ 12"

Figure Ex4.1-7
Negative flexure reinforcement for interior bays.

OK for Strength I negative moment in interior bays.

❑ Service I Limit State—Positive Moment Flexure Cracking

Load factors

$$\gamma_{DC} = 1 \qquad \gamma_{DW} = 1 \qquad \gamma_{LL} = 1$$

Positive service moment:

$$M_{POS} = \gamma_{DC}M_{DL\ slab} + \gamma_{DC}M_{DL\ parapet} + \gamma_{DW}M_{DL\ FSW} + \gamma_{LL}M_{LL}$$

$$= 1(0.097\ k/ft)3.48\ ft^2 + 1(0.46\ k/ft)0.28\ ft(1\ ft)$$

$$+ 1(0.029\ k/ft)3.48\ ft^2 + 1(6.19\ kft) = 6.76\ kft$$

Bar spacings is required to satisfy

5.7.3.4
$$s \le \frac{700\gamma_e}{\beta_s f_s} - 2d_c$$

where

5.7.3.4
$$\gamma_e = 1 \qquad \text{(class I exposure condition)}$$

$$d_c = C_{bottom} + \frac{\text{bar diameter}}{2} = 1\ in. + \frac{0.75\ in.}{2} = 1.38\ in.$$

$$h = t_{slab} = 8\ in.$$

$$\text{for}\ \beta_s = 1 + \frac{d_c}{0.7\,(h - d_c)} = 1 + \frac{1.38\ in.}{0.7(8\ in. - 1.38\ in.)} = 1.30$$

$$f_s = \frac{M_{POS}}{A_s j d_s}$$

where

$$A_s = 0.44\ in.^2$$

$$j = 1 - \frac{k}{3}$$

$$k = \sqrt{(\rho n)^2 + 2\rho n} - \rho n$$

$$\rho = \frac{A_s}{b d_s} = \frac{0.44\ in.^2}{12\ in.(6.13\ in.)} = 0.006$$

Therefore, for $n = 8$, $k = 0.27$, and $j = 0.91$,

$$f_s = \frac{6.76\ kft\,(12\ in./ft)}{0.44\ in.^2\,(0.91)\,6.13\ in.} = 33.05\ k/in.^2$$

$$s \le \frac{700\gamma_e}{\beta_s f_s} - 2d_c = \frac{700\,(1)}{1.3\,(33.05)}\,in. - 2\,(1.38\ in.) = 13.53\ in.$$

The selected spacing for strength I positive moment 12 in. < 13.53 in. Thus the strength I spacing should be used.

OK for Service I positive moment in interior bays.

☐ **Service I Limit State—Negative-Moment Flexure Cracking**

$$M_{NEG} = \gamma_{DC}M_{DL\ slab} + \gamma_{DC}M_{DL\ parapet} + \gamma_{DW}M_{DL\ FSW} + \gamma_{LL}M_{LL}$$

$$= 1(0.097\ k/ft)9.76 + 1(0.46\ k/ft)3.86$$

$$+ 1(0.029\ k/ft)9.76 + 1(5.11\ kft) = 8.1\ kft$$

The same requirement as for the positive moment needs to be satisfied.

$$d_c = C_{top} + \frac{bar\ diameter}{2} = 2.5\ in. + \frac{0.875\ in.}{2} = 2.94\ in.$$

$$\beta_s = 1 + \frac{d_c}{0.7\,(h - d_c)} = 1 + \frac{2.94\ in.}{0.7(8\ in. - 2.94\ in.)} = 1.83$$

$$A_s = 0.6\ in.^2(12\ in./8\ in.) = 0.9\ in.^2$$

$$\rho = \frac{A_s}{bd_s} = \frac{0.9\ in.^2}{(12\ in.)\left(7.5\ in. - 2.5\ in. - \dfrac{0.875}{2}\ in.\right)} = 0.016$$

For $n = 8$

$$k = \sqrt{(\rho n)^2 + 2\rho n} - \rho n$$

$$= \sqrt{.016^2(8^2) + 2\,(.016)\,8} - 0.016\,(8) = 0.39$$

$$j = 1 - \frac{k}{3} = 1 - \frac{0.39}{3} = 0.87$$

$$f_s = \frac{M_{NEG}}{A_s j d_s} = \frac{8.1\ kft\,(12\ in./ft)}{0.9\ in.^2\,(0.87)\,4.56\ in.} = 27.22\ k/in.^2$$

5.7.3.4 $$s \le \frac{700\gamma_e}{\beta_s f_s} - 2d_c = \frac{700\,(1)}{1.83\,(27.22)}\ in. - 2\,(2.94\ in.) = 8.17\ in.$$

Eight inches was selected for the strength I limit state < 8.17 in. **OK** for Service I negative moment in interior bays.

STRENGTH I LIMIT STATE FOR FLEXURAL CAPACITY
Both positive and negative moments are usually covered in this step of deck design, while shear is not explicitly checked. For that purpose, the strength I limit state is used to find the moment capacity requirement:

3.4.1   1.25 DC + 1.5 DW + 1.75 LL (1+IM)   (4.6-1)

where DC includes railing, deck self-weight, and concrete overlay if applicable; DW includes wearing surface and utilities if applicable; and LL refers to wheel live load with impact IM.

Under the strength I limit state, the reinforcement is designed accordingly. The conventional concrete flexural design equations may be used for this purpose as follows:

$$R_n = \frac{M}{\varphi_f b d_e^2}$$ (4.6-2)

where M = ultimate or factored moment from Strength I limit state load combination, Eq. 4.6-1
  ϕ_f = resistance factor for concrete in flexure, 0.9   *5.5.4.2.1*
  b = width of cross section (12 in. for the 1-ft strip)
  d_e = effective depth of cross section (total depth less cover, half of the bar diameter, and sacrificial layer depth if any)

The reinforcement ratio ρ is then calculated as

$$\rho = 0.85 \left(\frac{f_c'}{f_y}\right)\left(1 - \sqrt{1.0 - \frac{2R_n}{0.85f_c'}}\right)$$ (4.6-3)

where f_c' = concrete's 28-day compressive strength
  f_y = steel strength

The required reinforcement is then readily calculated as

$$A_s = \rho b d_e$$ (4.6-4)

Since b is typically 12 in., A_s is the required reinforcement per foot width of the deck. Note also that this approach can be applied to cover both positive and negative moments in the deck.

SERVICE I LIMIT STATE FOR CONTROLLING FLEXURAL CRACKING
The Service I limit state uses the following load combination:

3.4.1   1.0 DC + 1.0 DW + 1.0 LL (1+IM)   (4.6-5)

for the same DC, DW, and LL as in the strength I limit state. To control flexural cracking under the Service I limit state, the spacing of the main reinforcement determined above under the Strength I limit state needs to satisfy the following relation given in the AASHTO specifications:

5.7.3.4
$$s \leq \frac{700\gamma_e}{\beta_s f_s} - 2d_c \qquad (4.6\text{-}6)$$

where γ_e = exposure factor

   = 1 for Class I exposure condition

   = 0.75 for Class II exposure condition

The Class I exposure condition applies when cracks can be tolerated due to reduced concerns of appearance and/or corrosion. The Class II exposure condition applies to the transverse design of segmental concrete box girders for any loads applied prior to attaining full nominal concrete strength and when there is increased concern of appearance and/or corrosion:

$$d_c = \text{bottom cover} + \frac{\text{bar diameter}}{2} \qquad (4.6\text{-}7)$$

$$\beta_s = 1 + \frac{d_c}{0.7\,(h - d_c)} \qquad \text{with } h = \text{deck thickness} \qquad (4.6\text{-}8)$$

$$f_s = \frac{M}{A_s j d_s} \qquad (4.6\text{-}9)$$

where
$$j = 1 - \frac{k}{\frac{1}{3}} \qquad (4.6\text{-}10)$$

$$k = \sqrt{(\rho n)^2 + 2\rho n} - \rho n \qquad (4.6\text{-}11)$$

where n = modulus ratio of steel to concrete   (4.6-12)

$$\rho = \text{reinforcement ratio} = \frac{A_s}{bd_e} \qquad (4.6\text{-}13)$$

Since A_s is known as the result of Strength I limit state design, the requirement in Eq. 4.6-6 can be readily checked. In case A_s needs to be changed because it does not meet the requirement in Eq. 4.6-6, the practical approach to design is to first select a trial A_s and spacing s and then iterate the combination of A_s and s until Eq. 4.6-6 is satisfied.

DESIGN FOR BOTTOM LONGITUDINAL REINFORCEMENT
The AASHTO specifications give the following requirement for designing the bottom distribution reinforcement:

9.7.3.2

$$A_{\text{bottom}\%} = \begin{cases} \dfrac{220}{\sqrt{S_e}} \leq 67\% & \text{for primary reinfocement} \\ & \text{perpendicular to traffic} \end{cases} \qquad (4.6\text{-}14)$$

9.7.3.2

$$\begin{cases} \dfrac{100}{\sqrt{S_e}} \leq 50\% & \text{for primary reinfocement} \\ & \text{parallel to to traffic} \end{cases} \qquad (4.6\text{-}15)$$

where S_e in feet is the deck effective span length as defined in Figure 4.6-2.

DESIGN FOR TOP LONGITUDINAL REINFORCEMENT

According to the AASHTO specifications, reinforcement needs to be provided near surfaces of concrete exposed to daily temperature changes for shrinkage and temperature stresses. The top longitudinal steel in the concrete deck satisfies this requirement. In general, temperature and shrinkage reinforcement design ensures that the total reinforcement on exposed surfaces satisfies

5.10.8

$$A_s \geq \frac{1.3\, bh}{2\,(b+h)\,f_y} \qquad (4.6\text{-}16)$$

This value of A_s is limited to

5.10.8

$$0.11 \leq A_s \leq 0.60 \qquad (4.6\text{-}17)$$

where A_s = area of reinforcement in each direction and each face (in.²/ft)
  b = least width of component section (in.) see Figure 4.6-4.
  h = least thickness of component section (in.)
  f_y = yield strength of reinforcing bars ≤ 75 k/in.²

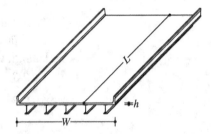

Figure 4.6-4
Parameters for temperature and shrinkage reinforcement calculation $b = \min(W, L)$.

When using the above equation, the calculated area of reinforcing steel must be equally distributed on both concrete faces. In addition, the maximum spacing of the temperature and shrinkage reinforcement must be the smaller of 3 times the deck thickness, or 18 in.

The empirical design method presented here is relatively simpler. Traditionally, concrete deck design uses an assumption that the failure mode is flexure, as detailed above. However, numerous lab and field test results have indicated that this assumption is not valid and the ultimate strength of concrete deck is controlled by the punching shear capacity. This observation has led to the empirical design method, which is not mechanistic but prescriptive. Example 4.2 presents a reinforced concrete deck design example using the empirical design method for interior bays.

4.6.6 Empirical Design Method
9.7.2

Example 4.2 Reinforced Concrete Deck Design (Empirical Design)

☐ **Design Requirement**
   Design a cast-in-place reinforced concrete deck for the steel beam bridge in Example 4.1 using the AASHTO empirical design method. The deck is supported by five steel girders spaced

at 8 ft 10 in. and an overhang 4 ft 5 in. wide, as shown in Figure Ex4.2-1. Assume $f_c' = 4$ k/in.^{2} and $f_y = 60$ k/in.^{2}

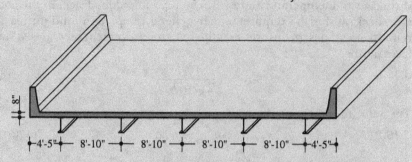

Figure Ex4.2-1
Superstructure dimensions.

4'-5"   8'-10"   8'-10"   8'-10"   8'-10"   4'-5"

The empirical deck design method is applicable to the interior bays, not the overhang.

☐ **Design Parameters**

5.12.3   Top cover $C_{top} = 2.5$ in. (exposure to deicing salts)

5.12.3   Bottom cover $C_{bottom} = 1$ in. (up to No. 11 bars)

Concrete compressive strength $f_c' \leq 5$ k/in.^{2}

9.7.1.1   Slab thickness $t_{slab} = 8$ in.   including sacrificial depth of 0.5 in.

Girder spacing $S = 8$ ft 10 in. $= 8.83$ ft

Traffic barriers are composite to the deck, and the deck is composite to the steel beams.
Full-depth diaphragms are used at the ends of the span.
There is no skew.

☐ **Effective Length**   9.7.2.3

8 ft 10 in. $-$ 12 in./2 $= 8.33$ ft

See Figure Ex4.2-2. The steel beam top flange is assumed to be 12 in. long.

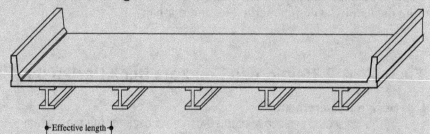

Figure Ex4.2-2
Effective length for deck on steel or concrete beams.

Effective length

□ **Check Design Conditions**   *9.7.2.4*

Cross frames or diaphragms are used throughout the cross section at lines of support. OK

The supporting components are made of steel and/or concrete. OK

The deck is fully cast in place and water cured. OK

The deck is of uniform depth except for hunches at girder flanges and other local thickening. OK

The ratio of effective length to design depth does not exceed 18.0 and is not less than 6.0.

$$8.33\text{ ft }(12\text{ in./ft})/(8\text{ in. }-0.5\text{ in.}) = 13.33 \quad \text{OK}$$

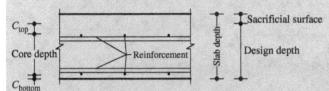

Figure Ex4.2-3
Design depth of reinforced concrete deck.

Core depth of the slab is not less than 4.0 in. OK

The effective length does not exceed 13.5 ft. OK

The minimum depth of the slab is not less than 7.0 in., excluding a sacrificial wearing surface. OK

There is an overhang beyond the centerline of the outside girder of at least 5.0 times the depth of the slab; this condition is satisfied if the overhang is at least 3.0 times the depth of the slab and a structurally continuous concrete barrier is made composite with the overhang. OK

The specified 28-day strength of the deck concrete is not less than 4.0 ksi. OK

The deck is made composite with the supporting structural components. OK

□ **Select Bottom Steel Reinforcement**

Assume No. 5 bars for both longitudinal and transverse directions. For the required reinforcement area of 0.27 in.²/ft, the corresponding spacing is

9.7.2.5 $S_{\text{reinforcement}} = 0.31\text{ in.}^2/0.27\text{ in.}^2/\text{ft} = 1.15\text{ ft}$

Use 1 ft, or 12 in.

□ **Select Top Steel Reinforcement**

Assume No. 4 bars for both longitudinal and transverse directions. For the required reinforcement area of 0.18 in.²/ft, the corresponding spacing is

9.7.2.5 $S_{reinforcement} = 0.20 \text{ in.}^2/0.18 \text{ in.}^2/\text{ft} = 1.11 \text{ ft}$

Use 1ft, or 12 in., as shown in Figure Ex4.2-4.

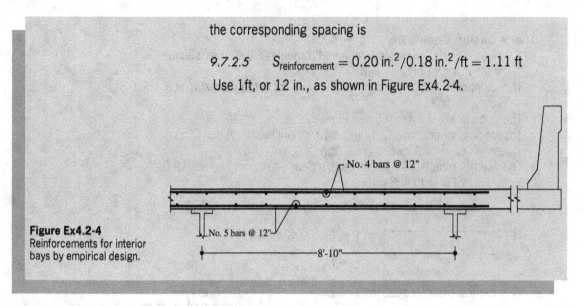

Figure Ex4.2-4
Reinforcements for interior bays by empirical design.

No. 4 bars @ 12"

No. 5 bars @ 12"

8'-10"

The empirical method prescribes a steel amount of 0.27 in.²/ft in each bottom layer (both the transverse and longitudinal layers) and 18 in.²/ft in each top layer if the conditions in the following section are met. The main reinforcement shall be placed between supporting beams. Spacing of steel shall not exceed 18 in. Reinforcing steel shall be Grade 60 or higher. All reinforcement shall be straight bars, except that hooks may be provided where required. The conditions that need to be satisfied for using this design are as follows:

❑ Cross frames or diaphragms are used throughout the cross section at lines of support.

❑ For cross sections involving torsionally stiff units, such as individual separated box beams, either intermediate diaphragms between the boxes are provided at a spacing not to exceed 25 ft or the need for supplemental reinforcement over the webs to accommodate transverse bending between the box units is investigated and reinforcement is provided if necessary.

❑ The supporting components are made of steel and/or concrete.

❑ The deck is fully cast in place and water cured.

❑ The deck is of uniform depth, except for haunches at girder flanges and other local thickening.

❑ The ratio of effective length S_e to design depth does not exceed 18 and is not less than 6, as shown in Figure 4.6-5.

❑ The core depth of the slab as indicated in Figure 4.6-5 is not less than 4 in.

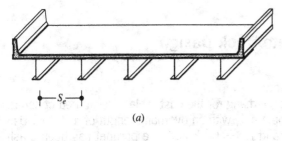

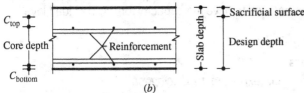

Figure 4.6-5
Geometric parameters: (a) deck effective length S_e, (b) design depth and core depth.

- ❑ The effective length S_e does not exceed 13.5 ft.
- ❑ The minimum depth of the slab is not less than 7 in., excluding a sacrificial wearing surface where applicable.
- ❑ There is an overhang beyond the centerline of the outside girder of at least 5 times the depth of the slab; this condition is satisfied if the overhang is at least 3 times the depth of the slab and a structurally continuous concrete barrier is made composite with the overhang.
- ❑ The specified 28-day strength of the deck concrete is not less than $4\,\mathrm{k/in.}^2$
- ❑ The deck is made composite with the supporting structural components.
- ❑ A minimum of two shear connectors at 24-in. centers shall be provided in the negative-moment region of continuous steel superstructures. For concrete girders, the use of stirrups extending into the deck shall be taken as sufficient to satisfy this requirement.

The deck overhang is required to be able to sustain not only the dead and live load as the deck in the interior bays but also truck collision load if it ever occurs. Accordingly, the specifications explicitly identify the following cases for overhang design (*A13.4*): Design Case 1 (Extreme Event II) for the horizontal force due to truck collision to the railing; Design Case 2 (Extreme Event II) for the vertical force due to truck collision to railing; and Design Case 3 (Strength I) for the dead and vehicular live loads to the overhang. Example 4.3 illustrates Design Case 1, and Example 4.4 demonstrates Design Cases 2 and 3.

4.6.7 Concepts for Deck Overhang Design

Example 4.3 Reinforced Concrete Deck Design (Overhang Design Case 1)

☐ **Design Requirement**

Design the deck overhang for the cast-in-place reinforced concrete deck in Example 4.1, with an overhang length of 4 ft 5 in. Use $f'_c = 4$ k/in.^{2} and $f_y = 60$ k/in.^{2} The parapet has been crash tested and is deemed to have the capacity for the TL-4 load level.
A13.3
Load modifier $\eta = 1.8$

☐ **Deck Parameters**

5.12.3 Top cover $C_{top} = 2.5$ in. (*exposure to deicing salts*)

5.12.3 Bottom cover $C_{bottom} = 1$ in.   (*up to No.11 bars*)

3.5.1 Concrete density $W_{concrete} = 0.145$ k/ft^3 (*Normal weight $f'_c \leq 5$ k/in^2*)

3.5.1 Future wearing surface (FWS) density $W_{FWS} = 0.14$ k/ft^3 (*bituminous*)
Future wearing surface thickness $t_{FWS} = 2.5$ in.

Slab thickness $t_{slab} = 8$ in.   with design depth of 7.5 in. excluding sacrificial surface of 0.5 in.

13.7.3.1.2 Deck overhang thickness $t_{overhang} = 9$ in.   with design depth of 8.5 in.
Girder spacing $S = 8$ ft 10 in. $= 8.83$ ft

☐ **Parapet Parameters**

Weight per foot of parapet: $W_{parapet} = 0.457$ k/ft

Base width $B_{base} = 1$ ft $3\frac{3}{4}$ in.

Parapet height $H_{parapet} = 42$ in. $= 3.5$ ft

Deck overhang needs to cover: Design Case 1 (Extreme II) – Horizontal force due to vehicle collision to railing; Design Case 2 (Extreme II) – Vertical force due to vehicle collision to railing; and Design Case 3 (Strength I): Dead and vehicular live loads to the overhang. This example addresses Design Case 1 and Design Cases 2 and 3 as well as several other items that are addressed in a later example.

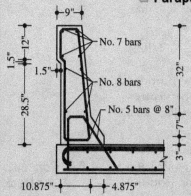

Figure Ex4.3-1
Railing reinforcement and dimension details.

☐ **Preparation for Design Case I: Bridge Railing Properties**

Refer to Figure Ex4.3-1 for railing details. Divide the railing into three portions for moment capacity calculations, as shown in Figure Ex4.3-2.

☐ **Flexural Resistance of Wall Beam about Vertical Axis, M_b**

For portion I defined in Figures Ex4.3-2 and 4.3-3:

$$\text{Area of portion I} = \frac{(9 \text{ in.} + 8.92 \text{ in.} + 1.5 \text{ in.}) \times (12 \text{ in.} + 1.5 \text{ in.})}{2}$$

$$- \left(\frac{1.5 \text{ in.} \times 1.5 \text{ in.}}{2} \right)$$

$$= 129.96 \text{ in.}^2$$

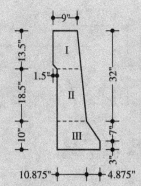

The cross section in Figure Ex4.3-3 for portion I is converted to an equivalent cross section in Figure Ex4.3-4, with the average depth $d_{average}$ computed here.

$$d_{average} = \frac{\text{area of portion I}}{\text{total height of portion I}} = \frac{129.96 \text{ in.}^2}{13.5 \text{ in.}} = 9.63 \text{ in.}$$

Figure Ex4.3-2
Division of parapet into three portions for calculating capacities M_b, M_w, and M_c.

$$d_s = d_{average} - \text{cover} - \frac{\text{bar diameter}}{2}$$

$$= 9.63 \text{ in.} - 1.5 \text{ in.} - \frac{1}{2} (0.88 \text{ in.}) = 7.69 \text{ in.}$$

$$A_s \text{ for two No. 7 bars} = 2 \left(0.6 \text{ in.}^2 \right) = 1.2 \text{ in.}^2$$

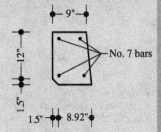

$$a = \frac{A_s F_y}{0.85 \, f'_c b} = \frac{1.2 \text{ in.}^2 \left(60 \text{ k/in.}^2 \right)}{0.85 \left(4 \text{ k/in.}^2 \right) 13.5 \text{ in.}} = 1.57 \text{ in.}$$

Figure Ex4.3-3
Portion I cross section for moment capacity M_b about vertical axis.

$$M_b = M_{b,\text{portion I}} = \varphi A_s F_y \left(d_s - \frac{a}{2} \right)$$

$$= 1 \left(1.2 \text{ in.}^2 \right) 60 \text{ k/in.}^2 \left(7.69 \text{ in.} - \frac{1.57 \text{ in.}}{2} \right) / 12 \text{ in./ft} = 41.43 \text{ kft}$$

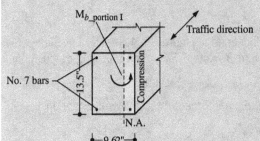

Figure Ex4.3-4
Equivalent cross section for portion I in Figure Ex4.3-3 for wall beam capacity M_b.

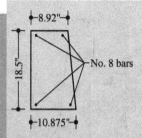

Figure Ex4.3-5
Portion II cross section for moment capacity M_b about vertical axis.

□ **Flexural Resistance of Wall about Vertical Axis, M_w**

For portion II defined in Figures Ex4.3-2 and 4.3-5:

The cross section in Figure Ex4.3-5 for portion II is converted to an equivalent cross section in Figure Ex4.3-6, with the average depth computed here.

$$d_{average} = \frac{(8.92 \text{ in.} + 10.88 \text{ in.})}{2} = 9.90 \text{ in.}$$

$$d_s = d_{average} - \text{cover} - \frac{\text{bar diameter}}{2}$$

$$= 9.90 \text{ in.} - 1.5 \text{ in.} - \frac{1 \text{ in.}}{2} = 7.90 \text{ in.}$$

$$A_S \text{ for two No. 8 bars} = 2 \left(0.79 \text{ in.}^2\right) = 1.57 \text{ in.}^2$$

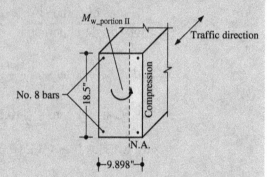

Figure Ex4.3-6
Equivalent cross section for portion II in Figure Ex4.3-5 for wall capacity M_w.

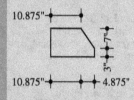

Figure Ex4.3-7
Portion III cross section for moment capacity M_w about vertical axis.

$$a = \frac{A_s F_y}{0.85 \, f'_c \, b} = \frac{1.57 \text{ in.}^2 \left(60 \text{ k/in.}^2\right)}{0.85 \left(4 \text{ k/in.}^2\right) 18.5 \text{ in.}} = 1.50 \text{ in.}$$

$$M_{w,\text{portion II}} = \varphi A_s F_y \left(d_s - \frac{a}{2}\right) = 1 \left(1.57 \text{ in.}^2\right)$$

$$\times 60 \text{ k/in.}^2 \left(7.90 \text{ in.} - \frac{1.50 \text{ in.}}{2}\right) /12 \text{ in./ft}$$

$$= 56.12 \text{ kft}$$

For portion III defined in Figures Ex4.3-2 and Ex4.3-8:
Use Figures Ex4.3-7 and 4.3-8 in the following calculations.

$$M_{w,\text{portion III}} = 0 \text{ kft}$$

There is no steel reinforcement in the cross section.

$$M_w = M_{w,\text{portion II}} + M_{w,\text{portion III}} = 56.12 \text{ kft}$$

$$+ \, 0 \text{ kft} = 56.12 \text{ kft}$$

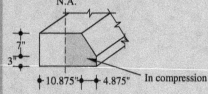

Figure Ex4.3-8
Equivalent cross section for portion III in Figure Ex4.3-7 for wall capacity M_w.

☐ Flexural Resistance of Wall about Horizontal Axis, M_c

For portion I defined in Figures Ex4.3-2 and Ex4.3-9:
Use Figures Ex4.3-9 and 4.3-10 in the following calculations.

$$d_{average} = 9.627 \text{ in.}$$

As portion I in M_b calculation.

$$d_s = d_{average} - \text{cover} - \frac{\text{bar diameter}}{2}$$

$$= 9.627 \text{ in.} - 1.5 \text{ in.} - \frac{0.625 \text{ in.}}{2}$$

$$= 7.815 \text{ in.}$$

$$A_S = \left(0.31 \text{ in.}^2\right) \left(\frac{12 \text{ in./ft}}{8 \text{ in.}}\right)$$

$$= 0.465 \text{ in.}^2/\text{ft}$$

$$a = \frac{A_s F_y}{0.85 f_c' b} = \frac{0.465 \text{ in.}^2 \left(60 \text{ k/in.}^2\right)}{0.85 \left(4 \text{ k/in.}^2\right) 12 \text{ in.}}$$

$$= 0.684 \text{ in.}$$

$$M_{c,\text{portion I}} = \varphi A_s F_y \left(d_s - \frac{a}{2}\right) = 1 \left(0.465 \text{ in.}^2/\text{ft}\right)$$

$$\times 60 \text{ k/in.}^2 \left(7.815 \text{ in.} - \frac{0.684 \text{ in.}}{2}\right) / 12 \text{ in./ft}$$

$$= 17.37 \text{ kft/ft}$$

For portion II defined in Figures Ex4.3-2 and Ex4.3-11:
Use Figures Ex4.3-11 and Ex4.3-12 in the following calculations.

$$d_{average} = 9.898 \text{ in.}$$

As in portion II of M_w calculation.

$$d_s = d_{average} - \text{cover} - \frac{\text{bar diameter}}{2} = d_{average}$$

$$= 9.898 \text{ in.} - 1.5 \text{ in.} - \frac{0.625 \text{ in.}}{2} = 8.086 \text{ in.}$$

Compression area depth a is the same as for portion I earlier.

$$M_{c,\text{portion II}} = \varphi A_s F_y \left(d_s - \frac{a}{2}\right) = 1 \left(0.465 \text{ in.}^2/\text{ft}\right)$$

$$\times 60 \text{ k/in.}^2 \left(8.086 \text{ in.} - \frac{0.684 \text{ in.}}{2}\right) / 12 \text{ in./ft}$$

$$= 18 \text{ kft/ft}$$

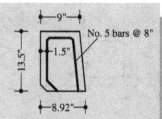

Figure Ex4.3-9
Portion I and steel for M_c about vertical axis.

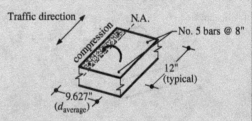

Figure Ex4.3-10
Equivalent cross section of portion I for M_c about vertical axis.

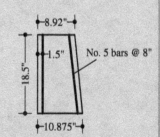

Figure Ex4.3-11
Portion II for M_c about horizontal axis.

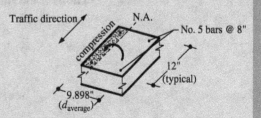

Figure Ex4.3-12
Equivalent cross section of portion II for M_c about horizontal axis.

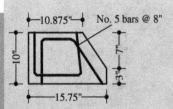

Figure Ex4.3-13
Portion III for M_c about horizontal axis.

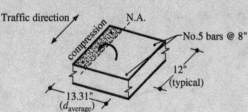

Figure Ex4.3-14
Equivalent cross section of portion III for M_c about horizontal axis.

For portion III is defined in Figures Ex4.3-2 and Ex4.3-13:
Use Figures Ex4.3-13 and 4.3-14 in the following calculations.

$$d_{average} = \frac{10.875 \text{ in.} + 15.75 \text{ in.}}{2} = 13.31 \text{ in.}$$

$$d_s = d_{average} - \text{cover} - \frac{\text{bar diameter}}{2} = 13.31 \text{ in.}$$

$$- 1.5 \text{ in.} - \frac{0.625 \text{ in.}}{2} = 11.5 \text{ in.}$$

Compression area depth a is the same as for portion I earlier.

$$M_{c,\text{portion III}} = \varphi A_s F_y \left(d_s - \frac{a}{2} \right)$$

$$= 1 \left(0.465 \text{ in.}^2/\text{ft} \right) 60 \text{ k/in.}^2$$

$$\times \left(11.5 \text{ in.} - \frac{0.684 \text{ in.}}{2} \right) /12 \text{ in./ft}$$

$$= 25.94 \text{ kft/ft}$$

Assume a failure mechanism including only portions I and II:

$$M_c = \frac{17.4 \text{ kft/ft} (13.5 \text{ in.}) + 18 \text{ kft/ft} (18.5 \text{ in.})}{32 \text{ in.}} = 17.73 \text{ kft/ft}$$

Assume a failure mechanism including the entire height:

$$M_c = \frac{17.37 \text{ kft/ft} (13.5 \text{ in.}) + 18.00 \text{ kft/ft} (18.5 \text{ in.}) + 25.94 \text{ kft/ft} (10 \text{ in.})}{42 \text{ in.}}$$

$$= 19.69 \text{ kft/ft}$$

☐ **Check for Impact within the Wall, Considering Failure Involving Portions I and II Only**

$$L_{c,\text{within wall I+II}} = \frac{L_t}{2} + \sqrt{\left(\frac{L_t}{2} \right)^2 + \frac{8H \left[M_b + M_w \right]}{M_c}} = \frac{3.5 \text{ ft}}{2}$$

A13.3

$$+ \sqrt{\left(\frac{3.5 \text{ ft}}{2} \right)^2 + \frac{8 (3.5 \text{ ft}) \left[41.43 \text{ kft} + 56.12 \text{ kft} \right]}{17.73 \text{ kft/ft}}}$$

$$= 14.28 \text{ ft}$$

and

$$R_{w,\text{within wall I+II}} = \left[\frac{2}{2L_c - L_t}\right]\left[8M_b + 8M_w + \frac{\left(M_c L_c^2\right)}{H}\right]$$

$$= \left[\frac{2}{2\,(14.28\text{ ft}) - 3.5\text{ ft}}\right]\left[8\,(41.43\text{ kft}) + 8\,(56.12\text{ kft})\right.$$

A13.2

$$\left. + \frac{(17.73\text{ kft/ft})\,(14.28\text{ ft})^2}{3.5\text{ ft}}\right] = 144.7\text{ k} > 54\text{ k}$$

OK for vehicle collision in the wall involving portions I and II only.
   A13.2

☐ **Check for Impact within the Wall, Considering Failure Involving Entire Height**

$$L_{c,\text{within wall I+II+III}} = \frac{L_t}{2} + \sqrt{\left(\frac{L_t}{2}\right)^2 + \frac{8H\,[M_b + M_w]}{M_c}} = \frac{3.5\text{ ft}}{2}$$

$$+ \sqrt{\left(\frac{3.5\text{ ft}}{2}\right)^2 + \frac{8\,(3.5\text{ ft})\,[41.43\text{ kft} + 56.12\text{ kft}]}{19.69\text{ kft/ft}}} = 13.66\text{ ft}$$

$$R_{w,\text{within wall I+II+III}} = \left[\frac{2}{2L_c - L_t}\right]\left[8M_b + 8M_w + \frac{\left(M_c L_c^2\right)}{H}\right]$$

$$= \left[\frac{2}{2\,(13.66\text{ ft}) - 3.5\text{ ft}}\right]\left[8\,(41.43\text{ kft}) + 8\,(56.12\text{ kft})\right.$$

$$\left. + \frac{19.69\text{ kft/ft}\,(13.66\text{ ft})^2}{3.5\text{ ft}}\right] = 153.66\text{ k} > 54\text{ k}$$

OK for vehicle collision in the wall involving entire parapet height.
   A13.2

☐ **Check for Impact at the End of a Wall or at a Joint, Considering Failure Involving Portions I and II Only**

$$L_{c,\text{wall end I+II}} = \frac{L_t}{2} + \sqrt{\left(\frac{L_t}{2}\right)^2 + H\frac{[M_b + M_w]}{M_c}} = \frac{3.5\text{ ft}}{2}$$

A 13.3.1-4

$$+ \sqrt{\left(\frac{3.5\text{ ft}}{2}\right)^2 + (3.5\text{ ft})\frac{[41.43\text{ kft} + 56.12\text{ kft}]}{17.73\text{ kft/ft}}}$$

$$= 6.47\text{ ft}$$

$$R_{w, \text{ wall end I+II}} = \left[\frac{2}{2L_c - L_t}\right]\left[M_b + M_w + \frac{(M_c L_c^2)}{H}\right]$$

$$= \left[\frac{2}{2\,(6.47\text{ ft}) - 3.5\text{ ft}}\right]\left[41.43\text{ kft} + 56.12\text{ kft}\right.$$

A 13.3.1-3

$$\left. + \frac{17.73\text{ kft/ft}\,(6.47\text{ft})^2}{3.5\text{ ft}}\right] = 65.59\text{ k} > 54\text{ k}$$

OK for vehicle collision at the end of a wall or at a joint involving portions I and II. *A13.2*

☐ **Check for Impact at the End of a Wall or at a Joint, Considering Failure Involving Entire Height**

$$L_{c, \text{ wall end I+II+III}} = \frac{L_t}{2} + \sqrt{\left(\frac{L_t}{2}\right)^2 + H\frac{[M_b + M_w]}{M_c}} = \frac{3.5\text{ ft}}{2}$$

$$+ \sqrt{\left(\frac{3.5\text{ ft}}{2}\right)^2 + (3.5\text{ ft})\frac{[41.43\text{ kft} + 56.12\text{ kft}]}{19.69\text{ kft/ft}}}$$

$$= 6.27\text{ ft}$$

$$R_{w, \text{ wall end I+II+III}} = \left[\frac{2}{2L_c - L_t}\right]\left[M_b + M_w + \frac{(M_c L_c^2)}{H}\right]$$

$$= \left[\frac{2}{2\,(6.27\text{ ft}) - 3.5\text{ ft}}\right]\left[41.43\text{ kft} + 56.12\text{ kft}\right.$$

$$\left. + \frac{19.69\text{ kft/ft}\,(6.27\text{ ft})^2}{3.5\text{ ft}}\right] = 70.51\text{ k} > 54\text{ k}$$

OK for vehicle collision at the end of a wall or at a joint involving entire parapet height. *A13.2*

The following design checks will use various free-body diagrams for static equilibrium shown in Figure Ex4.3-15.

☐ **Design Case 1 (Extreme-Event II Limit State) Check 1 at Parapet Face (Figure Ex4.3-16)**   *A13.4.1*

1.3.2.1	Resistance factor $\varphi_{\text{ext}} = 1$	
3.4.1	Component dead-load factor $\gamma_{p,DC} = 1.25$	
3.4.1	Vehicle collision load factor $\gamma_{CT} = 1$	

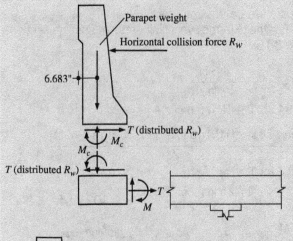

Figure Ex4.3-15
Free-body diagrams for design checking.

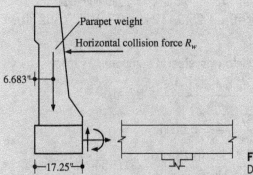

Figure Ex4.3-16
Design section of overhang at parapet face.

The axial tensile force is the maximum of the following values:

$$T_{\text{within wall I+II}} = \frac{R_{w,\text{within wall I+II}}}{L_{c,\text{within wall I+II}} + 2H} = \frac{144.72\ k}{14.28\ \text{ft} + 2\,(3.5\ \text{ft})}$$
$$= 6.80\ k/\text{ft}$$

$$T_{\text{within wall I+II+III}} = \frac{R_{w,\text{within wall I+II+III}}}{L_{c,\text{within wall I+II+III}} + 2H} = \frac{153.66\ k}{13.66\ \text{ft} + 2\,(3.5\ \text{ft})}$$
$$= 7.44\ k/\text{ft}$$

A13.4.2

$$T_{\text{wall end I+II}} = \frac{R_{w\ \text{wall end I+II}}}{L_{c,\ \text{wall end I+II}} + 2H} = \frac{65.59\ k}{6.47\ \text{ft} + 2\,(3.5\ \text{ft})}$$
$$= 4.87\ k/\text{ft}$$

$$T_{\text{wall end I+II+III}} = \frac{R_{w,\ \text{wall end I+II+III}}}{L_{c,\ \text{wall end I+II+III}} + 2H} = \frac{70.24\ k}{6.27\ \text{ft} + 2\,(3.5\ \text{ft})}$$
$$= 5.29\ k/\text{ft}$$

The worst condition to be designed for is $T = T_{\text{winthin wall I+II + III}}$ $= 7.44$ k/ft. For the extreme-event limit state and design section in Figure Ex4.3-16:

$$M_{\text{collision}} = M_{c,\text{ portion III}} + \frac{4.5 \text{ in.}}{12 \text{ in./ft}} T$$

$$= 25.94 \text{ kft/ft} + 0.375 \text{ ft} (7.44 \text{ k}) = 28.73 \text{ kft/ft}$$

$$M_{\text{DC deck}} = \frac{(9 \text{ in./12 in./ft}) \, 0.145 \text{ k/ft}^3 \, (17.25 \text{ in./12 in./ft})^2}{2}$$

$$= 0.112 \text{ kft/ft}$$

$$M_{\text{DC parapet}} = 0.457 \text{ k/ft} \frac{17.25 \text{ in.} - 6.683 \text{ in.}}{12 \text{ in./ft}} = 0.402 \text{ kft/ft}$$

$$M_u = \gamma_{CT} M_{\text{collision}} + \gamma_{p,DC} M_{\text{DC deck}} + \gamma_{p,DC} M_{\text{DC parapet}}$$

$$= 1(28.73 \text{ kft/ft}) + 1.25(0.112 \text{ kft/ft}) + 1.25(0.402 \text{ kft/ft})$$

$$= 29.37 \text{ kft/ft}$$

The required area of reinforcing steel is computed as follows:

$$d_e = t_{\text{overhang}} - C_{\text{top}} - \frac{\text{bar diameter}}{2} = 8.5 \text{ in.}$$

$$- 2.5 \text{ in.} - \frac{0.875 \text{ in.}}{2} = 5.56 \text{ in.}$$

$$R_n = \frac{M_u}{\varphi_{\text{ext}} b d_e^2} = \frac{(29.37 \text{ kft/ft}) \, 12 \text{ in./ft}}{(1) \, 12 \text{ in.}(5.56 \text{ in.})^2} = 0.95 \text{ k/in.}^2$$

$$\rho = 0.85 \left(\frac{f_c'}{f_y}\right) \left[1 - \sqrt{1 - \frac{2R_n}{0.85 f_c'}}\right]$$

$$= 0.85 \left(\frac{4 \text{ k/in.}^2}{60 \text{ k/in.}^2}\right) \left[1 - \sqrt{1 - \frac{2 \, (0.95 \text{ k/in.}^2)}{0.85 \, (4 \text{ k/in.}^2)}}\right] = 0.019$$

The above two equations are the general reinforced concrete design equations.

$$A_{s,\text{required}} = \rho b d_e = 0.019(12 \text{ in/ft})5.56 \text{ in.} = 1.27 \text{ in}^2/\text{ft}$$

Try one No. 5 bar and one No. 7 bundled at 8 in. One No. 7 bar at 8 in has been selected for the top reinforcement in the interior bays, which will extend into the overhang:

$$A_s = \left(0.6 \text{ in.}^2 + 0.31 \text{ in.}^2\right) \frac{12 \text{ in./ft}}{8 \text{ in.}} = 1.37 \text{ in.}^2/\text{ft} > A_{s,\text{required}}$$

$$= 1.27 \text{ in.}^2/\text{ft}$$

Check the provided resistance for the combined bending and tension, assuming an interaction relation:

$$\left(\frac{M_u}{\phi_b M_n}\right) + \left(\frac{P_u}{\phi_t P_n}\right) \le 1 \quad \text{or} \quad M_u \le \phi_b M_n \left(1 - \frac{P_u}{\phi_t P_n}\right)$$

$$a = \frac{A_s F_y}{0.85 f'_c b} = \frac{1.37 \text{ in.}^2 \left(60 \text{ k/in.}^2\right)}{0.85 \left(4 \text{ k/in.}^2\right) 12 \text{ in.}} = 2.01 \text{ in.}$$

$$1 - \frac{P_u}{\phi_t P_n} = 1 - \frac{7.44 \text{ k}}{1\left[(0.6 + 0.31) \text{ in.}^2 (12 \text{ in./8 in.}) + 0.44 \text{ in.}^2\right] 60 \text{ k/in.}^2} = 0.93$$

$$\phi_b M_n \left(1 - \frac{P_u}{\phi_t P_n}\right) = \phi_b A_s F_y \left(d_s - \frac{a}{2}\right)\left(1 - \frac{P_u}{\phi_t P_n}\right)$$

$$= 1(1.37 \text{ in.}^2)60 \text{ k/in.}^2 \left(8.5 \text{ in.} - 2.5 \text{ in.} - \frac{2.01 \text{ in.}}{2}\right)/12 \text{ in./ft } (0.93)$$

$$= 34.22 \text{ kft/ft}(0.93) = 31.82 \text{ kf/ft} > M_u = 29.37 \text{ kft/ft}$$

OK for vehicle collision at the parapet face.

☐ **Design Case 1 (Extreme-Event II limit state) Check 2 at Overhang (Figure Ex4.3-17).** See Figure Ex4.3-18 for moment load distribution.

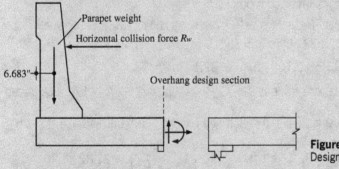

6.683″

Parapet weight

Horizontal collision force R_w

Overhang design section

Figure Ex4.3-17
Design section of overhang.

For the extreme-event limit state

$$M_{\text{collision}} = 28.73 \text{ kft/ft}$$

$$M_{\text{DC deck}} = \frac{\left(\frac{9 \text{ in.}}{12 \text{ in./ft}}\right) 0.145 \text{ k/ft}^3 \left(4.167 \text{ ft}\right)^2}{2} = 0.944 \text{ kft/ft}$$

$$M_{\text{DC parapet}} = 0.457 \text{ k/ft} \left(4.167 \text{ ft} - \frac{6.683 \text{ in.}}{1 \text{ in./ft}}\right) = 1.65 \text{ kft/ft}$$

$$M_{\text{DW FWS}} = \frac{(2.5 \text{ in./12 in./ft}) \, 0.14 \text{ k/ft}^3 \left(4.167 \text{ ft} - 1.438 \text{ ft}\right)^2}{2}$$

$$= 0.109 \text{ kft/ft}$$

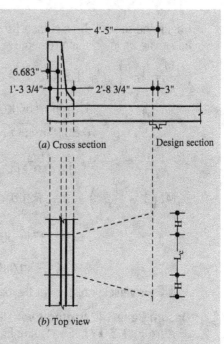

Figure Ex4.3-18
Collision moment load distribution in overhang.

(a) Cross section

Design section

(b) Top view

Check 2 total factored moment:

$$M_u = \gamma_{CT} M_{collision} + \gamma_{p,DC} M_{DC\ deck} + \gamma_{p,DC} M_{DC\ parapet} + \gamma_{p,DW} M_{DW\ FWS}$$

$$= 1(28.73\ \text{kft/ft}) + 1.25(0.944\ \text{kft/ft}) + 1.25(1.65\ \text{kft/ft})$$

$$+ 1.50(0.109\ \text{kft/ft}) = 32.14\ \text{kft/ft}$$

The axial tensile force is $T = T_{within\ wall\ I + II + III} = 7.44$ k/ft, calculated earlier.

Check provided flexural resistance for one No. 5 bar and one No. 7 bar bundled:

$$M_u = 32.14\ \text{kft/ft} > \phi_b M_n \left(1 - \frac{P_u}{\phi_t P_n}\right) = 31.82\ \text{kft/ft}$$

See check 1.

NG for vehicle collision at the overhang.

Change the negative-moment reinforcement in the overhang from No. 5 to No. 6.

$$R_n = \frac{M_u}{\varphi_{ext} b d_e^2} = \frac{(32.14\ \text{kft/ft})\ 12\ \text{in./ft}}{(1)\ 12\ \text{in.}\ (5.56\ \text{in.})^2} = 1.04\ \text{k/in.}^2$$

$$\rho = 0.85 \left(\frac{f_c'}{f_y}\right)\left[1 - \sqrt{1 - \frac{2R_n}{0.85 f_c'}}\right]$$

$$= 0.85 \left(\frac{4\ \text{k/in.}^2}{60\ \text{k/in.}^2}\right)\left[1 - \sqrt{1 - \frac{2\ (1.04\ \text{k/in.}^2)}{0.85\ (4\ \text{k/in.}^2)}}\right] = 0.021$$

$A_{s,required} = \rho b d_e = 0.021(12 \text{ in./ft})5.56 \text{ in.} = 1.4 \text{ in.}^2/\text{ft}$

$$A_s = (0.6 \text{ in.}^2 + 0.44 \text{ in.}^2)\frac{12 \text{ in./ft}}{8 \text{ in.}} = 1.56 \text{ in.}^2/\text{ft} > A_{s,\ required}$$

$$= 1.4 \text{ in.}^2/\text{ft}$$

Check the provided resistance for the combined bending and tension assuming an interaction relation:

$$\left(\frac{M_u}{\phi_b M_n}\right) + \left(\frac{P_u}{\phi_t P_n}\right) \le 1 \text{ or } M_u \le \phi_b M_n \left(1 - \frac{P_u}{\phi_t P_n}\right)$$

$$a = \frac{A_s F_y}{0.85 f'_c b} = \frac{1.56 \text{ in.}^2 \left(60 \text{ k/in.}^2\right)}{0.85 \left(4 \text{ k/in.}^2\right) 12 \text{ in.}} = 2.29 \text{ in.}$$

$$1 - \frac{P_u}{\phi_t P_n} = 1 - \frac{7.44 \text{ k}}{1\left[(0.6 + 0.44) \text{ in.}^2 (12 \text{ in./8 in.}) + 0.44 \text{ in.}^2\right] 60 \text{ k/in.}^2} = 0.94$$

$$\phi_b M_n \left(1 - \frac{P_u}{\phi_t P_n}\right) = \phi_b A_s F_y \left(d_s - \frac{a}{2}\right)\left(1 - \frac{P_u}{\phi_t P_n}\right)$$

$$= 1(1.56 \text{ in.}^2)60 \text{ k/in.}^2 \left(8.5 \text{ in.} - 2.5 \text{ in.} - \frac{2.29 \text{ in.}}{2}\right)/12 \text{ in./ft}(0.94)$$

$$= 37.87 \text{ kft/ft}(0.94) = 35.6 \text{ kf/ft} > M_u = 32.14 \text{ kft/ft}$$

`OK` for vehicle collision in overhang.

☐ **Design Case 1 (Extreme-Event II Limit State) Check 3 in first bay (Figure Ex4.3-19).**

This case will not control, compared with Check 2. This cross section is near that focused in Check 2. Thus, the total design moment is similar, as analogically seen in the moment distribution due to a unit parapet weight in Figure Ex4.3-20. The capacity of this section is the same as that in Check 2. Therefore this check will not control, while there is a significant reserve strength at the overhang section in Check 2 between the provided strength 32.16 kft/ft and the required 29.35 kft/ft.

`OK` for vehicle collision in the deck's first bay.

Use one No. 6 bar and one No. 7 bar bundled at 8 in. as shown in Figure Ex4.3-21.

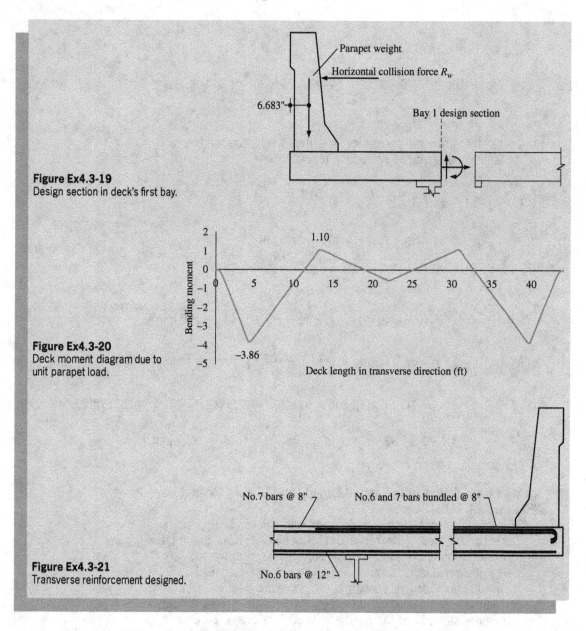

Figure Ex4.3-19
Design section in deck's first bay.

Figure Ex4.3-20
Deck moment diagram due to unit parapet load.

Figure Ex4.3-21
Transverse reinforcement designed.

The collision loads are much more unpredictable and more complex to analyze compared with the self-weight DC and truck load LL. The AASHTO specifications adopt a yield line method for estimating the forces that may be transferred to the deck overhang should such a collision take place. Then the design can proceed accordingly. The main philosophy here is to design the deck overhang stronger than the railing so that it will not fail before

Table 4.6-2
CT load magnitude depending on crash requirement for railing

Design Forces and Designations	Railing Test Levels					
	TL-1	TL-2	TL-3	TL-4	TL-5	TL-6
F_t Transverse (k)	13.5	27	54	54	124	175
F_L Longitudinal (k)	4.5	9	18	18	41	58
F_v Vertical (k) down	4.5	4.5	4.5	18	80	80
L_t and L_L (ft)	4	4	4	3.5	8	8
L_v (ft)	18	18	18	18	40	40
H_e (min) (in.)	18	20	24	32	42	56
Minimum height H of rail (in.)	27	27	27	32	42	90

Source: AASHTO LRFD Bridge Design Specifications, 2012. Used by permission.

the railing when a truck collision to the railing should occur. Table 4.6-2 displays the forces F_t, F_v, and F_L the railing is required to be able to sustain in the transverse, vertical, and longitudinal directions, respectively as indicated in Figure 4.6-6. While the vertical force F_v is directly applied to the deck overhang (to be addressed in Design Case 2) and the longitudinal force is resisted through the support of the span, only the transverse force F_t is transferred through the railing, addressed in Design Case 1. As seen, the force requirement depends on the railing test level for six different categories of truck traffic. In reality, the railing is often designed, constructed, and crash tested to have a higher capacity than that in Table 4.6-2 to be on the conservative side. Therefore, to design the deck overhang to be stronger

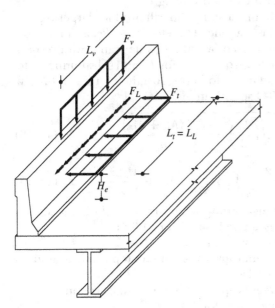

Figure 4.6-6
Forces in concrete parapet at truck collision.

than the railing, the force requirement in Table 4.6-2 should not be used directly. Rather, the capacity of the railing needs to be reasonably estimated, and then the deck overhang can be designed to a maximum between the railing capacity and the forces in Table 4.6-2 for which the railing has been crash tested. For proven capacity using a crash test, railings slightly modified from those physically crash tested are also acceptable to the AASHTO specifications.

To carry out this philosophy, the designer needs to determine the railing's capacity. The resulting deck overhang will then be designed to have a higher capacity than the railing so that it will not fail before the railing. The method given in the AASHTO specifications to estimate the load-carrying capacity of concrete parapet is briefly explained next, particularly with respect to the transverse load.

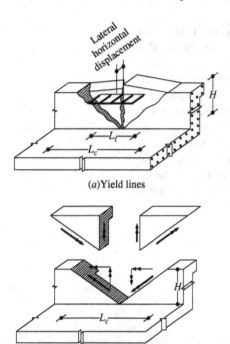

(a)Yield lines

(b) Yield line model

Figure 4.6-7
Yield lines of deck overhang failure under truck collision CT. *Source: AASHTO LRFD Bridge Design Specifications*, 2012. Used by permission.

Figure 4.6-7 shows the yield lines due to transverse collision load prescribed in the AASHTO specifications as the basis for further analysis for the capacity of the railing. It is seen that three yield lines are present. Free-body diagrams of the three pieces of the parapet separated by the yield lines are also shown in Figure 4.6-7. The free-body diagrams expose the internal forces at failure. These forces can be derived according to the reinforcement and the concrete cross section provided if the affected length L_c in the figure is given, which is apparently a function of the transverse force denoted as R_w. The parapet's capacity is thus quantified by R_w. The maximum between R_w and F_t in Table 4.6-2 will be the design load CT for the deck overhang.

The yield line assumption will produce larger R_w values than the real one. Therefore, the R_w as a function of L_c is minimized by setting its derivative with respect to L_c equal to zero. Solving for L_c and substituting L_c to find R_w lead to the following formulas for L_c and R_w in the AASHTO specifications:

$$L_c = \frac{L_t}{2} + \sqrt{\left(\frac{L_t}{2}\right)^2 + \frac{8H\left(M_b + M_w\right)}{M_c}} \qquad (4.6\text{-}18)$$

$$R_w = \left(\frac{2}{2L_c - L_t}\right)\left(8M_b + 8M_w + \frac{M_c L_c^2}{H}\right) \qquad (4.6\text{-}19)$$

where L_t = distance over which transverse collision load F_t is applied (see Figure 4.6-6)
  H = height of railing
  M_b = moment capacity of parapet top beam around vertical axis
  M_w = moment capacity of parapet wall around the same axis as for M_b

M_c = moment capacity of parapet wall around horizontal axis
L_c = maximum length of parapet over which yield lines are
formed (see Figure 4.6-6)

Figure 4.6-8 contains three free-body diagrams for designing three corresponding critical cross sections for overhang design. The forces include the dead load DC and the transverse collision load CT. According to the definition of Extreme–Event II limit state discussed in Chapter 3, for the first cross section at the parapet toe, no live load is possible to be applied. For the second and third cross section (outside and inside face of the first bay), the live load is also included. Depending on the overhang length, the live-load moment may or may not be negligible. As presented in Chapter 3, the Extreme–Event II load combination is as follows:

4.6.8 Design Case 1 Load (Transverse CT) under Extreme-Event II Limit State

3.4.1 $1.25\,DC + 1.50\,DW + 0.5\,LL\,(1 + IM) + 1.0\,CT$ (4.6-20)

This design case is illustrated in Example 4.3 for an RC deck on steel beams, which is actually valid for concrete beams with the same beam spacing as well.

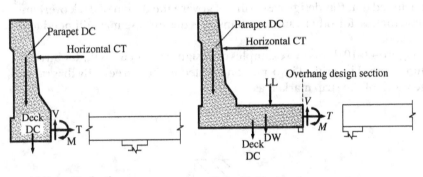

(a) Parapet toe section   (b) Outside bay 1 section

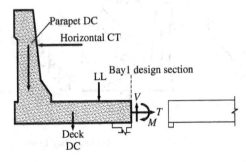

(c) Inside bay1 section

Figure 4.6-8
Loads in Design Case 1 under extreme-event II limit state.

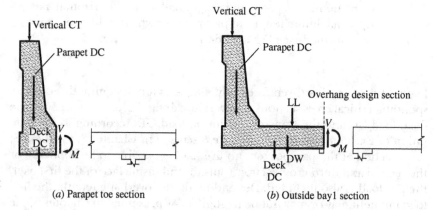

Figure 4.6-9
Loads of Design Case 2
under Extreme Event II limit
state.

(a) Parapet toe section   (b) Outside bay1 section

4.6.9 Design Case 2 Load (Vertical CT) under Extreme-Event II Limit State

Design Case 2 for deck overhang is under the same Extreme Event II limit state as Design Case 1 but covers the vertical CT load, not the transverse CT load. Figure 4.6-9 shows the same free-body diagrams for the same cross sections in Figure 4.6-8 for Design Case 1. It is seen that, by comparison of Figures 4.6-8 and 4.6-9, the resulting moment on the cross sections for the latter will be much smaller than for the former. Considering that the concrete parapet possesses a high strength along its length, it can be readily concluded that this design case will not govern the design of deck overhang. Nevertheless, for other railing systems, this design case may still need to be checked.

Figure 4.6-10 shows two examples of design cross sections under Strength I limit state. The critical information needed in these free-body diagrams is the width of the strip marked as

$$45 + 10X \tag{4.6-21}$$

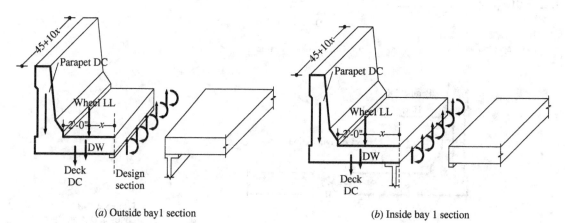

(a) Outside bay1 section   (b) Inside bay 1 section

Figure 4.6-10
Loads in Design Case 3 under Strength I limit state.

where X in feet is the distance between the wheel live load and the cross section being designed and the resulting value is in inches. The obtained width is considered to effectively participate in resisting the moment due to the live load. Namely, the wheel load–induced moment 16 k with an arm of X in feet is uniformly distributed over the width $45 + 10X$ in inches.

The dead-load moment is much easier to determine since it is uniformly distributed over the same length. Then the loads are combined under the Strength I limit state:

3.4.1 $$1.25\,\mathrm{DC} + 1.50\,\mathrm{DW} + 1.75\,\mathrm{LL}\,(1 + \mathrm{IM}) \qquad (4.6\text{-}22)$$

Note that LL also needs to include the multiple presence factor 1.2 for one lane load as the worst load.

This design case is illustrated in Example 4.4 for an RC deck on steel beams and is applicable to the same deck on concrete beams with the same beam spacing.

Example 4.4 Reinforced Concrete Deck Design (Overhang Design Cases 2 and 3, Other Checks)

❑ **Design Requirement**

Design the deck overhang in Examples 4.1 and 4.3 for Design Cases 2 and 3 and the longitudinal reinforcements and development lengths.

Load modifier $\eta = 1.0$

❑ **Deck Parameters**

5.12.3   Top cover $C_{top} = 2.5$ in.   (exposure to deicing salts)

5.12.3   Bottom cover $C_{bottom} = 1$ in.   (up to No. 11 bars)

9.7.1.1   Slab thickness $t_{slab} = 8$ in.   with design depth of 7.5 in.

13.7.3.1.2   Deck overhang thickness $t_{overhang} = 9$ in.   with design depth 8.5 in.

         Girder spacing $S = 8$ ft 10 in. $= 8.83$ ft

❑ **Design Case 2—Design for Vertical Collision Force**

This case of vertical force does not control for concrete parapets.

OK for Extreme-event II for vertical collision load on overhang.

❑ **Design Case 3—Design for Strength I Limit State: Check 4 at Design Section in Overhang**

Strip width over which live load is distributed for the distance between the wheel load and design section X, with X in feet and deck width in inches:

4.6.2.1.3 $W_{overhang} = 45.0 + 10.0X$

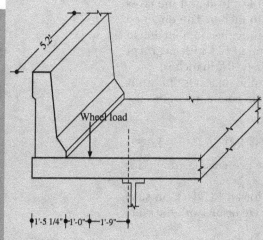

Figure Ex4.4-1
Design section of overhang.

For $X = 1.75$ ft as shown in Figure Ex4.4-1,

$$W_{overhang} = 45.0 + 10.0\,(1.75) = 62.5 \text{ in.}$$
$$= 5.2 \text{ ft}$$

$$M_{DC\,deck} = \frac{\left[(9 \text{ in.}/12 \text{ in.}/ft)\ 0.145 \text{ k/in.}^3 \times (4.167 \text{ ft})^2\right]}{2}$$
$$= 0.944 \text{ kft/ft}$$

$$M_{DC\,parapet} = 0.457 \text{ k/ft} \left(4.167 \text{ ft} - \frac{6.683 \text{ in.}}{12 \text{ in.}/ft}\right)$$
$$= 1.65 \text{ kft/ft}$$

$$M_{DC\,FWS} = \frac{\left[(2.5 \text{ in.}/12 \text{ in.}/ft)\ 0.14 \text{ k/ft}\,(4.167 \text{ ft} - 1.313 \text{ ft})^2\right]}{2}$$
$$= 0.119 \text{ kft/ft}$$

Use a multiple presence factor 1.2 for one lane load and an impact factor of 0.33.

$$M_{LL} = (1 + IM)\,1.2 \left(\frac{16 \text{ k}\,(1.75 \text{ ft})}{W_{overhang}}\right) 1$$

$$= (1.33)\,1.2 \left(\frac{16 \text{ k}\,(1.75 \text{ ft})}{5.2 \text{ ft}}\right) 1 = 8.59 \text{ kft/ft}$$

$$M_{total} = \gamma_{p,DC} M_{DL\,SLAB} + \gamma_{p,DC} M_{DL\,PARA} + \gamma_{p,DW} M_{DW\,FWS} + \gamma_{LL} M_{LL}$$

$$= 1.25\,(0.944 \text{ kft/ft}) + 1.25\,(1.65 \text{ kft/ft}) + 1.5\,(0.109 \text{ kft/ft})$$

$$+ 1.75\,(8.59 \text{ kft/ft}) = 18.44 \text{ kft/ft}$$

This total moment will not control, compared with Design Case 1 Check 2, where the design moment is 32.14 kft/ft in Example 4.3.
OK for Strength I moment in overhang.

☐ **Design Case 3—Design for Strength I Limit State: Check 5 at Design Section in First Bay**   *4.6.2.1.3*

$$X = 1 \text{ ft } 11\frac{3}{4} \text{ in.} = 1.98 \text{ ft}$$
$$W_{overhang} = 45 + 10\,(1.98) = 64.8 \text{ in.} = 5.4 \text{ ft}$$

Since this cross section is only about 6 in. apart from that used in Check 4 as shown in Figure Ex4.4-2, the total design moment will not increase significantly from 18.44 kft/ft. Thus, this cross section will not control.
OK for Strength I moment in first bay.

☐ Check Service I Limit State for Cracking in Overhang

In most deck overhang design cases, this check does not control. Therefore, the computations for cracking check are not shown here.
OK *for Service I for moment in overhang*

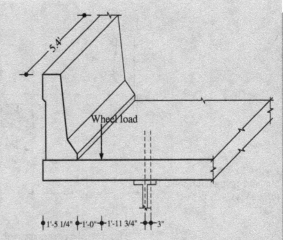

Figure Ex4.4-2
Design section of first bay.

☐ Design for Overhang Cutoff Length

The additional No. 6 bar required for extreme-event II limit state extends into the first bay. Its cutoff length is designed here to cover both the strength and the development of strength. The coverage for strength is give first.

The additional No. 6 bar is required to meet the strength requirement controlled in Design Case 1 Check 2 in the overhang design example. The corresponding total moment requirement is 32.14 kft/ft itemized as follows:

$$M_u = \gamma_{CT} M_{collision} + \gamma_{p,DC} M_{DC\ deck} + \gamma_{p,DC} M_{DC\ parapet} + \gamma_{p,DW} M_{DW\ FWS}$$

$$= 1(28.73\ kft/ft) + 1.25(0.944\ kft/ft) + 1.25(1.65\ kft/ft) + 1.50(0.109\ kft/ft)$$

$$= 28.73\ kft/ft + 1.18\ kft/ft + 2.06\ kft/ft + 0.16\ kft/ft = 32.14\ kft/ft$$

The No. 7 bar alone without the additional No. 6 bar is required by Strength I limit state for the interior bay's negative moment. The corresponding total moment requirement is 12.76 kft/ft, also itemized as follows:

$$M_{NEG} = \gamma_{p,DC} M_{DL\ slab} + \gamma_{p,DC} M_{DL\ parapet} + \gamma_{p,DW} M_{DW\ FWS} + \gamma_{LL} M_{LL}$$

$$= 1.25(0.097\ k/ft)9.76\ ft(1\ ft) + 1.25(0.457\ k/ft)3.86\ ft^2$$

$$+ 1.5(0.029\ k/ft)9.76\ ft(1\ ft) + 1.75(5.11\ kft)$$

$$= 1.18\ kft/ft + 2.21\ kft/ft + 0.42\ kft/ft + 8.94\ kft/ft = 12.76\ kft/ft$$

Design for overhang cutoff length now is equivalent to designing a length X indicated in Figure Ex4.4-3 at which the extreme-Event

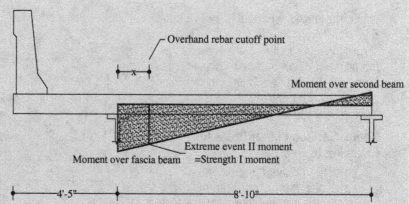

Figure Ex4.4-3
Moment diagram for collision
force in first bay of deck.

II limit state requirement becomes the same as the strength I limit state. Since the moments for the deck, future wearing surface, and parapet in the Extreme-Event II check are not greater than the corresponding items in the strength I check, when the collision moment 28.73 kft/ft reduces to a distance X to 8.94 kft/ft for truck live load, the extreme-Event II requirement will become the same as strength I. That X will represent the cutoff point where the additional No. 5 bar will not be needed, considering strength. A length for strength development will need to be added to determine the cutoff point for satisfying both strength and development requirements. We use this idea to determine X as follows. The distribution of the collision moment is assumed to follow the same trend as the moment diagram of the parapet load moment given in Figure Ex4.4-4. For convenience, this distribution is expressed with X in feet:

$$-3.86 + 0.562X$$

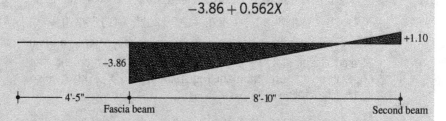

Figure Ex4.4-4
Assumed collision moment diagram resembling moment of parapet weight.

Based on the assumed similarity, the following relation is established to solve for X:

$$\frac{28.73 \text{ kft/ft}}{-3.86} = \frac{8.94 \text{ kft/ft}}{-3.86 + 0.562X}$$

Find $X = 4.7$ ft to satisfy this required similarity relation, as the theoretical cutoff length for the No. 6 bar considering the strength requirement.

The general additional length to extend beyond this theoretical cutoff point is the maximum of the following three requirements: *5.11.1.2*

Effective depth of the member: $d_e = 7.5$ in. $- 2.5$ in. $- \dfrac{0.625 \text{ in.}}{2}$
$= 4.69$ in.

15 times the nominal bar diameter: $15(0.625 \text{ in.}) = 9.4$ in.

$\frac{1}{20}$ of the clear span: $\frac{1}{20}$ (8.83 ft) 12 in./ft $= 5.298$ in.

Use 10 in. The total required length beyond the centerline of the fascia beam into the first bay is

$$\text{Cutoff}_{total} = 4.7 \text{ ft } (12 \text{ in./ft}) + 10 \text{ in.} = 67 \text{ in.}$$

☐ **Check for Development Length Requirements**
The basic development length is the larger of the following three values:

5.11.2.1 $\dfrac{1.25 \text{ (bar area) } f_y}{\sqrt{f_c'}}$ or 0.4(bar diameter) f_y or 12 in.

$$\frac{1.25 \left(0.44 \text{ in.}^2\right)\left(60 \text{ k/in.}^2\right)}{\sqrt{4 \text{ k/in.}^2}} = 16.51 \text{ in.}$$

$$0.4 \left(0.75 \text{ in.}\right)\left(60 \text{ k/in.}^2\right) = 18 \text{ in.}$$

Thus, use $l_d = 18$ in. with the following modification factors:

For epoxy coated bars: 1.2   *5.11.2.1*
For two bundled bars: 1.0   *5.11.2.3*

Use $l_d = 18$ in.(1.2)1=21.6 in. The required length past the cneterline of the fascia girder is

$$3 \text{ in.} + l_d = 24.6 \text{ in.}$$
$$\text{Use } 4.7 \text{ ft } (12 \text{ in./ft}) + 24.6 \text{ in.} = 81 \text{ in.} > 67 \text{ in.}$$

OK for overhang reinforcement cutoff length.

☐ **Design Bottom Longitudinal Distribution Reinforcement**
The bottom longitudinal distribution reinforcement is required as a minimum percentage of the primary (transverse) reinforcement and based on whether the primary reinforcement is parallel or perpendicular to traffic.
Requirement:

$$A_{bottom\%} = \frac{220}{\sqrt{S_e}} = \frac{220}{\sqrt{8 \text{ ft } 10 \text{ in.} - 12 \text{ in.} + 12 \text{ in.}/2}}$$

9.7.3.2

$$= \frac{220}{\sqrt{8.33 \text{ ft}}} = 76\% \qquad \text{required} \leq 67\%$$

Use $A_{\text{bottom\%}} = 67\%$. For No. 6 bars at 12 in. selected for primary positive moment in interior bays,

$$A_{s,\text{bottom longitudinal}} = A_{\text{bottom\%}} A_s = 0.67\left(0.44 \text{ in.}^2/\text{ft}\right) = 0.29 \text{ in.}^2/\text{ft}$$

Required spacing using No. 6 bars in the longitudinal direction in the bottom face:

$$\text{Spacing} = \frac{\text{bar area}}{A_{s,\text{bottom longitudinal}}} = \frac{0.44 \text{ in.}^2}{0.29 \text{ in.}^2/\text{ft}} \, 12 \text{ in.}/\text{ft} = 18.2 \text{ in.}$$

Use No. 6 bars at 18-in. spacing for the bottom longitudinal reinforcement, as shown in Figure Ex4.4-5.
OK for bottom longitudinal reinforcement.

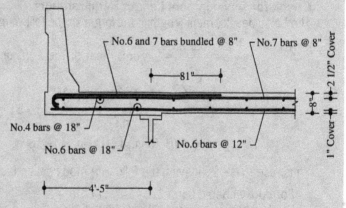

Figure Ex4.4-5
Reinforcement details.

☐ **Design Top Longitudinal Distribution Reinforcement**
Requirements for longitudinal temperature and shrinkage reinforcement for surfaces exposed to daily temperature variation:

5.10.8 $$A_s = \frac{1.3\,bh}{2\,(b+h)\,f_y} = \frac{1.3\,(106)\,8}{2\,(106+8)\,60} = 0.062 \text{ in.}^2/\text{ft}$$

and

5.10.8 $$0.11 \leq A_s \leq 0.60$$

The maximum spacing of the temperature and shrinkage reinforcement shall not exceed 3.0 times the deck thickness or 18 in.
Check No. 4 bars at 18-in. spacing:

$$0.11 < \text{provided } A_s = 0.13 \text{ in.}^2/\text{ft} < 0.60$$

Use No. 4 bars at 18-in. spacing for the top longitudinal temperature and shrinkage reinforcement.
OK for top longitudinal reinforcement.
All reinforcements are shown in Figure Ex4.4-5.

4.7 Design of Steel I Beams

Figure 4.7-1 displays a typical design procedure for a steel beam superstructure in accordance with the AASHTO specifications. Examples 4.5 to 4.8 illustrate the design of a steel I-beam superstructure with rolled beams. Examples 4.9 to 4.11 are for a steel I-beam superstructures with plate girders.

Load effect estimation may need to be carried out separately for exterior beams and interior beams if they obviously carry different portions of the deck. The exterior beams have no other beams to share the load with at one of the two sides, but the interior beams do have other beams on both sides to distribute the load. Therefore, these two groups are often treated differently in design. Of course they may be made identical particularly when future widening is foreseeable.

4.7.1 Dead-Load Effects

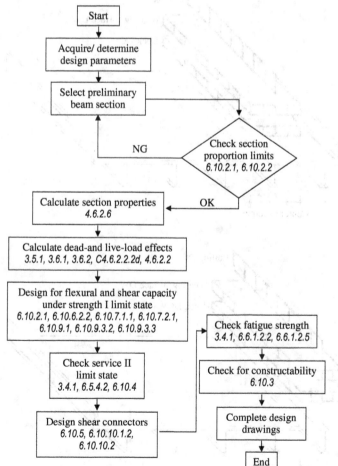

Figure 4.7-1
Recommended steel composite I beam design procedure for short and medium spans.

Dead-load effect calculation for primary beams in beam bridges is relatively simple since the concept of tributary area can be used for most dead loads if not all. Figures 4.7-2 and 4.7-3 show dead loads on an interior and an exterior beam, respectively, based on this concept. The railings may be the most complex items in dead-load distribution, since they are carried by more than the exterior beams. However, it is very popularly practiced that the railing's weight is distributed evenly among all the beams in the cross section because the deck is considered to be rigid enough. Note that if the deck were infinitely rigid the beams would indeed share loads on the deck evenly. For overwhelmingly used reinforced concrete decks cast in place, this approach is considered to be acceptable. Of course, 3D analysis can be performed to find more accurate distribution of the railing load to each

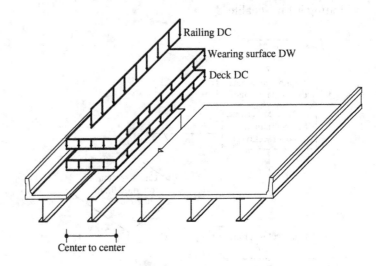

Figure 4.7-2
Dead load on interior beam.

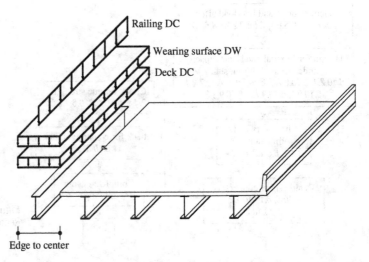

Figure 4.7-3
Dead load on exterior beam.

beam. However, since the deck usually occupies the majority of the dead load on a beam, inaccuracy in estimating the portion of the railing for DC carried by a particular beam may not be that critical anyway. As indicated qualitatively in Figures 4.7-2 and 4.7-3 the railings' gravity force is evenly distributed over all the beams in the cross section.

Examples 4.5 and 4.9 illustrate typical calculations for estimating dead load effects for design.

Live-load distribution is much more complex than dead-load distribution, because the position of live load on the bridge span can be a significant variable in both the longitudinal and transverse directions. Apparently, a 3D structural analysis may be applied to find how much load is carried by each beam in the cross section for the worst situation of each beam. However, the AASHTO specifications also offer a simplified beamline analysis method. Following this method, the user needs to perform structural analysis of the bridge as a beam (a simple span as a simply supported beam or continuous spans as a continuous beam) for one lane of truck live load positioned longitudinally at the worst location, depending on the load effect of interest (moment or shear). Then the designer needs to multiply the resulting load effect by the number of lanes available along with the multiple presence factor (Chapter 3) to find the total load effect of the bridge, then multiply that with a code specified load distribution factor to distribute the total load effect to one beam to be designed. This process eliminates complex and time-consuming refined analysis (often 3D analysis using numerical methods implemented in computer software programs).

This method includes tables of the so-called lateral load distribution factor for a variety of commonly used superstructure cross sections (such as concrete deck slab on steel beams, on prestressed concrete I beams, and on prestressed multibox beams). Two different load effects are covered in these tables: moment and shear. The specifications also provide simplified correction factors for skewed spans. These distribution factors and skew correction factors are derived from more complex 3D analysis of many bridge spans. Due to the scope of the analysis cases, the following ranges are given in the AASHTO specifications as the conditions for applying the simplified beamline analysis method:

4.7.2 Live-Load Effects

- ☐ The deck width is constant.
- ☐ The number of beams is not less than 4.
- ☐ The beams are parallel and have approximately the same stiffness.
- ☐ The overhang width minus the barrier width is less than 3.0 ft.
- ☐ Curvature in the plan is zero.
- ☐ The following applicability requirements are met: (i) $3.5 \text{ ft} \leq S \leq 16 \text{ ft}$, where S is the beam spacing; (ii) $4.5 \text{ in.} \leq t \leq 12 \text{ in.}$, where t is the deck slab thickness; (iii) $20 \text{ ft} \leq L \leq 240 \text{ ft}$, where L is the beam span length; (iv) $10,000 \text{ in.}^4 \leq K_g \leq 17,000,000 \text{ in.}^4$, where K_g is a stiffness factor to de defined below; (v) $-1 \text{ ft} \leq d_e \leq 5.5 \text{ ft}$, where d_e is the transverse

distance between the inside web face of the exterior girder and the toe of the curb in feet.

For steel beams supporting a reinforced concrete deck, the distribution factors DF are given here, taken from the AASHTO specifications.

Distribution Factors DF for Interior Steel Beams

❑ DF for moment for one lane loaded:

$$4.6.2.2 \quad \mathrm{DF}_{\text{moment interior } 1} = 0.06 + \left(\frac{S}{14}\right)^{0.4} \left(\frac{S}{L}\right)^{0.3} \left[\frac{K_g}{12\,L\,t^3}\right]^{0.1} \quad (4.7\text{-}1)$$

where S = beam spacing (ft)
L = beam span length (ft)
t = concrete deck slab thickness (in.)
K_g = longtudinal stiffness parameter (in.^{4}) = $n(I + A\,e_g^2)$
4.6.2.2

where $n = E_b/E_d$, ratio of Young's modulus of beam and deck slab *4.6.2.2*
I = moment of inertia of noncomposite beam (in.^{4})
A = cross-sectional area of noncomposite beam (in.^{2})
e_g = distance between centers of gravity of noncomposite beam and deck (in.)

Note that, with the bridge owner's concurrence, $\left[K_g/\left(12\,L\,t_s^3\right)\right]^{0.1} = 1.02$ is allowed in the specifications as a first estimate. This is particularly helpful when the cross section has not been finalized yet, which is often the case because DF is needed to find the design load effects to complete design. Until then, the cross section and K_g are unknown. It also needs to be noted that the multiple presence factor of 1.2 in Table 3.3-1 for one lane has been included in Eq. 4.7-1 and similar formulas in the specifications (but not those referring to the so-called lever rule, to be seen below).

❑ DF for moment for two or more lanes loaded:

$$4.6.2.2 \quad \mathrm{DF}_{\text{moment interior } 2} = 0.075 + \left(\frac{S}{9.5}\right)^{0.6} \left(\frac{S}{L}\right)^{0.2} \left[\frac{K_g}{12\,L\,t^3}\right]^{0.1} \quad (4.7\text{-}2)$$

Between the DF values for one lane and multiple lanes loaded with the multiple presence factor already included, whichever is larger will control the capacity of the beam. This concept is applicable in designing shear and other beams. Again, with the bridge owner's concurrence, $\left[K_g / \left(12\,L\,t_s^3\right)\right]^{0.1} = 1.02$ is allowed in the specifications.

❑ DF for shear for one lane loaded:

4.6.2.2 $\mathrm{DF}_{\text{shear interior 1}} = 0.36 + \dfrac{S}{25}$ (4.7-3)

❑ DF for shear for two or more lanes loaded:

4.6.2.2 $\mathrm{DF}_{\text{shear interior 2}} = 0.2 + \left(\dfrac{S}{12}\right) - \left(\dfrac{S}{35}\right)^2$ (4.7-4)

Again, whichever is larger between one lane and multiple lanes loaded shall be used in the design to distribute the load to the beam. The multiple presence factor has been included in both equations.

Distribution Factors DF for Exterior Steel Beams

❑ DF for moment for one lane loaded: Use the lever rule along with the multiple presence factor of 1.2 in Table 3.3-1. *3.6.1.1.2*

The so-called lever rule is referred to many times in the AASHTO specifications for structural analysis and load effect distribution. It is a simple mechanics model thought to be analogical to load distribution here. It can be easily and best understood using the simple graph in Figure 4.7-4. Under a unit load 1 consisting of two wheel loads each at 0.5, the reaction at the support is taken as the distribution factor of that support beam. The unit load 1 represents one truck load and the two 0.5 loads indicate two wheel lines making up the truck load.

Apparently, this approach does not include consideration to random multiple-lane simultaneous occupancy by trucks. Thus, the 1.2 multiple presence factor in Table 3.3-1 needs to be explicitly applied to the resulting distribution factor if one lane of loading is considered. If more lanes are loaded, the corresponding multiple presence factors should be included by multiplying the corresponding 1.0 to the resulting load effect.

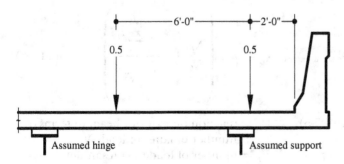

Figure 4.7-4
Lever rule.

❑ DF for moment for two or more lanes loaded:

4.6.2.2 $DF_{\text{moment exterior 2}} = eDF_{\text{moment interior 2}}$

$$e = 0.77 + \frac{d_e}{9.1} \qquad\qquad (4.7\text{-}5)$$

Distance d_e is between the web of the exterior girder and the interior edge of the curb in feet. If the former is inside the latter, d_e is positive; otherwise it is negative. Also d_e needs to satisfy $-1 \leq d_e \leq 5.5$. If the calculated d_e is below the minimum value of -1, then it is set at -1, and if it is above the maximum value of 5.5, then the 5.5 is used for d_e.

❑ DF for shear for one lane loaded: Use the lever rule, that is,

4.6.2.2 $DF_{\text{shear exterior 1}} = DF_{\text{moment exterior 1}}$ (4.7-6)

Note again that the multiple presence factor 1.2 needs to be multiplied with the analysis result because the lever rule method is used here, not a formula given in the specifications.

❑ DF for shear for two or more lanes loaded:

4.6.2.2 $DF_{\text{shear exterior 2}} = eDF_{\text{shear interior 2}}$ $e = 0.6 + \frac{d_e}{10}$ (4.7-7)

Note that, in Eq. 4.7-4, $DF_{\text{shear interior 2}}$ has the multiple presence factor imbedded.

Additional Investigation for Distribution Factor
In beam bridges, the cross section is usually braced using diaphragms or cross frames, as seen earlier in this chapter. The distribution factor for the exterior beam shall not be less than what would be obtained by assuming the cross section to deflect and rotate as a rigid-body cross section. This is required in the AASHTO specifications. To satisfy this code requirement, the following distribution factor for shear along with the multiple presence factor in Table 3.3-1 needs to be computed and compared with the corresponding DF values discussed above. The larger DF shall be used in beam design.

C4.6.2.2.2d $DF_{\text{shear exterior, } N_L} = m_{N_L} \left(\dfrac{N_L}{N_b} + \dfrac{x_{\text{ext}} \sum\limits_{i=1}^{N_L} e_i}{\sum\limits_{j=1}^{N_b} x_j^2} \right)$ (4.7-8)

where (see Figure 4.7-5)   DF = reaction on exterior beam in terms of number of lanes of load, N_L

N_L = number of loaded lanes under consideration

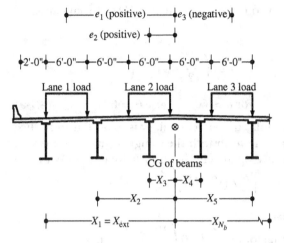

Figure 4.7-5
Truck placement in additional investigation for exterior beam distribution factor.

$$e_i = \text{eccentricity of design truck or lane load in lane } i \text{ from center of gravity of beams}$$

$$x_j = \text{horizontal distance from center of gravity of beams to beam } j$$

$$x_{\text{ext}} = \text{horizontal distance from center of gravity of beams to exterior beam}$$

$$N_b = \text{number of beams}$$

$$m_{N_L} = \text{multiple presence factor in Table 3.3-1 for } N_L \text{ lanes}$$

This additional investigation is required because the distribution factor for beams in a multibeam cross section was determined without consideration to diaphragm or cross frames. This check is an interim provision until research provides a better solution.

Correction for Effect of Skew
Skewed support is commonly used in highway bridges, particularly beam bridges, because it is relatively simpler to provide for beam bridges compared with other superstructure types, such as truss, arch, cable stayed, and suspension bridges. As discussed earlier in this chapter, skewed supports may save the cost of additional acquisition of right of way when the road carried by the bridge intersects at a nonstraight angle with a roadway, waterway, railway, and so on. Figure 4.7-6 shows the relation of the beamlines and the support lines and the definition of the skew angle θ to be used in the correction formulas in the AASHTO specifications given below.

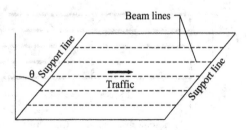

Figure 4.7-6
Definition of skew angle θ in horizontal plane.

The skew correction factor to DF given in the AASHTO specifications is as follows:

❑ For moment:

$$\text{Correction factor} = 1 - 0.25 \left(\frac{K_g}{12\,Lt^3}\right)^{0.25} \left(\frac{S}{L}\right)^{0.5} \tan^{1.5}\theta$$

4.6.2.2

$$30° \le \theta \le 60° \qquad \text{for } \theta > 60° \qquad \text{use } \theta = 60° \qquad (4.7\text{-}9)$$

As seen in the above formula, the effect of skew is expected to reduce the design moment, because the term with skew angle θ has a minus sign and thus reduces the effective span length when θ increases. With the bridge owner's concurrence, $\left[K_g/\left(12\,Lt^3\right)\right]^{0.25} = 1.03$ can be used for simplification.

❑ For obtuse corner shear:

$$\text{Correction factor} = 1 + 0.20 \left(\frac{12\,Lt^3}{K_g}\right)^{0.3} \tan\theta$$

4.6.2.2

$$0° \le \theta \le 60° \qquad (4.7\text{-}10)$$

As seen, the effect of skew is to increase the obtuse shear when the skew angle θ increases. This effect is induced because skew makes adjacent beams behave differently since their respective effective span lengths differ from each other under the load. This effect induces additional torsion requiring an extra shear force to maintain equilibrium. Note also that with the bridge owner's concurrence, $\left(12\,Lt^3/K_g\right)^{0.3} = 0.97$ can be used for simplification.

Example 4.5 shows a typical load effect analysis for a simple span highway bridge superstructure made of multiple steel I beams. Example 4.9 illustrates the same but for a superstructure of multiple steel plate girders of I section.

Example 4.5 Steel Rolled Beam Bridge Design (Load Effects)

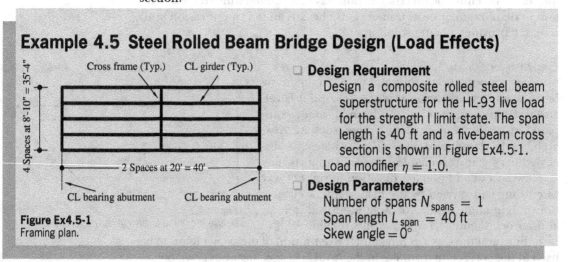

Figure Ex4.5-1
Framing plan.

❑ **Design Requirement**
Design a composite rolled steel beam superstructure for the HL-93 live load for the strength I limit state. The span length is 40 ft and a five-beam cross section is shown in Figure Ex4.5-1. Load modifier $\eta = 1.0$.

❑ **Design Parameters**
Number of spans $N_{spans} = 1$
Span length $L_{span} = 40$ ft
Skew angle $= 0°$

Number of girders $N_{girders} = 5$
Girder spacing $S = 8$ ft 10 in. $= 8.83$ ft
Deck overhang length $S_{overhang} = 3$ ft 11 in. $= 3.92$ ft
Cross-frame spacing $L_b = 20$ ft
Web yield strength $F_{yw} =$ Flange yield strength $F_{yf} = 50$ k/in.^{2}
Concrete 28-day compressive strength $f_c' = 4.5$ k/in.^{2}
Reinforcement strength $f_y = 60$ k/in.^{2}
Total deck thickness $t_{slab} = 9.0$ in.
Deck design thickness $t_{design\_slab} = 8.5$ in.
Steel density $W_{steel} = 0.490$ k/ft^3
Concrete density $W_{concrete} = 0.145$ k/ft^3
Miscellaneous structural steel dead load (per beam) $W_{miscellaneous}$
    $= 0.02$ k/ft
Stay-in-place deck form weight $W_{deck\ forms} = 14.98$ lb/ft $= 0.015$ k/ft
Parapet weight (each) $W_{parapet} = 0.53$ k/ft
Future wearing surface density $W_{FWS} = 0.14$ k/ft^3
Future wearing surface thickness $t_{FWS} = 2.5$ in.
Deck width $w_{deck} = 43$ ft 2 in. $= 43.20$ ft
Roadway width $w_{roadway} = 40$ ft 4 in. $= 40.33$ ft
Haunch depth $d_{haunch} = 1$ in.

☐ **Load Factors (see Table Ex4.5-1)**

Table Ex4.5-1
Load factors

Limit State	Load Factors			
	DC	**DW**	**LL**	**IM**
Strength I	1.25	1.5	1.75	1.75
Service II	1	1	1.3	1.3

☐ **Resistance Factors**
For flexural resistance $\phi_f = 1$, shear resistance $\phi_v = 1$.

☐ **Effective Flange Width**

4.6.2.6.1

$w_{effective\ interior} = 8$ ft 10 in. $= 8.83$ ft

$w_{effective\ exterior} = 4$ ft 5 in. $+ 3$ ft 11 in. $= 8.33$ ft

☐ **Dead Loads for Interior Beams**

1. Concrete deck:

$W_{concrete} = 0.145$ k/ft^3 $S = 8$ ft 10 in. $t_{slab} = 9$ in.

$DL_{deck} = W_{concrete} S T_{slab} = 0.145$ k/ft^3 (8 ft 10 in.)9 in./(12 in./ft)

$\qquad = 0.96$ k/ft

2. Stay-in-place forms, dead load per unit length:

$$W_{deck\ forms} = 0.015\ k/ft^2 \qquad S = 8\ ft\ 10\ in \qquad w_{top\ flange} = 14\ in.$$

The top flange width is assumed.

$$DL_{deck\ forms} = W_{deck\ forms}\left(S - w_{top\ flange}\right) = 0.015\ k/ft^2\ (8\ ft\ 10\ in. - 14\ in.)$$

$$= 0.12\ k/ft$$

3. Miscellaneous:

$$DL_{miscellaneous} = 0.02\ k/ft$$

The miscellaneous weight is for the weight of traffic signs, illumination, and so on.

4. Concrete parapet:

$$W_{parapet} = 0.457\ k/ft \qquad N_{girders} = 5$$
$$DL_{parapet} = 2W_{parapet}/N_{girders} = 0.183\ k/ft$$

For the concrete parapets, the dead load per unit length is computed assuming that the superimposed dead load of the two parapets is distributed uniformly among all the girders.

5. Future wearing surface:

$$W_{road\ way} = 40\ ft\ 4\ in. \qquad N_{girders} = 5$$

$$DL_{FWS} = \frac{W_{FWS}\ (t_{FWS}/12\ in./ft)\ w_{road\ way}}{N_{girders}}$$

$$= \frac{0.140\ k/f^3\ (2.5\ in./12\ in./ft)\ 40.33\ ft}{5} = 0.24\ k/ft$$

6. Steel:

$$DL_{steel} = 0.194\ k/ft$$

Assume a W 27 × 194 section.

7. Total dead load from beam (for noncomposite cross section):

$$DL_{beam} = DL_{steel} + DL_{deck} + DL_{deck\ forms} + DL_{miscellaneous}$$
$$= 0.194\ k/ft + 0.96\ k/ft + 0.12\ k/ft + 0.02\ k/ft = 1.29\ k/ft$$

☐ **Dead-Load Moments and Shears for Design of Interior Beams**

(a) Total beam:

$$M_{beam} = 1.29\ k/ft\ (40\ ft)^2/8 = 258\ kft$$
$$V_{beam} = 1.29\ k/ft\ (40\ ft)/2 = 25.8\ k$$

(b) Parapet:
$$M_{parapet} = 0.183 \text{ k/ft } (40 \text{ ft})^2/8 = 36.6 \text{ kft}$$
$$V_{parapet} = 0.183 \text{ k/ft } (40 \text{ ft})/2 = 3.66 \text{ k}$$

(c) Future wearing surface:
$$M_{FWS} = 0.24 \text{ k/ft } (40 \text{ ft})^2/8 = 48 \text{ kft}$$
$$V_{FWS} = 0.24 \text{ k/ft } (40 \text{ ft})/2 = 4.8 \text{ k}$$

Dead Loads for Exterior Beams

8. Concrete deck:

$$W_{concrete} = 0.145 \text{ k/ft}^3 \qquad \text{Width} = 8 \text{ ft 4 in.} \qquad t_{slab} = 9 \text{ in.}$$

$$\begin{aligned} DL_{deck} &= W_{concrete} \, St_{slab} = 0.145 \text{ k/ft}^3 \, (8 \text{ ft 4 in}) \, 9 \text{ in.}/(12 \text{ in./ft}) \\ &= 0.91 \text{ k/ft} \end{aligned}$$

9. Stay-in-place forms, dead load per unit length:

$$W_{deck \, forms} = 0.015 \text{ k/ft}^2 \qquad \text{Width} = 4 \text{ ft 5 in.} \qquad w_{top \, flange} = 14 \text{ in.}$$

Overhang will not use stay-in-place forms. The top flange width is assumed.

$$DL_{deck \, forms} = W_{deck \, forms} \left(\text{width} - w_{top \, flange}/2 \right) = 0.015 \text{ k/ft}^2$$

$$\times \, (4 \text{ ft 5 in.} - 7 \text{ in.}) = 0.16 \text{ k/ft}$$

10. Miscellaneous:
$$DL_{miscellaneous} = 0.02 \text{ k/ft}$$

11. Concrete parapet:

$$W_{parapet} = 0.457 \text{ k/ft} \qquad N_{girders} = 5$$
$$DL_{parapet} = 2W_{parapet}/N_{girders} = 0.183 \text{ k/ft}$$

For the concrete parapets, the dead load per unit length is computed assuming that the superimposed dead load of the two parapets is distributed uniformly among all the girders

12. Future wearing surface:

$$w_{road \, way} = 40 \text{ ft 4 in.} \qquad N_{girders} = 5$$

$$\begin{aligned} DL_{FWS} &= \frac{W_{FWS} \left(t_{FWS}/12 \text{ in./ft} \right) w_{road \, way}}{N_{girders}} \\ &= \frac{0.140 \text{ k/f}^3 \, (2.5 \text{ in.}/12 \text{ in./ft}) \, 40.33 \text{ ft}}{5} = 0.24 \text{ k/ft} \end{aligned}$$

13. Steel:

$$DL_{steel} = 0.194 \text{ k/ft}$$

Assume a W 27 × 194 section.

14. Total dead load from beam (for noncomposite cross section):

$$DL_{beam} = DL_{steel} + DL_{deck} + DL_{deck\ forms} + DL_{miscellaneous}$$
$$= 0.194 \text{ k/ft} + 0.91 \text{ k/ft} + 0.06 \text{ k/ft} + 0.02 \text{ k/ft} = 1.18 \text{ k/ft}$$

☐ **Dead-Load Moments and Shears for Design of Exterior Beams**

(d) Total beam:

$$M_{beam} = 1.18 \text{ k/ft} \ (40 \text{ ft})^2/8 = 236 \text{ kft}$$
$$V_{beam} = 1.18 \text{ k/ft} (40 \text{ ft})/2 = 23.6 \text{ k}$$

(e) Parapet:

$$M_{parapet} = 0.183 \text{ k/ft} \ (40 \text{ ft})^2/8 = 36.6 \text{ kft}$$
$$V_{parapet} = 0.183 \text{ k/ft} \ (40 \text{ ft})/2 = 3.66 \text{ k}$$

(f) Future wearing surface:

$$M_{FWS} = 0.24 \text{ k/ft} \ (40 \text{ ft})^2/8 = 48 \text{ kft}$$
$$V_{FWS} = 0.24 \text{ k/ft} (40 \text{ ft})/2 = 4.8 \text{ k}$$

☐ **Live-Load Effects for Design**

In calculating the live-load effects for strength I limit state, the following two cases need to be checked:

1. Truck load (with IM) + lane load
2. Tandem load (with IM) + lane load

The larger value of the two cases will control. These load effects need to include the respective distribution factors and the dynamic allowance IM.

☐ **Live-Load Distribution Factors**

The AASHTO distribution factors can be used since the following conditions are met: (1) The deck width is constant. (2) The number of beams is not less than 4. (3) The beams are parallel and have approximately the same stiffness. (4) The overhang width minus the barrier width is less than 3.0 ft. (5) Curvature in the plan is zero. In addition, the following quantitative range applicability requirements are met: (i) $3.5 \text{ ft} \le S \le 16 \text{ ft}$; (ii) $4.5 \text{ in.} \le t_{deck} \le 12 \text{ in.}$; (iii) $20 \text{ ft} \le L_{design} \le 240 \text{ ft}$; (iv) $10{,}000 \text{ in.}^4 \le K_g \le 17{,}000{,}000 \text{ in.}^4$; (v) $-1 \text{ ft} \le d_e \le 5.5 \text{ ft}$.

Interior Beams

1. Distribution Factor for Moment—One Lane Loaded *4.6.2.2.2*
   Assume $K_g/12\,Lt_s^3 = 1$ for preliminary design:

$$DF_{\text{moment interior 1 lane}} = 0.06 + \left(\frac{S}{14}\right)^{0.4}\left(\frac{S}{L}\right)^{0.3}\left[\frac{K_g}{12\,L\,t_s^3}\right]^{0.1}$$

$$= 0.06 + \left(\frac{8.83}{14}\right)^{0.4}\left(\frac{8.83}{40}\right)^{0.3} 1^{0.1} = 0.59$$

*Multiple presence factor for one lane, 1.2, has been included.
3.6.1.1.2*

2. Distribution Factor for Moment—Two or More Lanes Loaded

$$DF_{\text{moment interior 2 lane}} = 0.075 + \left(\frac{S}{9.5}\right)^{0.6}\left(\frac{S}{L}\right)^{0.2}\left[\frac{K_g}{12\,L\,t_s^3}\right]^{0.1}$$

$$= 0.075 + \left(\frac{8.83}{9.5}\right)^{0.6}\left(\frac{8.83}{40}\right)^{0.2} 1^{0.1} = 0.78$$

For the interior beam design, the controlling moment distribution factor
is 0.78.

3. Distribution Factor for Shear—One Lane Loaded *4.6.2.2.2*

$$DF_{\text{shear interior 1 lane}} = 0.36 + \frac{S}{25.0} = 0.36 + \frac{8.83}{25.0} = 0.71$$

4. Distribution Factor for Shear—Two or More Lanes Loaded

$$DF_{\text{shear interior 2 lane}} = 0.2 + \left(\frac{S}{12}\right) - \left(\frac{S}{35}\right)^{2.0}$$

$$= 0.2 + \left(\frac{8.83}{12}\right) - \left(\frac{8}{40}\right)^{2.0} = 0.87$$

For the interior beam design, the controlling shear distribution factor
is 0.87.

Exterior Beams

Distance d_e is between the web of the exterior girder and the interior edge
of the curb:

$$d_e = 2.5\,\text{ft} \quad \text{satisfying} \; -1 \le d_e \le 5.5$$

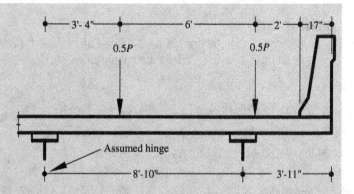

Figure Ex4.5-2
Lever rule.

1. Distribution Factor for Moment—One Lane Loaded Use the lever rule as shown in Figure Ex4.5-2:

   3.6.1.1.2
   $$DF_{moment\ exterior\ 1\ lane} = \frac{0.5\,(3.33\ ft) + 0.5\,(9.33\ ft)}{8.83\ ft} \quad (1.2)$$
   $$= 0.72\,(1.2) = 0.86\ lane$$

   Multiple presence factor for one lane 1.2, is used.

2. Distribution Factor for Moment—Two or More Lanes Loaded

   4.6.2.2.2
   $$e = 0.77 + \frac{d_e}{9.1} = 0.77 + \left(\frac{2.5}{9.1}\right) = 1.04$$

   $$DF_{moment\ exterior\ 2\ lane} = e(DF_{moment\ interior\ 2\ lane}) = 1.04\,(0.78)$$
   $$= 0.81\ lane$$

   For the exterior beam design, the controlling moment distribution factor is 0.86.

3. Distribution Factor for Shear—One Lane Loaded Use the lever rule as in the moment calculation:

   $$DF_{shear\ exterior\ 1\ lane} = DF_{moment\ exterior\ 1\ lane} = 0.86\ lane$$

4. Distribution Factor for Shear—Two or More Lanes Loaded

   $$e = 0.6 + \frac{d_e}{10} = 0.6 + \frac{2.5}{10} = 0.85$$

   $$DF_{shear\ exterior\ 2\ lane} = e(DF_{shear\ interior\ 2\ lane}) = 0.85\,(0.87) = 0.74\ lane$$

   For the exterior beam design, the controlling shear distribution factor is 0.86.

Maximum Live-Load Shear

Use Figures Ex4.5-3 and Ex4.5-4 to find corresponding shear force to compare and determine which one
governs.

Unfactored shears:

$$V_{truck} = (1 + IM) \text{ (static truck load shear)}$$

$$= 1.33 \left[32 + \frac{32\,(26)}{40} + \frac{8\,(12)}{40} \right]$$

$$= 1.33(55.2) = 73.4 \text{ k}$$

$$V_{lane} = \frac{0.64\,(40)}{2} = 12.8 \text{ k}$$

Total interior beam design shear:

$$V_{interior} = (12.8 + 73.4)\, DF_{shear\ interior\ 2\ lane}$$

$$= 86.2\,(0.87) = 75.0 \text{ k}$$

Total exterior beam design shear:

$$V_{exterior} = (12.8 + 73.42)\, DF_{shear\ exterior\ 1\ lane}$$

$$= 86.2\,(0.86) = 74.1 \text{ k}$$

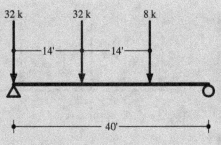

Figure Ex4.5-3
HL-93 truck load placement for maximum shear.

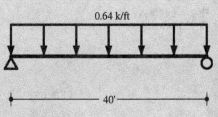

Figure Ex4.5-4
HL-93 lane load placement for maximum shear.

Maximum Live-Load Moment

Truck Load

First, we locate the center of gravity (CG) of the HL-93
truck to be used later (see Figure Ex4.5-5 and
Table Ex4.5-2.) Take moments about the 32-k axle
to the right.

$$Y' = \frac{\text{sum of moments}}{\text{sum of forces}} = \frac{672 \text{ kft}}{72 \text{ kft}} = 9 \text{ ft 4 in.}$$

$$Y = 14 - Y' = 4.67 \text{ ft} = 4 \text{ ft 8 in.}$$

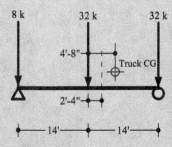

Figure Ex4.5-5
Center of gravity of HL-93 truck.

Table Ex4.5-2
Calculation for center of gravity of HL-93 truck

Force	Arm	Moment
8 k	28 ft	224 kft
32 k	14 ft	448 kft
32 k	0 ft	0 kft
total		672 kft

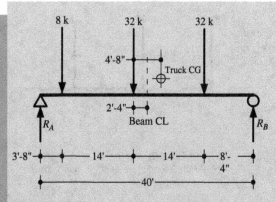

Figure Ex4.5-6
HL-93 truck placement for maximum moment.

To find the maximum moment of the HL-93 truck load, the midspan point needs to bisect the distance between the center of gravity and the nearest 32-k load, as shown in Figure Ex4.5-6.

Reactions:

$$R_B = \frac{1}{40}\left[8\,\text{k}(3.67\,\text{ft}) + 32\,\text{k}\,(17.67\,\text{ft})\right.$$

$$\left. + 32\,\text{k}\,(31.67\,\text{ft})\right] = 40.2\,\text{k}$$

$$R_A = (8\,\text{k} + 32\,\text{k} + 32\,\text{k}) - 40.2\,\text{k} = 31.8\,\text{k}$$

Unfactored truck load moment including IM:

$$M_{\text{truck}} = \left[R_A\,(17.67\,\text{ft}) - 8\,\text{k}\,(14\,\text{ft})\right](1 + \text{IM})$$

$$= (449.7\,\text{kft})\,1.33 = 598.1\,\text{kft}$$

Unfactored lane load moment:

$$M_{\text{lane}} = \frac{0.64\,\text{k/ft}\,(40\,\text{ft})^2}{8} = 128.0\,\text{kft}$$

Total moment:

$$M_{\text{LL}} = M_{\text{truck}} + M_{\text{lane}} = 598.14\,\text{kft} + 128\,\text{kft} = 726.1\,\text{kft}$$

Total moment in interior beam:

$$M_{\text{LL interior}} = M_{\text{LL}}DF_{\text{moment linterior 2 lane}} = (726.14\,\text{kft})\,0.78 = 566.4\,\text{kft}$$

Total moment in exterior beam:

$$M_{\text{LL exterior}} = M_{\text{LL}}DF_{\text{moment exterior 1 lane}} = (726.14\,\text{kft})\,0.86 = 624.5\,\text{kft}$$

Tandem load

To determine the maximum moment of tandem load, the two 25-k axles need to be bisected by the center of the span, as shown in Figure Ex4.5-7. The maximum tandem load moment is given as

$$M_{\text{tandem}} = 25\,\text{k}\,(18\,\text{ft}) = 450.0\,\text{kft}$$

This is approximately the same moment as truck load (TL) and thus TL moment M_{TL} without IM is used.

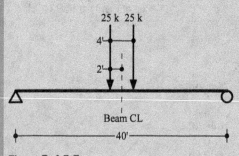

Figure Ex4.5-7
Tandem loading placement for maximum moment.

Tandem Live-Load Shear

Use Figure Ex4.5-8 to calculate the maximum shear for the tandem load. The maximum tandem load shear is given as

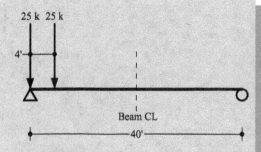

$$V_{tandem} = 25 \text{ k} \left(\frac{36 \text{ ft}}{40 \text{ ft}} \right) + 25 \text{ k}$$

$$= 22.5 \text{ k} + 25 \text{ k} = 47.5 \text{ k}$$

Figure Ex4.5-8
Tandem loading placement for maximum shear.

This maximum shear is less than the truck load maximum shear, 55.2 k, and thus the truck load controls.

Design Load Effects

Based on the dead- and live-load effects calculated, the design load effects for the strength I limit state are summarized here:

Load modifier $\eta = 1.0$

Factored moments:

$$M_{DC} = 1.25 \left(M_{beam} + M_{parapet} \right) = 1.25 \left(258 \text{ kft} + 42 \text{ k} \right)$$

$$= 1.25 \left(300 \text{ kft} \right) = 375.0 \text{ kft}$$

$$M_{DW} = 1.5 \left(48 \text{ kft} \right) = 72 \text{ kft}$$

$$M_{LL} = \begin{cases} 1.75 \left(566.4 \text{ kft} \right) = 991.2 \text{ kft} & \text{for interior beams} \\ 1.75 \left(624.48 \text{ kft} \right) = 1092.8 \text{ kft} & \text{for exterior beams} \end{cases}$$

Total factored moment:

$$M_{total} = M_u = \begin{cases} M_{DC} + M_{DW} + M_{LL} = 375 \text{ kft} + 72 \text{ kft} & \text{for interior beams} \\ \quad +991.2 \text{ kft} = 1438.2 \text{ kft} \\ 375 \text{ kft} + 72 \text{ kft} + 1092.8 \text{ kft} & \text{for exterior beams} \\ \quad = 1540 \text{ kft} \end{cases}$$

Factored shears:

$$V_{DC} = 1.25 \left(V_{DL \text{ beam}} + V_{DL \text{ parapet}} \right) = 1.25 \left(25.8 \text{ k} + 4.2 \text{ k} \right)$$

$$= 1.25 \left(30.0 \text{ k} \right) = 37.5 \text{ k}$$

$$V_{DW} = 1.5 \left(4.8 \text{ k} \right) = 7.2 \text{ k}$$

$$V_{LL} = \begin{cases} 1.75\,(75.0\,\text{k}) = 131.2\,\text{k} & \text{for interior beams} \\ 1.75\,(74.1\,\text{k}) = 129.7\,\text{k} & \text{for exterior beams} \end{cases}$$

Total factored shear:

$$V_{\text{total}} = V_u = \begin{cases} V_{DC} + V_{DW} + V_{LL} = 37.5\,\text{k} + 7.2\,\text{k} & \text{for interior beams} \\ \quad +131.2\,\text{k} = 175.9\,\text{k} \\ 37.5\,\text{k} + 7.2\,\text{k} + 129.7\,\text{k} = 174.4\,\text{k} & \text{for exterior beams} \end{cases}$$

Examples 4.6 and 4.7 demonstrate the beam design considering the strength and service limits for the same superstructure in Example 4.5, as discussed below in Sections 4.7.3 through 4.7.7. Example 4.8 continues the same design problem but addresses the design of shear studs, which is explained in Section 4.7.9.

Example 4.6 Steel Rolled Beam Bridge (Strength I and Service II Limit States for Interior Beams)

Figure Ex4.6-1
W 24 × 84
dimensions.

□ **Design Requirement**

Design a rolled beam section for the bridge in Example 4.5 using the design load effect results obtained there.

Load modifier $\eta = 1.0$

□ **Trial Section**

Try Section W 24 × 84 (see Figure Ex4.6-1):

$I_x = 2370\,\text{in.}^4 \quad I_y = 94.4\,\text{in.}^4 \quad A = 24.7\,\text{in.}^2 \quad Z_x = 224\,\text{in.}^3$

$S_x = 196\,\text{in.}^3 \quad b_f = 9.02\,\text{in.} \quad t_f = 0.77\,\text{in.}$

$t_w = 0.47\,\text{in.} \quad d = 24.1\,\text{in.}$

□ **Composite Section Properties for Interior Beams**

For dead loads, use only the steel cross-sectional properties given in Table Ex4.6-1. For superimposed dead loads, use $3(n) = 3(8) = 24$:

$$w_{\text{eff}} = 8\,\text{ft}\,10\,\text{in.}$$

$$b_{\text{eff}} = \frac{w_{\text{eff}}}{3\,(n)} = \frac{(8\,\text{ft}\,10\,\text{in.})}{3\,(8)} = 4.42\,\text{in.}$$

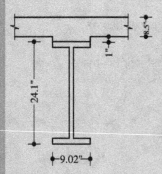

Figure Ex4.6-2
Composite section dimensions for superimposed dead load.

Based on Figure Ex4.6-2, Table Ex4.6-1 summarises calcutations for the section properties.

Table Ex4.6-1
Superimposed dead-load section properties

Component	A (in.^{2})	Y (in.)	AY (in.)	AY^2 (in.^{4})	I_0 (in.^{4})
W 24 × 84	24.7	24.1/2 = 12.05	297.64	3,586.56	2,370
Slab (3n)	4.42(8.5) = 37.57	8.5/2 + 1 + 24.1 = 29.35	1,102.68	32,363.64	4.42(8.5)^3/12 = 226.2
Sum	62.27		1,400.32	35,950.2	2,596.2

$$I_z = \sum I_0 + \sum Ay^2$$

$$= 2596.2 \text{ in.}^4 + 35{,}950.2 \text{ in.}^4 = 38{,}546.4 \text{ in.}^4$$

$$Y' = \frac{\sum Ay}{\sum A} = \frac{1400.32 \text{ in.}^3}{62.27 \text{ in.}^2} = 22.49 \text{ in.}$$

$$Y_t = 24.1 \text{ in.} + 1 \text{ in.} + 8.5 \text{ in.} - 22.49 \text{ in.} = 11.11 \text{ in.}$$

$$I_x = I_z - \left(\sum A\right)\left(Y'\right)^2 = 38{,}546.4 \text{ in}^4 - 62.27 \text{ in.}^2 (22.49 \text{ in.})^2$$

$$= 7050.23 \text{ in.}^4$$

$$S_{t,\text{SDL}} = \frac{7050.23 \text{ in.}^4}{11.11 \text{ in.}} = 634.58 \text{ in.}^3$$

$$S_{b,\text{SDL}} = \frac{7050.23 \text{ in.}^4}{22.49 \text{ in.}} = 313.48 \text{ in.}^3$$

For live loads, also use $n = 8$:

$$w_{\text{eff}} = 8 \text{ ft } 10 \text{ in.}$$

$$b_{\text{eff}} = \frac{w_{\text{eff}}}{n} = \frac{8 \text{ ft } 10 \text{ in.}}{8} = 13.25 \text{ in.}$$

$$I_{\text{slab}} = \frac{13.25 \text{ in.} (8.5 \text{ in.})^3}{12} = 678.10 \text{ in.}^4$$

The live load cross section propeties are calculated in Table Ex4.6-2.

$$I_z = \sum I_0 + \sum Ay^2 = 3048.1 \text{ in.}^4 + 100{,}608.56 \text{ in.}^4$$

$$= 103{,}656.67 \text{ in.}^4$$

$$Y' = \frac{\sum Ay}{\sum A} = \frac{3603.33 \text{ in.}^3}{137.33 \text{ in.}^2} = 26.24 \text{ in.}$$

Table Ex4.6-2
Live-load section properties

Component	A (in²)	Y (in)	AY (in)	AY² (in⁴)	I_0 (in.⁴)
W 24 × 84	24.7	24.1/2 = 12.05	297.64	3,586.56	2,370
Slab (n)	13.25(8.5) = 112.63	8.5/2 + (1 + 24.1) = 29.35	3,305.69	97,022	13.25(8.5)³/12 = 678.10
Sum	137.33		3,603.33	100,608.56	3,048.1

$$Y_t = 24.1 \text{ in.} + 1 \text{ in.} + 8.5 \text{ in.} - 26.24 \text{ in.} = 7.36 \text{ in.}$$

$$I_x = I_z - \left(\sum A \right) \left(Y' \right)^2 = 103,656.67 \text{ in.}^4$$

$$- \left(137.33 \text{ in.}^2 \right) (26.24 \text{ in.})^2 = 9,099.8 \text{ in.}^4$$

$$S_{t,LL} = \frac{9099.8 \text{ in.}^4}{7.36 \text{ in.}} = 1236.39 \text{ in.}^3$$

$$S_{b,LL} = \frac{9099.8 \text{ in.}^4}{26.24 \text{ in.}} = 346.79 \text{ in.}^3$$

☐ **Check Member Properties**   *A 6.10.2*
Refer to Figure Ex4.6-3 for the following checks.

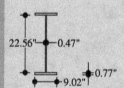

22.56" 0.47"

0.77"

9.02"

Figure Ex4.6-3
W 24 × 84 dimension details.

$$\frac{D}{t_w} = \frac{22.56}{0.47} = 48 \leq 150 \qquad \frac{b_f}{2t_f} = \frac{9.02}{2(0.77)} = 5.86 \leq 12$$

$$b_f = 9.02 \text{ in.} \geq \frac{D}{6} = \frac{22.56}{6} = 3.76 \text{ in.}$$

$$t_f = 0.77 \text{ in.} \geq 1.1 \, t_w = 1.1(0.47) = 0.52 \text{ in.}; \qquad \frac{I_{compression\,flange}}{I_{tension\,flange}} = 1$$

$$\text{satisfying} \qquad 0.1 \leq \frac{I_{compression\,flange}}{I_{tension\,flange}} \leq 10$$

☐ **Check Moment Capacity for Strength I Limit State**
Plastic neutral axis location $a = A F_y /(0.85 f_c' b_{eff}) = 24.7 (50) / [0.85 (4.5) 106] = 3.05$ in. is in the concrete deck. Thus, $D_{cp} = 0$ since no web is under compression, satisfying

$$2 (D_{cp}/t_w) = 0 \leq 3.76 \sqrt{E/F_{yc}}$$

In addition, $D/t_w = 22.56/0.47 = 48 < 150$ and $F_{yw} = 50$ k/in.² < 70 k/in.² Therefore the section is compact, and

$$D_p = a = 3.05 \text{ in.} \qquad D_t = 24.1 + 1 + 8.5 = 33.6 \text{ in.}$$

Thus $D_p \le 0.1\, D_t$ and the ductility requirement is satisfied as

$$\frac{D_p}{D_t} = \frac{3.05}{33.6} = 0.09 \le 0.42$$

and thus

$$M_n = M_p = AF_y \left(d - \frac{a}{2} \right)$$

$$= (24.7 \text{ in.}^2)50 \text{ k/in.}^2 \left(\frac{24.1 \text{ in.}}{2} + 1 \text{ in.} + 8.5 \text{ in.} - \frac{3.05 \text{ in.}}{2} \right)$$

$$\times \frac{1}{12 \text{ in./ft}} = 2061 \text{ kft}$$

The beam is not subjected to lateral bending, and thus $f_l = 0$:

$$M_u + f_l\, S_{xt} = M_u = 1438 \text{ kft} \le \varphi_f M_n = 1\,(2061 \text{ kft}) = 2061 \text{ kft}$$

OK for the interior beams. See Example 4.5.

☐ Check Shear Capacity for Strength I Limit State
Assume $k = 5$ for this unstiffened beam:

$$\frac{D}{t_w} = \frac{22.56}{0.47} = 48 \le 1.12 \sqrt{\frac{kE}{F_y}} = 1.12 \sqrt{\frac{5\,(29,000)}{50}} = 60.3$$

and thus $C = 1.0$

$$V_n = CV_p = C\,(0.58\, F_y\, Dt_w) = 1\,(0.58)\,50\,(22.56)\,0.47 = 307 \text{ k}$$
$$V_u = 175.94 \text{ k} \le \varphi_v V_n = 1\,(307) = 307 \text{ k}$$

OK for the interior beams. See Example 4.5.

☐ Service II Limit State for Steel   6.10.4

$$f_{DL} = \frac{M_{DL \text{ beam}}}{S_{NC}} = \frac{258 \text{ kft}\,(12 \text{ in./ft})}{196 \text{ in.}^3} = 15.79 \text{ k/in.}^2$$

$$f_{SDL} = \frac{M_{DL \text{ parapet}} + M_{DL \text{ FWS}}}{S_{b,SDL}} = \frac{(48 \text{ kft} + 42 \text{ kft})\,12 \text{ in./ft}}{313.48 \text{ in.}^3} = 3.45 \text{ k/in.}^2$$

$$f_{LL+I} = \frac{M_{LL}}{S_{b,LL}} = \frac{556.4 \text{ kft}\,(12 \text{ in./ft})}{346.79 \text{ in.}^3} = 19.25 \text{ k/in.}^2$$

$$1.0\,DC + 1.0\,DW + 1.3\,LL = 1.0(15.79) + 1.0(3.45) + 1.3(19.25)$$
$$= 44.3 \text{ k/in.}^2 \le \varphi_y F_y = 0.95(50) = 47.5 \text{ k/in.}^2$$

OK for the interior beams.

Example 4.7 Steel Rolled Beam Bridge Design (Strength I and Service II Limit States for Exterior Beams)

Figure Ex4.7-1
W 24 × 84
dimensions.

❑ **Design Requirement**

Design the exterior beams for the bridge in Example 4.5 using the design load effect results obtained there.

Load modifier $\eta = 1.0$

❑ **Trial Section**

Try section W 24 × 84 (see Figure Ex4.7-1):

$$I_x = 2370 \text{ in.}^4 \quad I_y = 94.4 \text{ in.}^4 \quad A = 24.7 \text{ in.}^2 \quad Z_x = 224 \text{ in.}^3$$

$$S_x = 196 \text{ in.}^3 \quad b_f = 9.02 \text{ in.} \quad t_f = 0.77 \text{ in.} \quad t_w = 0.47 \text{ in.}$$

$$d = 24.1 \text{ in.}$$

❑ **Composite Section Properties for Exterior Beams**

For dead loads, use only the steel cross-sectional properties given in Table Ex4.7-1. For superimposed dead loads, use $3(n) = 3(8) = 24$:

$$w_{eff} = 4 \text{ ft } 5 \text{ in.} \ + 3 \text{ ft } 11 \text{ in.} = 8 \text{ ft } 4 \text{ in.}$$

$$b_{eff} = \frac{w_{eff}}{3\,(n)} = \frac{8 \text{ ft } 4 \text{ in.}}{3\,(8)} = 4.17 \text{ in.}$$

Table Ex4.7-1
Superimposed dead-load section properties

Component	A (in.²)	Y (in.)	AY (in.³)	AY² (in.⁴)	I_0(in.⁴)
W 24 × 84	24.7	24.1/2 = 12.05	297.64	3,586.56	2370
Slab (3n)	4.17(8.5) = 35.42	8.5/2 + (1 + 24.1) = 29.35	1039.6	30,512	4.42(8.5)³/12 = 213.4
Sum	60.1		1337	34,099	2583

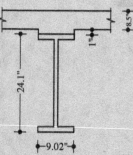

Figure Ex4.7-2
Composite section dimensions
for superimposed dead load.

Table Ex4.7-1 demonstrates calculations of section properties for superimposed dead load based on Figure Ex4.7-2.

$$I_z = \sum I_0 + \sum Ay^2$$

$$= 2583 \text{ in.}^4 + 34,099 \text{ in.}^4 = 36,682 \text{ in.}^4$$

$$Y' = \frac{\sum Ay}{\sum A} = \frac{1337 \text{ in.}^3}{60.1 \text{ in.}^2} = 22.25 \text{ in.}$$

$$Y_t = 24.1 \text{ in.} + 1 \text{ in.} + 8.5 \text{ in.} - 22.25 \text{ in.} = 11.35 \text{ in.}$$

$$I_x = I_z - \left(\sum A \right) \left(Y' \right)^2 = 36{,}682 \text{ in.}^4$$

$$- 60.1 \text{ in.}^2 (22.25 \text{ in.})^2 = 6929 \text{ in.}^4$$

$$S_{t,\text{SDL}} = \frac{6929 \text{ in.}^4}{11.35 \text{ in.}} = 610.5 \text{ in.}^3$$

$$S_{b,\text{SDL}} = \frac{6929 \text{ in.}^4}{22.25 \text{ in.}} = 313.4 \text{ in.}^3$$

For live loads, also use $n = 8$:

$$w_{\text{eff}} = 8 \text{ ft } 4 \text{ in.}$$

$$b_{\text{eff}} = \frac{w_{\text{eff}}}{n} = \frac{8 \text{ ft } 4 \text{ in.}}{8} = 12.5 \text{ in.}$$

$$I_{\text{slab}} = \frac{12.5 \text{ in.} (8.5 \text{ in.})^3}{12} = 639.71 \text{ in.}^4$$

Table Ex4.7-2
Live-Load Section Properties

Component	$A(\text{in}^2)$	$Y(\text{in})$	$AY(\text{in})$	$AY^2(\text{in}^4)$	$I_0(\text{in.}^4)$
W 24 × 84	24.7	24.1/2 $= 12.05$	297.64	3586.56	2,370
Slab (n)	12.5(8.5) $= 106.25$	8.5/2 + $(1 + 24.1)$ $= 29.35$	3118.44	91,526	12.5(8.5)³/12 $= 639.71$
Sum	130.95		3416	95,113.7	3010

The results in Table Ex4.7-2 are used below to find the section modula for live load.

$$I_z = \sum I_0 + \sum Ay^2 = 3010 \text{ in.}^4 + 95{,}113 \text{ in.}^4 = 98{,}113 \text{ in.}^4$$

$$Y' = \frac{\sum Ay}{\sum A} = \frac{3416 \text{ in}^3}{131 \text{ in}^2} = 26.09 \text{ in.}$$

$$Y_t = 24.1 \text{ in.} + 1 \text{ in.} + 8.5 \text{ in.} - 26.09 \text{ in.} = 7.51 \text{ in.}$$

$$I_x = I_z - \left(\sum A \right) \left(Y' \right)^2 = 98{,}123 \text{ in.}^4 - \left(131 \text{ in.}^2 \right) (26.09 \text{ in.})^2$$

$$= 8953 \text{ in.}^4$$

$$S_{t,LL} = \frac{8953 \text{ in.}^4}{7.51 \text{ in.}} = 1192 \text{ in.}^3$$

$$S_{b,LL} = \frac{8953 \text{ in.}^4}{26.09 \text{ in.}} = 344.2 \text{ in.}^3$$

☐ Check Member Properties   A6.10.2

Refer to Figure Ex4.7-3 in the following checks.

$$\frac{D}{t_w} = \frac{22.56}{0.47} = 48 \leq 150 \qquad \frac{b_f}{2t_f} = \frac{9.02}{2\,(0.77)} = 5.86 \leq 12$$

$$b_f = 9.02 \text{ in.} \geq \frac{D}{6} = \frac{22.56}{6} = 3.76 \text{ in.}$$

Figure Ex4.7-3
W 24 × 84 dimension details.

☐ Check Moment Capacity for Strength I Limit State

The plastic neutral axis location $a = A F_y / \left(0.85 f_c' b_{eff}\right) = 24.7\,(50) /$ $[0.85\,(4.5)\,100] = 3.23$ in. is in the concrete deck. Thus, $D_{cp} = 0$ since no web is under compression, satisfying

$$2\frac{D_{cp}}{t_w} = 0 \leq 3.76\sqrt{\frac{E}{F_{yc}}}$$

In addition, $D/t_w = 22.56/0.47 = 48 < 150$ and $F_{yw} = 50$ k/in.^{2} < 70 k/in.^{2}

Therefore the section is compact, and

$$D_p = a = 3.23 \text{ in.} \qquad D_t = 24.1 + 1 + 8.5 = 33.6 \text{ in.}$$

Thus $D_p \leq 0.1 D_t$ and the ductility requirement is satisfied as

$$\frac{D_p}{D_t} = \frac{3.23}{33.6} = 0.10 \leq 0.42$$

and thus

$$M_n = M_p = AF_y \left(d - \frac{a}{2}\right)$$

$$= (24.7 \text{ in.}^2)50 \text{ k/in.}^2 \left(\frac{24.1 \text{ in.}}{2} + 1 \text{ in.} + 8.5 \text{ in.} - \frac{3.23 \text{ in.}}{2}\right)$$

$$\times \frac{1}{12 \text{ in./ft}} = 2051 \text{ kft}$$

The beam is not subjected to lateral bending, and thus $f_l = 0$:

$$M_u + f_l\, S_{xt} = M_u = 1534 \text{ kft} \leq \varphi_f M_n = 1\,(2051 \text{ kft}) = 2051 \text{ kft}$$

OK for the exterior beams.

☐ **Check Shear Capacity for Strength I Limit State**

Assume $k = 5$ for this unstiffened beam:

$$\frac{D}{t_w} = \frac{22.56}{0.47} = 48 \leq 1.12\sqrt{\frac{kE}{F_y}} = 1.12\sqrt{\frac{5\,(29{,}000)}{50}} = 60.3 \quad \text{and } C = 1.0$$

$$V_n = CV_p = C\,(0.58\,F_y\,Dt_w) = 1\,(0.58)\,50\,(22.56)\,0.47 = 307\,k$$

$$V_u = 174.4k \leq \phi_v V_n = 1\,(307) = 307\,k$$

OK for the exterior beams.

☐ **Service II limit State for Steel Yielding**

$$f_{DL} = \frac{M_{DL\ beam}}{S_{NC}} = \frac{236\ \text{kft}\,(12\ \text{in./ft})}{196\ \text{in.}^3} = 14.44\ \text{k/in.}^2$$

$$f_{SDL} = \frac{M_{DL\ parapet} + M_{DL\ FWS}}{S_{b,SDL}} = \frac{(36.6\ \text{kft} + 48\ \text{kft})\,12\ \text{in./ft}}{313.4\ \text{in.}^3} = 3.24\ \text{k/in.}^2$$

$$f_{LL+I} = \frac{M_{LL}}{S_{b,LL}} = \frac{624.5\ \text{kft}\,(12\ \text{in./ft})}{344.2\ \text{in.}^3} = 21.76\ \text{k/in.}^2$$

$$1.0DC + 1.0DW + 1.3LL = 1.0(14.44) + 1.0(3.24) + 1.3(21.76)$$

$$= 46.0\ \text{k/in.}^2 \leq \varphi_y F_y = 0.95(50) = 47.5\ \text{k/in.}^2$$

OK for the exterior beams.

Example 4.8 Steel Rolled Beam Bridge Design (Shear Studs)

☐ **Design Requirement**

Design the shear studs for the bridge in Example 4.5 for the following parameters:

Total deck thickness $t_{slab} = 9.0$ in.
Deck design thickness $t_{design\ slab} = 8.5$ in.
Average daily truck traffic (ADTT) = 900 trucks/day for driving lane
Load modifier $\eta = 1.0$

☐ **Fatigue Limit State**

Shear Range

Envelope shear ranges at various points along the span are calculated as the strength requirement for fatigue limit state design. Conventionally, the span is divided at every 10th point where the shear ranges are computed. For the 40-ft symmetric simple span in this example, only half of the span is analyzed as follows. One direction of truck traffic is analyzed assuming that the bridge carries one direction of traffic. The largest shear force ranges at the specified points on the span are taken as the envelope values for design. Each shear range value is also to be factored by the dynamic impact factor 1.15, load factor γ_L, and live load distribution factor with the multiple presence factor excluded, $DF_{SHEAR\ INT\ 1\ lane}/1.2$. For $DF_{shear\ interior\ 1\ lane}$ found as 0.71 in Example 4.5:

$$\text{Total factor TF} = (1 + IM)\,\gamma_L \left(\frac{DF_{shear\ interior\ 1\ lane}}{1.2}\right)$$

$$= (1.15)\left(\frac{0.71}{1.2}\right)\gamma_L = 0.68\gamma_L$$

Enveloping Shear Forces at 0 ft from Support

Positive maximum shear is found with the fatigue truck placed on the span as shown in Figures Ex4.8-1 and the corresponding shear diagram in Ex4.8-2

$$V_{0-P} = \left[32\,k + 32\,k\left(\frac{10\,ft}{40\,ft}\right)\right](TF) = 40\,k\,(TF)$$

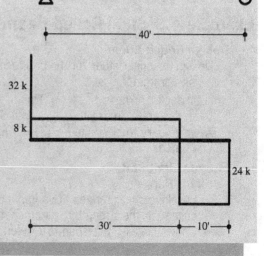

Figure Ex4.8-1
Fatigue truck placement for positive shear at supports.

Figure Ex4.8-2
Shear diagram for fatigue truck in Figure Ex4.8-1.

Negative maximum shear is given as
$$V_{0\text{-}N} = 0\,\text{k}$$

Shear range V_f:
$$V_f = V_{0\text{-}P} - V_{0\text{-}N} = (40\,\text{k} + 0\,\text{k})\,\text{TF} = 40\,\text{K (TF)}$$

Enveloping Shear Forces at 4 ft from Support

Positive maximum shear is found with the fatigue truck placed on the span as shown in Figures Ex4.8-3 and corresponding shear diagram in Ex4.8-4

$$V_{4\text{-}p} = \left[32\,\text{k}\left(\frac{36\,\text{ft}}{40\,\text{ft}}\right) + 32\,\text{k}\left(\frac{6\,\text{ft}}{40\,\text{ft}}\right) \right]\text{(TF)}$$
$$= 33.6\,\text{k (TF)}$$

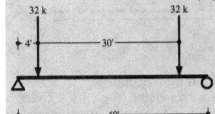

Figure Ex4.8-3
Fatigue truck placement for positive shear at 4 ft from support.

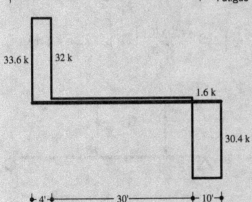

Figure Ex4.8-4
Shear diagram for fatigue truck in Figure Ex4.8-3.

Negative maximum shear is found with the fatigue truck placed on the span as shown in Figures Ex4.8-5 and corresponding shear diagram in Ex4.8-6

$$V_{4\text{-}N} = \left[8\,\text{k}\left(\frac{36\,\text{ft}}{40\,\text{ft}}\right) - 8\,\text{k} \right]\text{(TF)} = -0.8\,\text{k (TF)}$$

Shear range V_f:
$$V_f = V_{4\text{-}P} - V_{4\text{-}N} = (33.6\,\text{k} + 0.8\,\text{k})\,\text{TF} = 344\,\text{k (TF)}$$

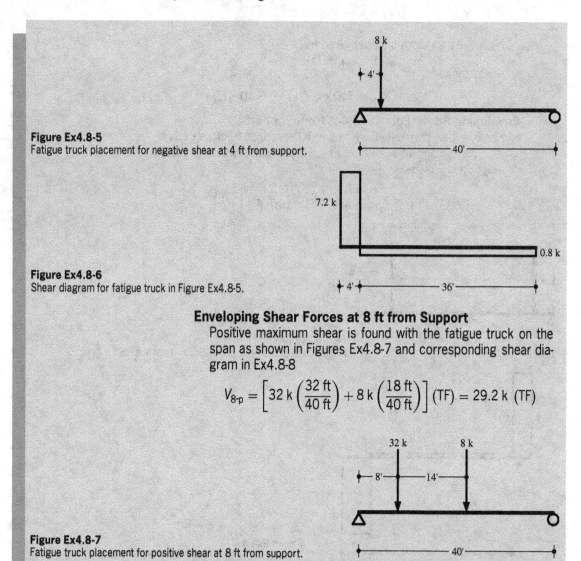

Figure Ex4.8-5
Fatigue truck placement for negative shear at 4 ft from support.

Figure Ex4.8-6
Shear diagram for fatigue truck in Figure Ex4.8-5.

Enveloping Shear Forces at 8 ft from Support

Positive maximum shear is found with the fatigue truck on the span as shown in Figures Ex4.8-7 and corresponding shear diagram in Ex4.8-8

$$V_{8\text{-}p} = \left[32\,k \left(\frac{32\,ft}{40\,ft} \right) + 8\,k \left(\frac{18\,ft}{40\,ft} \right) \right] (TF) = 29.2\,k\ (TF)$$

Figure Ex4.8-7
Fatigue truck placement for positive shear at 8 ft from support.

Figure Ex4.8-8
Shear diagram for fatigue truck in Figure Ex4.8-7.

Negative maximum shear is found with the fatigue truck on the span as shown in Figures. Ex4.8-9 and corresponding shear diagram in Ex4.8-10

$$V_{8\text{-}N} = \left[32\ k \left(\frac{32\ ft}{40\ ft} \right) + 32\ k \left(\frac{38\ ft}{40\ ft} \right) \right] (TF) = -4.8\ k\ (TF)$$

Shear range V_f:

$$V_f = V_{8\text{-}P} - V_{8\text{-}N} = (29.2\ k + 4.8\ k)\ TF = 34\ k\ (TF)$$

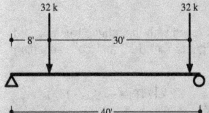

Figure Ex4.8-9
Fatigue truck placement for negative shear at 8 ft from support.

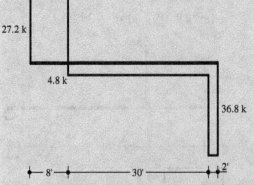

Figure Ex4.8-10
Shear diagram for fatigue truck in Figure Ex4.8-9.

Enveloping Shear Forces at 12 ft from Support

Positive maximum shear is found with the fatigue truck on the span as shown in Figures Ex4.8-11 and corresponding shear diagram in Ex4.8-12

$$V_{12\text{-}p} = \left[32\ k \left(\frac{28\ ft}{40\ ft} \right) + 8\ k \left(\frac{14\ ft}{40\ ft} \right) \right] (TF) = 25.2\ k\ (TF)$$

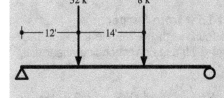

Figure Ex4.8-11
Fatigue truck placement for positive shear at 12 ft from support.

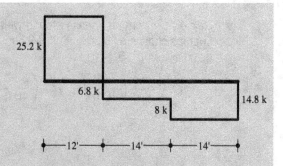

Figure Ex4.8-12
Shear diagram for fatigue truck in Figure Ex4.8-11.

Negative maximum shear is found with the fatigue truck placed on the span as shown in Figures Ex4.8-13 and corresponding shear diagram in Ex4.8-14

$$V_{12\text{-N}} = \left[32\,k\left(\frac{28\,ft}{40\,ft}\right) - 32\,k\right](TF) = -9.6\,k\,(TF)$$

Shear range V_f:

$$V_f = V_{12\text{-P}} - V_{12\text{-N}} = (25.2\,k + 9.6\,k)\,TF = 34.8\,k\,(TF)$$

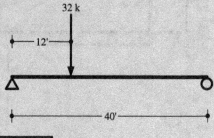

Figure Ex4.8-13
Fatigue truck placement for negative shear at 12 ft from support.

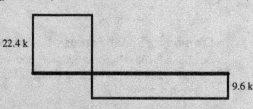

Figure Ex4.8-14
Shear diagram for fatigue truck in Figure Ex4.8-13.

Enveloping Shear Forces at 16 ft from Support

Positive maximum shear is found with the fatigue truck on the span as shown in Figures Ex4.8-15 and corresponding shear diagram in Ex4.8-16.

$$V_{16\text{-p}} = \left[32\,k\left(\frac{24\,ft}{40\,ft}\right) + 8\,k\left(\frac{10\,ft}{40\,ft}\right)\right](TF) = 21.2\,k\,(TF)$$

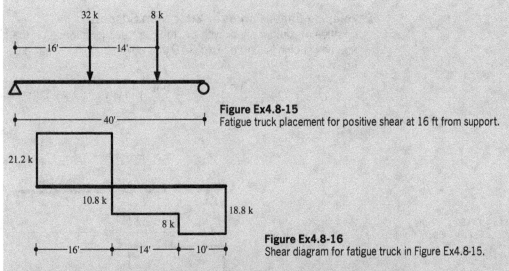

Figure Ex4.8-15
Fatigue truck placement for positive shear at 16 ft from support.

Figure Ex4.8-16
Shear diagram for fatigue truck in Figure Ex4.8-15.

Negative maximum shear is found with the fatigue truck on the span as shown in Figures Ex4.8-17 and corresponding shear diagram in Ex4.8-18.

$$V_{16\text{-}N} = \left[32\,\text{k} \left(\frac{24\,\text{ft}}{40\,\text{ft}} \right) - 32\,\text{k} \right] (\text{TF}) = -12.8\,\text{k} \ (\text{TF})$$

Shear range V_f:

$$V_f = V_{16\text{-}P} - V_{16\text{-}N} = (21.2\,\text{k} + 12.8\,\text{k})\,\text{TF} = 34\,\text{k} \ (\text{TF})$$

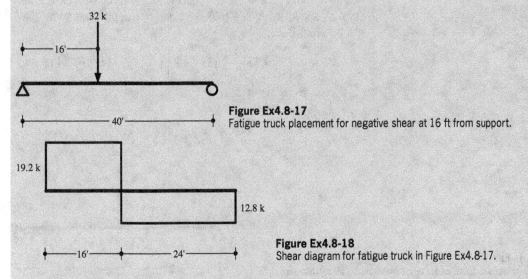

Figure Ex4.8-17
Fatigue truck placement for negative shear at 16 ft from support.

Figure Ex4.8-18
Shear diagram for fatigue truck in Figure Ex4.8-17.

Enveloping Shear Forces at 20 ft from Support

Positive maximum shear with the fatigue truck positioned on the span as shown in Figures Ex4.8-19 and corresponding shear diagram in Ex4.8-20:

$$V_{20\text{-}P} = \left[32\,k\left(\frac{20\,ft}{40\,ft}\right) + 8\,k\left(\frac{6\,ft}{40\,ft}\right)\right](TF) = 17.2\,k\,(TF)$$

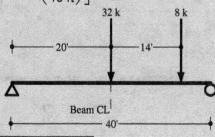

Figure Ex4.8-19
Fatigue truck placement for positive shear at 20 ft from support.

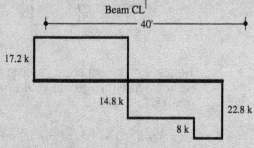

Figure Ex4.8-20
Shear diagram for fatigue truck in Figure Ex4.8-19.

Negative maximum shear is found with the fatigue truck on the span as shown in Figures Ex4.8-21 and corresponding shear diagram in Ex4.8-22.

$$V_{20\text{-}N} = \left[32\,k\left(\frac{20\,ft}{40\,ft}\right) - 32\,k\right](TF) = -16\,k\,(0.51)$$
$$= -8.16\,k$$

Shear range V_f:

$$V_f = V_{20\text{-}P} - V_{20\text{-}N} = (17.2\,k + 16\,k)\,TF = 33.2\,k\,(TF)$$

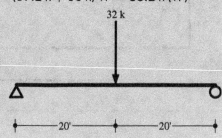

Figure Ex4.8-21
Fatigue truck placement for negative shear at 20 ft from support.

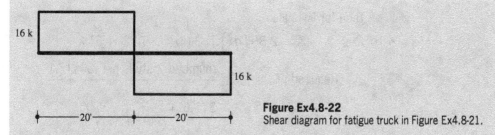

16 k

16 k

|←——— 20' ———→|←——— 20' ———→|

Figure Ex4.8-22
Shear diagram for fatigue truck in Figure Ex4.8-21.

The above calculated shear force values are summarized in Table Ex4.8-1 for the shear ranges to be used later as the load.

Table Ex4.8-1
Shear ranges as fatigue load

Location (ft)	Positive Shear (k TF)	Negative Shear (k TF)	Shear Range, V_f (k TF)
0	40	0	40
4	33.6	−0.8	34.4
8	29.2	−4.8	34
12	25.2	−9.6	34.8
16	21.2	−12.8	34
20	17.2	−16	33.2

Shear Stud Design for Fatigue Limit State   *6.10.10.2*

The section properties are as follows:

$I_x = 9,099.8$ in.^{4}

$Q =$ (area of girder) (distance from neutral axis to stringer's center of gravity)

$$= 24.7 \text{ in.}^2 \left(26.24 \text{ in.} - \frac{24.1 \text{ in.}}{2} \right) = 350.49 \text{ in.}^3$$

See Example 4.6 calculations for composite section properties for live load.
Use 3/4-in. shear studs:

Transverse stud spacing $s > 4(d) = 4(0.75 \text{ in.}) = 3 \text{ in.}$ $b_f = 9.02 \text{ in.}$

Use three studs per transverse row:

$$s = \frac{b_f - 2 \text{ (edge distance)}}{2 \text{ (spaces)}} = \frac{9.02 \text{ in.} - 2(1.375 \text{ in.})}{2} = 3.14 \text{ in.} > 3 \text{ in.}$$

$N = $ ADTT (365 days/year/lane) (75 years)

6.10.10.2 $= 900(365)75 = 24.6 \times 10^6$ cycles for driving lane

$\alpha = 34.5 - 4.28 \log N = 34.5 - (4.28) \log \left(24.6 \times 10^6 \right) = 2.86$

$\alpha > 5.5/2$

Then for finite life:

6.10.10.2 $Z_r = 2.86 \left(d^2\right) = 2.86 \left(0.75\right)^2 = 1.61 \text{ k}$

Required pitch $p \le \dfrac{(\text{number of studs per row})\,(Z_r)}{V_{sr}}$

6.10.10.1 $= \dfrac{3\,(1.61)}{V_{sr}}$ with $V_{sr} = \dfrac{V_f Q}{I_x}$

For Fatigue II Limit State, TF $= 0.68\ \gamma_L = 0.68\,(0.75) = 0.51$. Based on these calculations, the required stud pitch p is calculated in Table Ex4.8-2.

Table Ex4.8-2
Required shear stud pitches for interior beams

Location (ft)	Shear Range, V_f (k)	V_{sr} (k/in.)	Required Pitch, p (in.)
0	20.4	0.79	6.11
4	17.54	0.71	6.80
8	17.34	0.67	7.21
12	17.75	0.63	7.10
16	17.34	0.68	7.21
20	16.93	0.65	7.43

Use 15 spaced at 6 in. and then 21 spaced at 7 in. with a total of $3(15+21)=108$ studs on each half of the beam for the interior beams.

☐ **For Exterior Beams**

Total factor TF $= (1 + \text{IM})\,\gamma_L \left(\dfrac{\text{DF}_{\text{shear exterior 1 lane}}}{1.2} \right)$

$= (1.15)\,0.75 \left(\dfrac{0.86}{1.2} \right) = 0.51\,(1.21) = 0.62$

Q/I_x is similar between the internal and external beams. The only difference is between their respective TF values and their ratio is 1.21. For the exterior beams, the shear ranges will be 1.21 times larger. The pitch needs to be 1.21 times smaller, as shown in Table Ex4.8-3. Use 20 spaces at 5 in. and then 23 spaces at 6 in. with a total of $3(20+23)=129$ studs on each half of the beam for the exterior beams.

☐ **Strength Limit State**   *6.10.10.4*

Determine the Design Force: The force that the steel beam can carry is given as

6.10.10.4 $P_{2p} = A_s F_y = 24.7 \text{ in.}^2 \left(50 \text{ k/in.}^2\right) = 1235 \text{ k}$

Table Ex4.8-3
Required shear stud pitches for exterior beams

Location (ft)	Shear Range, V_f (k)	V_{sr} (k/in.)	Required Pitch, p (in.)
0	24.7	0.95	5.05
4	21.22	0.86	5.62
8	20.98	0.81	5.96
12	21.48	0.83	5.87
16	20.98	0.81	5.96
20	20.49	0.79	6.14

the force that the concrete slab can carry as

$$P_{1p} = 0.85f_c' \, w_{eff}t = 0.85 \left(4.5 \text{ k/in.}^2\right) 8.83 \text{ ft} \,(8.5 \text{ in.}) \, 12 \text{ in./ft} = 3445 \text{ k}$$

and the design force as the minimum of the two forces:

$$P = \min\left(P_{1p}, P_{2p}\right) = 1235 \text{ k}$$

Ultimate strength of stud:

6.10.10.4.3
$$Q_n = 0.5A_{sc}\sqrt{f_c' E_{sc}} \le A_{sc}F_u$$
$$A_{sc} = \frac{\pi \,(0.75)^2}{4} = 0.44 \text{ in.}^2$$

C5.4.2.4
$$E_c = 1820\sqrt{f_c'} = 1820\sqrt{4.5} = 3860 \text{ k/in.}^2$$

Thus

$$Q_n = 0.5 \left(0.44 \text{ in.}^2\right) \sqrt{4.5 \text{ k/in.}^2 \,(3860 \text{ k/in.}^2)} = 28.99 \text{ k}$$

and Q_n needs to be $\le 0.44 \text{ in.}^2 \left(60 \text{ k/in.}^2\right) = 26.4\text{k}$.

Number of studs required for strength limit state:

6.10.10.4.1
$$\text{Required number of studs} = \frac{P}{\varphi_{sc}\,(Q_n)} = \frac{1235 \text{ k}}{0.85\,(26.4 \text{ k})} = 55.04$$

This number is smaller than that required by fatigue limit state 108. Use 108 studs for each half of the span in the interior beams and 129 in the exterior beams.
OK for studs for both fatigue and strength limit states.

Example 4.9 Steel Plate Girder Bridge Design (Load Effects)

❏ **Design Requirement**

Design a composite steel plate girder superstructure for the HL-93 live load for the strength I limit state. The span is 115 ft long and needs to carry two lanes of vehicle traffic. The framing plan is shown in Figure Ex4.9-1.

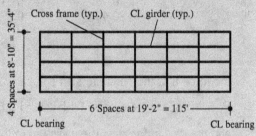

Figure Ex4.9-1
Plan view of framing.

Load modifier $\eta = 1.0$

❏ **Design Parameters**

Span number $N_{spans} = 1$

Span length $L_{span} = 115$ ft

Skew angle $= 0°$

Girder number $N_{girders} = 5$

Girder spacing $S = 8$ ft 10 in. $= 8.83$ ft

Deck overhang length $S_{overhang} = 3$ ft 11 in.

Cross-frame spacing $L_b = 19$ ft 2 in. $= 19.17$ ft

Web yield strength $F_{yw} = 50$ k/in.^{2}

Flange yield strength $F_{yf} = 50$ k/in.^{2}

Concrete 28-day compressive strength $f'_c = 4.0$ k/in.^{2}

Reinforcement strength $f_y = 60$ k/in.^{2}

Total deck thickness $=$ deck effective thickness $t_{slab} = 8.5$ in.

Total overhang deck thickness $t_{slab\ overhang} = 9.5$ in.

Steel density $W_{steel} = 0.490$ k/ft^3 *3.5.1*

Concrete density $W_{concrete} = 0.145$ k/ft^3 *3.5.1*

Additional miscellaneous dead load (per girder) $DL_{miscellaneous}$
    $= 0.015$ k/ft

Stay-in-place deck form weight $W_{deck\ forms} = 14.98$ lb/ft
    $= 0.015$ k/ft

Parapet weight (each) $W_{parapet} = 0.457$ k/ft

Wearing surface weight $W_{WS} = 0.14$ k/ft^3 *3.5.1*

Wearing surface thickness $t_{WS} = 2.5$ in.

Deck width $w_{deck} = 43$ ft 2 in.

Roadway width $w_{roadway} = 40$ ft 4 in. $= 40.33$ ft

Haunch depth $d_{haunch} = 1$ in.
Average daily truck traffic (single-lane) ADTT= 5000 trucks/day

☐ **Trial Girder Section**
As a first step the girder section is assumed as in Figure Ex4.9-2.

☐ **Section Properties**
The noncomposite dead loads are applied to the steel girder-only cross section. The superimposed dead loads are applied to the composite cross section based on a modular ratio of $3n$ or n, whichever gives the higher stress.

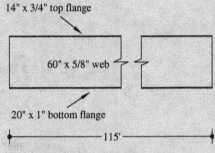

14" x 3/4" top flange

60" x 5/8" web

20" x 1" bottom flange

115'

Figure Ex4.9-2
Plate girder cross-sectional details.

$$E_c = 33,000\left(W_c^{1.5}\right)\sqrt{f_c'}$$

$$= 33,000\left(0.145 \text{ k/ft}^3\right)^{1.5}\sqrt{4 \text{ k/in.}^2} = 3644 \text{ k/in.}^2$$

$$n = E_s/E_c = 29,000/3644 = 7.95 \qquad \text{Use } n = 8$$

☐ **Effective Flange Width**

$$4.6.2.6.1 \qquad W_{eff} = \begin{cases} 8.83 \text{ ft} = 8 \text{ ft } 10 \text{ in.} & \text{for interior girders.} \\ \dfrac{8 \text{ ft } 10 \text{ in.}}{2} + 3 \text{ ft } 11 \text{ in.} = 8 \text{ ft } 4 \text{ in.} & \text{for exterior girders} \end{cases}$$

☐ **Dead-Load Effects for Interior Girders**
1. Concrete deck:

$$DL_{deck} = W_{concrete} \, St_{slab} = \left(0.145 \text{ k/ft}^3\right) 8.83 \text{ ft } (8.5 \text{ in.})/(12 \text{ in./ft})$$
$$= 0.907 \text{ k/ft}$$

2. Haunch:

$$DL_{haunch} = W_c \, b_{f,top} \, t_{haunch} = 0.145 \text{ k/ft}^3 \, (1.17 \text{ ft}) \, 1 \text{ in.}/(12 \text{ in./ft})$$
$$= 0.014 \text{ k/ft}$$

3. Stay-in-place forms:

$$DL_{deck\,forms} = W_{deck\,forms}\,(S - t_{f,\,top}) = 0.015 \text{ k/ft}^2 \left(8.83 \text{ ft} - \frac{14 \text{ in.}}{12 \text{ in./ft}}\right)$$
$$= 0.115 \text{ k/ft}$$

4. Concrete parapet: *Superimposed dead loads are evenly distributed among all girders.*

$$DL_{pararapet} = 2W_{parapet}/N_{girders} = 2(0.457 \text{ k/ft})/5 = 0.183 \text{ k/ft}$$

5. Wearing surface:

$$DL_{FWS} = \frac{W_{WS}\,(t_{WS}/12\text{ in./ft})\,w_{roadway}}{N_{girders}}$$

$$= \frac{0.140\text{ k/ft}^3\,(2.5\text{ in./12 in./ft})\,40.33\text{ ft}}{5} = 0.235\text{ k/ft}$$

6. Dead load of steel:

$$DL_{steel} = (W_{steel})\,(\text{cross-sectional area}) = 0.49\text{ k/ft}^3\,(68\text{ in.}^2)\,/$$

$$(12\text{ in./ft})^2 = 0.231\text{ k/ft}$$

7. Miscellaneous items:

$$DL_{miscellaneous} = 0.015 \quad \text{k/ft}$$

☐ Dead-Load Effects for Exterior Girders

1. Concrete deck:

$$DL_{deck} = W_{concrete}\left[\frac{S}{1\frac{1}{2}}\,t_{slap} + (3.2\text{ ft})\,t_{slap\ overhanf}\right]$$

$$= \left(0.145\text{ k/ft}^3\right)6.23\text{ ft}^2 = 0.903\text{ k/ft}$$

2. Haunch:

$$DL_{haunch} = W_c\,b_{f,\,top}\,t_{haunch} = 0.145\text{ k/ft}^3\,(1.17\text{ ft})\,1\text{ in.}/$$

$$(12\text{ in./ft}) = 0.014\text{ k/ft}$$

3. Stay-in-place forms:

$$DL_{deck\ forms} = W_{deck\ forms}\left(\frac{S - t_{f,\,top}}{2}\right)$$

$$= 0.015\text{ k/ft}^2\left(\frac{8.83\text{ ft} - 14\text{ in./12 in./ft}}{2}\right) = 0.057\text{ k/ft}$$

4. Concrete parapet:
   Superimposed dead loads are evenly distributed among all girders.

$$DL_{pararapet} = 2W_{parapet}/N_{girders} = 2\,(0.457\text{ k/ft})\,/5 = 0.183\text{ k/ft}$$

5. Wearing surface:

$$DL_{FWS} = \frac{W_{WS}\,(t_{WS}/12\text{ in./ft})\,w_{roadway}}{N_{girders}}$$

$$= \frac{0.140\text{ k/ft}^3\,(2.5\text{ in./12 in./ft})\,40.33\text{ ft}}{5} = 0.235\text{ k/ft}$$

6. Dead load of steel:

$$DL_{steel} = (W_{steel})\,(\text{cross-sectional area}) = 0.49\ \text{k/ft}^3\left(68\ \text{in.}^2\right)/$$

$$(12\ \text{in./ft})^2 = 0.231\ \text{k/ft}$$

7. Miscellaneous items:

$$DL_{miscellaneous} = 0.015\ \text{k/ft}$$

❑ **Dead Load for Noncomposite Section for Interior Girders**

$$DL = DL_{deck} + DL_{haunch} + DL_{deck\ forms} + DL_{steel} + DL_{miscellaneous}$$
$$= 0.907 + 0.014 + 0.115 + 0.231 + 0.015 = 1.282\ \text{k/ft}$$

❑ **Dead Load for Noncomposite Section for Exterior Girders**

$$DL = DL_{deck} + DL_{haunch} + DL_{deck\ forms} + DL_{steel} + DL_{misecellaneous}$$
$$= 0.903 + 0.014 + 0.057 + 0.231 + 0.015 = 1.220\ \text{k/ft}$$

❑ **Dead Load for Composite Section for Interior and Exterior Girders**

$$DC = DL_{parapet} = 0.183\ \text{k/ft}$$

$$DW = DL_{FWS} = 0.235\ \text{k/ft}$$

❑ **Dead-Load Moments and Shears**

1. Total girder—Moment:

$$M_{DL\ girder\ interior} = 1.282\,(115^2)/8 = 2119\ \text{k-ft}$$

$$M_{DL\ girder\ exterior} = 1.220\,(115^2)/8 = 2017\ \text{k-ft}$$

Shear:

$$V_{DL\ girder\ interior} = 1.282\,(115)/2 = 73.72\ \text{k}$$

$$V_{DL\ girder\ exterior} = 1.220\,(115)/2 = 70.15\ \text{k}$$

2. Parapet—Moment:

$$M_{DL\ parapet} = 0.183\left(115^2\right)/8 = 302.5\ \text{k-ft}$$

Shear:

$$V_{DL\ parapet} = 0.183\,(115)/2 = 10.5\ \text{k-ft}$$

3. Future wearing surface—Moment

$$M_{DL\ WS} = 0.235\left(115^2\right)/8 = 388.9\ \text{k-ft}$$

Shear:

$$V_{DL\ WS} = 0.235\,(115)/2 = 13.50\ \text{k}$$

◻ **Longitudinal Stiffness Parameter:**

4.6.2.2.1 $K_g = n\left[I + A\left(e_g^2\right)\right]$

is calculated in Table Ex4.9-1.

Table Ex4.9-1
Calculation for longitudinal stiffness parameter K_g

n	8
I (in.^{4})	38,305
A (in.^{2})	68
e_g(in.)	40.28
K_g(in.^{4})	1,189,068

◻ **Live-Load Distribution Factors for Interior Girders**
This example is identified as case "a" in 4.6.2.2.1. Check the range of applicability:

For $3.5 \leq S \leq 16.0$, $S = 8.83$ ft.   OK
For $4.5 \leq t_s \leq 12.0$, $t_s = t_{slab} = 8.5$ in.   OK
For $20 \leq L \leq 240$, $L = L_{span} = 115$ ft.   OK
For $N_b \geq 4$, $N_b = N_{girders} = 5$.   OK
For $10,000 \leq K_g \leq 7,000,000$ in.^{4}, $K_g = 1,189,068$ in.^{4}
     OK

1. For moment—One lane loaded 4.6.2.2

$$DF_{\text{interior moment 1}} = 0.06 + \left(\frac{S}{14}\right)^{0.4}\left(\frac{S}{L}\right)^{0.3}\left[\frac{K_g}{12\,Lt_s^3}\right]^{0.1}$$

$$= 0.06 + \left(\frac{8.83\,\text{ft}}{14}\right)^{0.4}\left(\frac{8.83\,\text{ft}}{115\,\text{ft}}\right)^{0.3}$$

$$\times \left[\frac{1,189,068\,\text{in.}^4}{(12)\,115\,\text{ft}\,(8.5\,\text{in.})^3}\right]^{0.1} = 0.46\,\text{lane}$$

2. For moment—Two or more lanes loaded

$$DF_{\text{interior moment 2}} = 0.075 + \left(\frac{S}{9.5}\right)^{0.6}\left(\frac{S}{L}\right)^{0.2}\left[\frac{K_g}{12\,L\,t_s^3}\right]^{0.1}$$

$$= 0.075 + \left(\frac{8.83\,\text{ft}}{9.5}\right)^{0.6}\left(\frac{8.83\,\text{ft}}{115\,\text{ft}}\right)^{0.2}$$

$$\times \left[\frac{1,189,068\,\text{in.}^4}{12.0\,(115\,\text{ft})\,8.5^3}\right]^{0.1} = 0.67\,\text{lane}$$

For interior girder moment, the controlling distributor factor is 0.67.

3. For shear—One lane loaded $4.6.2.2$

$$DF_{\text{interior shear } 1} = 0.36 + \frac{S}{25} = 0.36 + \frac{8.83\,\text{ft}}{25} = 0.71 \text{ lane}$$

4. For shear—Two or more lanes loaded

$$DF_{\text{interior shear } 2} = 0.2 + \left(\frac{S}{12}\right) - \left(\frac{S}{35}\right)^{2.0}$$

$$= 0.2 + \left(\frac{8.83\,\text{ft}}{12}\right) - \left(\frac{8.83\,\text{ft}}{35}\right)^{2.0} = 0.87 \text{ lane}$$

For interior girder shear, the controlling distributor factor is 0.87.

☐ **Live-Load Distribution Factors for Exterior Girders**

$$d_e = 2.5\,\text{ft} \quad \text{satisfying} - 1.0 \le d_e \le 5.5$$

1. For moment—One lane loaded

Use the lever rule per $4.6.2.2.2$ as shown in Figure Ex4.9-3.

$$DF_{\text{exterior moment } 1} = \frac{0.5\,(40\,\text{in.}) + 0.5\,(112\,\text{in.})}{106\,\text{in.}}\,(1.2) = 0.86 \text{ lane}$$

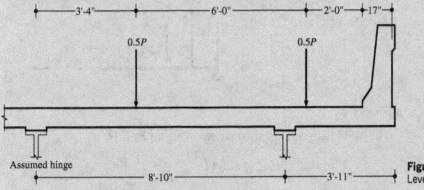

Figure Ex4.9-3
Lever rule.

Multiple presence factor 1.20 has been included.

2. For moment—Two or more lanes loaded

$4.6.2.2$

$$e = 0.77 + \left(\frac{d_e}{9.1}\right) = 0.77 + \left(\frac{2.52\,\text{ft}}{9.1}\right) = 1.05$$

$$DF_{\text{exterior moment } 2} = e\,(DF_{\text{interior moment } 2 \text{ lane}}) = 1.05\,(0.67) = 0.70 \text{ lane}$$

For exterior girder moment, the controlling distributor factor is 0.86.

3. For shear—One lane loaded

For one design lane loaded, the lever rule is used per 4.6.2.2 illustrated in Figure Ex4.9-3. A multiple presence factor of 1.2 also needs to be applied when the lever rule is used.

$$DF_{\text{exterior shear 1}} = \frac{0.5(40 \text{ in.}) + 0.5(112 \text{ in.})}{106 \text{ in.}} (1.2) = 0.86 \text{ lane}$$

4. For shear—Two or more lanes loaded

$$e = 0.6 + \left(\frac{d_e}{10}\right) = 0.6 + \left(\frac{2.5 \text{ ft}}{10}\right) = 0.85$$

$$DF_{\text{exterior shear 2}} = e \left(DF_{\text{interior shear 2 lane}}\right) = 0.85 \, (0.87) = 0.74 \text{ lane}$$

For exterior girder shear, the controlling distributor factor is 0.86.

☐ **Live-Load Design Shears**

For HL93 truck load on span shown in Figure Ex4.9-4:

$$V_{\text{truck}} = \left[32 + 32 \left(\frac{101}{115}\right) + 8 \left(\frac{87}{115}\right)\right] (1 + IM) = 66.16 \text{ k} \, (1.33)$$
$$= 87.99 \text{ k}$$

This shear governs the shear due to the HL93 tandem load.

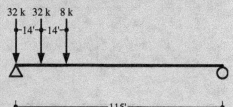

Figure Ex4.9-4
HL93 truck load placement for maximum shear.

For HL93 lane load on span shown in Figure Ex4.9-5

$$V_{\text{lane}} = 0.64 \left(\frac{115}{2}\right) = 36.8 \text{ k}$$

Figure Ex4.9-5
HL93 lane load placement for maximum shear.

Unfactored design shears:

$$(87.99 \text{ k} + 36.8 \text{ k}) \, 0.87 = 108.6 \text{ k} \quad \text{for interior girders}$$
$$(87.99 \text{ k} + 36.8 \text{ k}) \, 0.86 = 107.3 \text{ k} \quad \text{for exterior girders}$$

Live-Load Design moments
The midspan bisects between truck CG and closest heavier axle load
32 k for maximum moment, shown in Figure Ex4.9-6.

$$R_B = \frac{41.17\,(8) + 55.17\,(32) + 69.17\,(32)}{115} = 37.46\text{ k}$$

$$R_A = 34.54\text{ k}$$

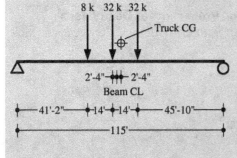

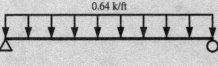

Figure Ex4.9-6
HL93 truck load for maximum moment.

Figure Ex4.9-7
HL93 lane load for maximum moment.

$$M_{truck} = [R_A\,(55.17) - 8\,(14)]\,(1 + \text{IM}) = 1794\text{ kft }(1.33) = 2385\text{ kft}$$

This moment governs the moment due to the HL93 tandem load.

$$M_{lane} = 0.64\left(115^2/8\right) = 1{,}058\text{ kft}$$

Unfactored design moments:

$$(2385\text{ kft} + 1058\text{ kft})\,0.67 = 2307\text{ kft}\quad\text{for interior girders}$$

$$(2385\text{ kft} + 1058\text{ kft})\,0.86 = 2961\text{ kft}\quad\text{for exterior girders}$$

Load Effect Combination for Strength I Limit State
Load effect combination for Strength I Limit State.
For moment:

$$M_{total\ strength\ I} = \gamma_{DC}(M_{DL\ beam} + M_{DL\ parapet}) + \gamma_{DW}M_{DL\ WS} + \gamma_{LL}M_{LL}$$

$$= 1.25(2119 + 302.5) + 1.5(388.9) + 1.75(2307)$$

$$= 7647\text{ kft}\quad\text{for interior girders}$$

$$M_{total\ strength\ I} = 1.25(2119 + 302.5) + 1.5(388.9) + 1.75(2961)$$

$$= 8792\text{ kft}\quad\text{for exterior girders}$$

For shear

$$V_{\text{total strength I}} = \gamma_{DC}(V_{\text{DL beam}} + V_{\text{DL parapet}}) + \gamma_{DW} V_{\text{DL WS}} + \gamma_{LL} V_{LL}$$
$$= 1.25(73.72 + 10.5) + 1.5(13.5) + 1.75(108.6)$$
$$= 315.6 \text{ k for interior girders}$$
$$V_{\text{total strength I}} = 1.25(73.72 + 10.5) + 1.5(13.5) + 1.75(107.3)$$
$$= 313.3 \text{ k for exterior girders}$$

☐ **Steel Noncomposite Section Properties for Dead Loads**
The distances Y are measured from the bottom of the section in Table Ex4.9-2.

$$I_z = \sum I_0 + \sum Ay^2 = 11{,}252 \text{ in.}^4 + 75{,}602 \text{ in.}^4$$
$$= 86{,}854 \text{ in.}^4$$

$$Y' = \frac{\sum Ay}{\sum A} = \frac{1817 \text{ in.}^3}{68 \text{ in.}^2} = 26.72 \text{ in.}$$

$$Y_{\text{top girder}} = 61.75 \text{ in.} - 26.72 \text{ in.} = 35.03 \text{ in.}$$

$$I_x = I_z - \left(\sum A\right)\left(Y'\right)^2 = 86{,}854 \text{ in.}^4 - 68 \text{ in.}^2 (26.72 \text{ in.})^2$$
$$= 38{,}305 \text{ in.}^4$$

$$S_{\text{top girder}} = \frac{38{,}305 \text{ in.}^4}{35.03 \text{ in.}} = 1093 \text{ in.}^3$$

$$S_{\text{bottom girder}} = \frac{38{,}305 \text{ in}^4}{26.72 \text{ in}} = 1434 \text{ in.}^3$$

Figure Ex4.9-8
Preliminary section dimensions.

Table Ex4.9-2
Noncomposite Section Properties for Dead Load

Component	Width (in.)	Height (in.)	A (in.²)	Y (in.)	AY (in.³)	AY² (in.⁴)	I_0 (in.⁴)
Top flange	14	0.75	10.5	61.38	644.5	39,559	0.492
Web	0.625	60	37.5	31	1,162.5	36,038	11,250
Bottom flange	20	1	20	0.5	10	5	1.667
Sum		61.75	68	92.88	1,817	75,602	11,252

Example 4.10 highlights the beam design for the same steel superstructure in Example 4.9, with some details omitted because they have been covered in Examples 4.6 through 4.8. Instead, Example 4.11 touches upon steel beam fatigue design, presented in Section 4.7.8. Fatigue is often a

relevant issue to plate girders since more steel members are welded together to construct the structural system, as opposed to the rolled beam superstructure with much fewer pieces welded together.

Example 4.10 Steel Plate Girder Bridge Design (Strength I and Service II Limit States)

☐ Design Requirement

Check strength I and service II limit states for the composite steel plate girder superstructure in Example 4.9.
Load modifier $\eta = 1.0$

☐ Composite Sectional Properties

Steel Noncomposite Section for Dead Loads as shown in Table Ex4.10-1

Distances Y are measured from the bottom of the section.

$$I_z = \sum I_0 + \sum Ay^2 = 11{,}252 \text{ in.}^4 + 75{,}602 \text{ in.}^4$$

$$= 86{,}854 \text{ in.}^4$$

$$Y' = \frac{\sum Ay}{\sum A} = \frac{1817 \text{ in.}^3}{68 \text{ in.}^2} = 26.72 \text{ in.}$$

$$Y_{top\ girder} = 61.75 \text{ in.} - 26.72 \text{ in.} = 35.03 \text{ in.}$$

$$I_x = I_z - \left(\sum A\right)\left(Y'\right)^2 = 86{,}854 \text{ in.}^4 - 68 \text{ in.}^2\,(26.72 \text{ in.})^2$$

$$= 38{,}305 \text{ in.}^4$$

$$S_{top\ girder} = \frac{38{,}305 \text{ in.}^4}{35.03 \text{ in.}} = 1093 \text{ in.}^3$$

$$S_{bottom\ girder} = \frac{38{,}305 \text{ in.}^4}{26.72 \text{ in.}} = 1434 \text{ in.}^3$$

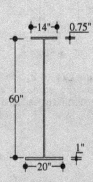

Figure Ex4.10-1
Preliminary section dimensions.

Table Ex4.10-1
Noncomposite section properties for dead load

Component	Width (in.)	Height (in.)	A (in.²)	Y (in.)	AY (in.³)	AY² (in.⁴)	I_0(in.⁴)
Top flange	14	0.75	10.5	61.38	644.5	39,559	0.492
Web	0.625	60	37.5	31	1,162.5	36,038	11,250
Bottom flange	20	1	20	0.5	10	5	1.667
Sum		61.75	68	92.88	1,817	75,602	11,252

Note: The distances are measured from the bottom of section.

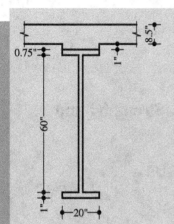

Figure Ex4.10-2
Composite section dimensions for superimposed dead load.

Composite Section for Superimposed Dead Loads as shown in Figure Ex4.10-2

For superimposed dead loads, use $3(n) = 3(8) = 24$:

$$b_{eff} = \frac{w_{eff}}{3(n)} = \frac{(8.83)\ 12\ \text{in./ft}}{3(8)} = 4.42\ \text{in.}$$

$$I_{slab} = \frac{4.42\ \text{in.}\ (8.5\ \text{in.})^3}{12} = 226.2\ \text{in.}^4$$

$$I_z = \sum I_0 + \sum Ay^2 = 38{,}531\ \text{in.}^4 + 217{,}201\ \text{in.}^4$$
$$= 255{,}732\ \text{in.}^4$$

$$Y' = \frac{\sum Ay}{\sum A} = \frac{4{,}334\ \text{in.}^3}{105.6\ \text{in.}^2} = 41.04\ \text{in.}$$

$$Y_{top\ slab} = 61.75\ \text{in.} + 1\ \text{in.} + 8.5\ \text{in.} - 41.041\ \text{in.} = 30.21\ \text{in.}$$

$$Y_{top\ girder} = 61.75\ \text{in.} - 41.04\ \text{in.} = 20.71\ \text{in.}$$

$$I_x = I_z - \left(\sum A\right)\left(Y'\right)^2 = 255{,}732\ \text{in.}^4$$
$$- 105.6\ \text{in.}^2\ (41.04\ \text{in.})^2 = 77{,}872\ \text{in.}^4$$

$$S_{bottom\ girder\ SDL} = \frac{77{,}872\ \text{in.}^4}{41.04\ \text{in.}} = 1{,}897\ \text{in.}^3$$

$$S_{top\ slab\ SDL} = \frac{77{,}872\ \text{in.}^4}{30.21\ \text{in.}} = 2{,}577\ \text{in.}^3$$

$$S_{top\ girder\ SDL} = \frac{77{,}872\ \text{in.}^4}{20.71\ \text{in.}} = 3{,}760\ \text{in.}^3$$

Table Ex4.10-2
Composite section properties for superimposed dead load

Component	A (in.^{2})	Y (in.)	AY (in.)	AY^2 (in.^{4})	I_0 (in.^{4})
Steel	68	26.72	1,817	48,549	38,305
Slab (3n)	37.57	67	2,517	168,652	226.2
Sum	105.6		4,334	217,201	38,531

Composite Section for Live Loads

For superimposed dead loads, use $n = 8$:

$$b_{eff} = \frac{w_{eff}}{n} = \frac{106\ \text{in.}}{8} = 13.25\ \text{in.}$$

$$I_{slab} = \frac{13.25 \text{ in.} (8.5 \text{ in.})^3}{12} = 678.1 \text{ in.}^4$$

$$I_z = \sum I_0 + \sum Ay^2 = 38,983 \text{ in.}^4 + 554,010 \text{ in.}^4 = 592,993 \text{ in.}^4$$

$$Y' = \frac{\sum Ay}{\sum A} = \frac{9361 \text{ in.}^3}{180.6 \text{ in.}^2} = 51.83 \text{ in.}$$

$$Y_{top \text{ slab}} = 61.75 \text{ in.} + 1 \text{ in.} + 8.5 \text{ in.} - 51.83 \text{ in.} = 19.42 \text{ in.}$$

$$Y_{top \text{ girder}} = 61.75 \text{ in.} - 51.83 \text{ in.} = 9.92 \text{ in.}$$

$$I_x = I_z - \left(\sum A\right)\left(Y'\right)^2 = 592,993 \text{ in.}^4 - 180.6 \text{ in.}^2 (51.83 \text{ in.})^2$$

$$= 107,838 \text{ in.}^4$$

$$S_{bottom \text{ girder LL}} = \frac{107,838 \text{ in.}^4}{51.83 \text{ in.}} = 2,081 \text{ in.}^3$$

$$S_{top \text{ slab LL}} = \frac{107,838 \text{ in.}^4}{19.42 \text{ in.}} = 5,553 \text{ in.}^3$$

$$S_{top \text{ girder LL}} = \frac{107,838 \text{ in.}^4}{9.92 \text{ in.}} = 10,871 \text{ in.}^3$$

Table Ex4.10-3
Composite section properties for live load

Component	A (in²)	Y (in)	AY (in)	AY² (in⁴)	I_0 (in.⁴)
Steel	68	26.72	1,817	48,549	38,305
Slab (n)	112.6	67	7,544	505,461	678.1
Sum	180.6		9,361	554,010	38,983

☐ **Load Effect Combination for Service II Limit State**   *6.10.4*
   Girder Top Stresses

Dead-load stress on noncomposite section:

$$f_{non \text{ composite DL}} = \frac{M_{non \text{ composite DL}}}{S_{top \text{ girder}}} = \frac{2119 \text{ kft} (12 \text{ in./ft})}{1093 \text{ in.}^3}$$

$$= 23.26 \text{ k/in.}^2 \text{ (compressive)}$$

Parapet dead-load stress on 3n-composite section:

$$f_{parapet} = \frac{M_{parapet}}{S_{top \text{ girder SDL}}} = \frac{302.5 \text{ kft} (12 \text{ in./ft})}{3760 \text{ in.}^3}$$

$$= 0.965 \text{ k/in.}^2 \text{ (compressive)}$$

Wearing surface dead-load stress on $3n$-composite section:

$$f_{WS} = \frac{M_{WS}}{S_{\text{top girder SDL}}} = \frac{388.9 \text{ kft} (12 \text{ in./ft})}{3{,}760 \text{ in.}^3}$$

$$= 1.24 \text{ k/in.}^2 \quad \text{(compressive)}$$

Live-load stress on n-composite section:

$$f_{LL} = \frac{M_{LL}}{S_{\text{top girder LL}}} = \frac{2307 \text{ kft} (12 \text{ in./ft})}{10{,}871 \text{ in.}^3}$$

$$= 2.55 \text{ k/in.}^2 \quad \text{(compressive)}$$

Service II limit state checking: For interior girders

$$f_{\text{total service II}} = \gamma_{DC}(f_{\text{noncomposite DL}} + f_{\text{parapet}}) + \gamma_{DW}\, f_{WS} + \gamma_{LL}\, f_{LL}$$

$$= 1.0(23.26 + 0.965) + 1.0(1.24) + 1.3(2.55)$$

$$= 28.8 \text{ k/in.}^2 < \phi_b F_y = 0.95(50) = 47.5 \text{ k./in.}^2$$

OK for yielding in interior girders.

For exterior girders, using $M_{LL} = 2961$ kft,

$$f_{\text{total service II}} = 1.0(23.26 + 0.965) + 1.0(1.24) + 1.3(3.39)$$

$$= 29.8 \text{ k/in.}^2 < \phi_b F_y = 0.95(50) = 47.5 \text{ k/in}^2$$

OK for yielding in exterior girders.

Girder Bottom Stresses

Dead-Load stress on noncomposite section:

$$f_{\text{noncomposite DL}} = \frac{M_{\text{noncomposite DL}}}{S_{\text{bottom girder}}} = \frac{2119 \text{ kft} (12 \text{ in./ft})}{1434 \text{ in.}^3}$$

$$= 17.73 \text{ k/in.}^2 \quad \text{(tensile)}$$

Parapet dead-load stress on $3n$-composite section:

$$f_{\text{parapet}} = \frac{M_{\text{parapet}}}{S_{\text{bottom girder SDL}}} = \frac{(302.5 \text{ kft}) (12 \text{ in./ft})}{1897 \text{ in.}^3}$$

$$= 1.91 \text{ k/in.}^2 \quad \text{(tensile)}$$

Wearing surface dead load stress on $3n$-composite section:

$$f_{WS} = \frac{M_{WS}}{S_{\text{bottom girder SDL}}} = \frac{(388.9 \text{ kft}) (12 \text{ in./ft})}{1897 \text{ in.}^3}$$

$$= 2.46 \text{ k/in.}^2 \quad \text{(tensile)}$$

Live-load stress on n-composite section:

$$f_{LL} = \frac{M_{LL}}{S_{\text{bottom girder LL}}} = \frac{(2307 \text{ kft})(12 \text{ in./ft})}{2081 \text{ in.}^3}$$
$$= 13.3 \text{ k/in.}^2 \text{ (tensile)}$$

Service II limit state checking: For interior girders

$$f_{\text{total service II}} = \gamma_{DC}(f_{\text{noncomposite DL}} + f_{\text{parapet}}) + \gamma_{DW} f_{WS} + \gamma_{LL} f_{LL}$$
$$= 1.0(17.73 + 1.91) + 1.0(2.46) + 1.3(13.3)$$
$$= 39.4 \text{ k/in.}^2 < \phi_b F_y = 0.95(50) = 47.5 \text{ k/in.}^2$$

OK for yielding in interior girders.

For exterior girders, using $M_{LL} = 2961$ kft

$$f_{\text{total service II}} = 1.0(17.73 + 1.91) + 1.0(2.46) + 1.3(17.07)$$
$$= 44.3 \text{ k/in.}^2 < \phi_b F_y = 0.95(50) = 47.5 \text{ k/in.}^2$$

OK for yielding in exterior girders.

☐ **Check Section Proportion Limits for Moment Design under Strength I Limit State**   *6.10.2*

Compression flange moment of inertia about y-axis:

$$I_{yc} = \frac{0.75\left(14^3\right)}{12} = 171.5 \text{ in.}^4$$

Tension flange moment of inertia about y-axis:

$$I_{yt} = \frac{1\left(20^3\right)}{12} = 666.7 \text{ in.}^4$$

Thus,

$$0.1 \leq \frac{I_{yc}}{I_{yt}} = \frac{171.5}{666.7} = 0.26 \leq 10 \qquad \boxed{\text{OK}}$$

The next section proportion check relates to the web slenderness. For a section without longitudinal stiffeners, the web must be proportioned such that:

$$\frac{D}{t_w} = \frac{60 \text{ in.}}{0.625 \text{ in.}} = 96 \leq 150 \qquad \boxed{\text{OK}}$$

For the top flange,

$$\frac{b_f}{2t_f} = \frac{14 \text{ in.}}{2(0.75 \text{ in.})} = 9.33 \leq 12 \qquad \boxed{\text{OK}}$$

$$b_f = 14 \text{ in.} \geq D/6 = 60/6 = 10 \text{ in.} \qquad \boxed{\text{OK}}$$

$$t_f = 0.75 \text{ in.} \geq 1.1 t_w = 1.1(0.625) = 0.69 \text{ in.} \qquad \boxed{\text{OK}}$$

For the bottom flange,

$$\frac{b_f}{2t_f} = \frac{20 \text{ in.}}{2(1 \text{ in.})} = 10 \leq 12 \qquad \boxed{\text{OK}}$$

$$b_f = 20 \text{ in.} \geq D/6 = 60/6 = 10 \text{ in.} \qquad \boxed{\text{OK}}$$

$$t_f = 1 \text{ in.} \geq 1.1 t_w = 1.1(0.625) = 0.69 \text{ in.} \qquad \boxed{\text{OK}}$$

Plastic Moment Capacity   D6.1

Total force in the yielded tension (bottom) flange:

$$P_t = F_{yt} b_t t_t = 50 \text{ k/in.}^2 (20 \text{ in.})1 \text{ in.} = 1000 \text{ k}$$

Total force in the yielded web:

$$P_w = F_{yw} D_w t_w = 50 \text{ k/in.}^2 (60 \text{ in.})0.625 \text{ in.} = 1875 \text{ k}$$

Total force in the yielded compression (top) flange:

$$P_c = F_{yc} b_c t_c = 50 \text{ k/in.}^2 (14 \text{ in.})0.75 \text{ in.} = 525 \text{ k}$$

Total force in the ultimate concrete slab:

$$P_s = 0.85 f_c b_s t_s = 0.85 \left(4 \text{ k/in.}^2 \right) 106 \text{ in.}(8.5 \text{ in.}) = 3063 \text{ k}$$

The forces in the longitudinal reinforcement may be conservatively neglected.

Determine the location of the plastic neutral axis:

$$P_t + P_w = 1000 \text{ k} + 1875 \text{ k} = 2875 \text{ k} < P_s = 3063 \text{ k}$$

$$P_t + P_w + P_c = 1000 \text{ k} + 1875 \text{ k} + 525 \text{ k} = 3400 \text{ k} > P_s = 3063 \text{ k}$$

Therefore, the plastic neutral axis is located within the compression (top) flange, its distance from girder top is

D6.1

$$\overline{Y} = \left(\frac{t_c}{2} \right) \left[\frac{P_t + P_w - P_s}{P_c} + 1 \right] = \left(\frac{0.75 \text{ in.}}{2} \right)$$
$$\times \left[\frac{1000 \text{ k} + 1875 \text{ k} - 3063 \text{ k}}{525 \text{ k}} + 1 \right] = 0.241 \text{ in.}$$

and the plastic moment M_p is given as

D6.1 $\quad M_P = \left(\dfrac{P_c}{2t_c}\right)\left[\overline{Y}^2 + \left(t_c - \overline{Y}\right)^2\right] + P_s d_s + P_w d_w + P_t d_t$

where

$d_s = \dfrac{t_s}{2} + t_{hunch} + \overline{Y} = \dfrac{8.5\text{ in.}}{2} + 1\text{ in.} + 0.241\text{ in.} = 5.49\text{ in.}$

$d_w = \dfrac{D_w}{2} + t_c - \overline{Y} = \dfrac{60\text{ in.}}{2} + 0.75\text{ in.}\ 0.241\text{ in.} = 30.51\text{ in.}$

$d_t = \dfrac{t_t}{2} + D_w + t_c - \overline{Y} = \dfrac{1\text{ in.}}{2} + 60\text{ in.} + 0.75\text{ in.} - 0.241\text{ in.}$

$\qquad = 61.01\text{ in.}$

Thus,

$M_P = \left(\dfrac{525\text{ k}}{2\,(0.75\text{ in.})}\right)\left[(0.241\text{ in.})^2 + (0.75\text{ in.} - 0.241\text{ in.})^2\right]$

$\qquad + (3063\text{ k})\,(5.491\text{ in.}) + (1875\text{ k})\,(30.51\text{ in.}) + (1000\text{ k})\,(61.01\text{ in.})$

$\qquad = 135{,}146\text{ k in.} = 11{,}262\text{ kft}$

☐ **Determine If Section Is Compact or Noncompact.**
Where the specified minimum yield strength does not exceed 70 k/in.^{2}, the girder has a constant depth, and the girder does not have longitudinal stiffeners or holes in the tension flange, the first step is to check the compact-section web slenderness provisions.

6.10.6.2 $\qquad\qquad 2\dfrac{D_{cp}}{t_w} = 0 \le 3.76\sqrt{\dfrac{E}{F_{yc}}}$

Since the plastic neutral axis is located within the compression flange, $D_{cp} = 0$.
Therefore the web is deemed compact. Since this is a composite section in positive flexure, the flexural resistance is computed as defined by the composite compact-section positive flexural resistance in *6.10.6.2.2.*

☐ **Check for Flexure—Strength I Limit State** $\qquad$ *6.10.7.1*
For checking $D_p < 0.1\,D_t$

$D_p = t_s + t_{haunch} + \overline{Y} = 8.5\text{ in.} + 1\text{ in.} + 0.241\text{ in.} = 9.74\text{ in.}$

$D_t = t_s + t_{haunch} + t_c + D_w + t_t = 8.5\text{ in.} + 1\text{ in.}$

$\qquad + 0.75\text{ in.} + 60\text{ in.} + 1\text{ in.} = 71.3\text{ in.}$

Because $D_p > 0.1 \, D_t$, the nominal flexural resistance is

$$M_n = M_p\left(1.07 - 0.7\frac{D_p}{D_t}\right) = 11{,}262 \text{ kft}\left[1.07 - 0.7\left(\frac{9.74 \text{ in.}}{71.3 \text{ in.}}\right)\right]$$
$$= 10{,}973 \text{ kft}$$

Then flexural resistance at this design section is checked with $f_l = 0$:
For the interior girders,

$$M_u + \frac{1}{3}f_l \, S_{xt} = M_u = 1.25(2119 + 302.5)$$

6.10.7.1 $$\qquad\qquad + 1.5(388.9) + 1.75(2307) = 7647 \text{ kft}$$
$$\leq \varphi_f M_n = 1.0(10{,}973) = 10{,}973 \text{ kft}$$

OK for the interior girders.
For the exterior girders,

$$M_u = 1.25\,(2119 + 302.5) + 1.5\,(388.9) + 1.75\,(2961)$$
$$= 8792 \text{ kft} \leq \varphi_f \, M_n = 10{,}973 \text{ kft}$$

OK for the exterior girders.

☐ **Check for Shear—Strength I Limit State**
Shear must be checked at each section of the girder. For this design example, shear is maximum at the bridge end. The first step in the design for shear is to check if the web must be stiffened. The nominal shear resistance of unstiffened webs of hybrid and homogenous girders is

6.10.9.2 $\quad V_n = CV_p$
6.10.9.2 $\quad V_p = 0.58F_{yw}D_w t_w = 0.58\,(50)\,(60)\,(0.625) = 1088 \text{ k}$

$$1.4\sqrt{\frac{Ek}{F_{yw}}} = 1.4\sqrt{\frac{29{,}000\,(5)}{50}} = 75.4 < \frac{D}{t_w} = \frac{60 \text{ in.}}{0.625 \text{ in.}} = 96$$

Then

$$C = \frac{1.57}{(D_w/t_w)^2}\left(\frac{Ek}{F_{yw}}\right) = \frac{1.57}{96^2}\frac{29{,}000\,(5)}{50} = 0.494$$
$$V_n = CV_p = 0.494\,(1088) = 537.5 \text{ k}$$

For the interior girders

$$V_u = 1.25(V_{\text{DL girder}} + V_{\text{DL parapet}}) + 1.50(V_{\text{DL WS}}) + 1.75\,(V_{\text{LL+IM}})$$
$$= 1.25(73.72 + 10.5) + 1.50(13.5) + 1.75(108.6)$$
$$= 315.6 \text{ k} < \phi_v V_n = 1.0(537.5) = 537.5 \text{ k}$$

OK for the interior girders.

For the exterior girders

$$V_u = 1.25(73.72 + 10.5) + 1.50(13.5) + 1.75(107.3) = 313.3 \text{ k} < \phi_v V_n$$
$$= 537.5 \text{ k}$$

OK for the exterior girders.
No transverse intermediate stiffeners are needed and only bearing stiffeners are needed.

- **Design for Bearing Stiffeners—Strength I Limit State**
Try $b_t = 5$ in. and $t_p = \frac{3}{4}$ in. (See Figures Ex4.10-3 and Ex4.10-4):

6.10.11.2.2 $\quad b_t = 5 \text{ in.} \le 0.48\, t_p \sqrt{\dfrac{E}{F_{ys}}} = 0.48\,(0.75)\sqrt{\dfrac{29{,}000}{50}} = 8.67 \text{ in. OK}$

Check Bearing Resistance

6.10.11.2.2 $\quad (R_{sb})_r = \phi_b\,(R_{sb})_n = 1.40 A_{pn} F_{ys} = 1\,(1.4)\,2\,(5-1)\,\dfrac{3}{4}\,(50)$
$$= 420 \text{ k} > 315.6 \text{ k}$$

See Figure Ex4.10-3 for area A_{pn} of the projected elements of the stiffener.

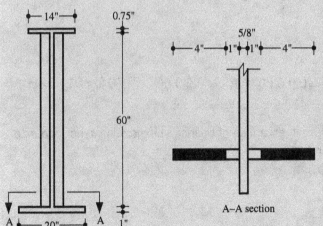

A–A section

Figure Ex4.10-3
Bearing stiffener details.

OK for bearing resistance of bearing stiffeners in the interior and exterior girders.

Check Axial Resistance

Calculate the following cross-sectional properties according to Figure Ex4.10-4:

$$r_s = \sqrt{\frac{I_s}{A_s}} = \sqrt{\frac{(5.625 - 0.75)\,0.625^3 + (0.75)\,10.625^3}{12\,[0.625\,(5.625) + 2\,(5)\,0.75]}} = 2.61 \text{ in}$$

$$\lambda = \frac{KL}{r_s \pi}\frac{F_y}{E} = \frac{0.75\,(60 \text{ in})}{(2.61 \text{ in})\,\pi}\frac{50 \text{ k/in}^2}{29{,}000 \text{ k/in}^2} = 0.0095 \le 2.25$$

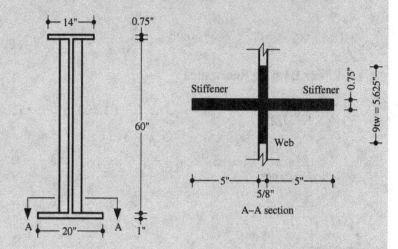

Figure Ex4.10-4
Bearing stiffener cross section for axial resistance.

Thus

$$\varphi_c P_n = 0.9(0.66^\lambda\,F_y A_s) = 0.9(0.66^{0.0095})\,50 \text{ k/in}^2(11.02 \text{ in}^2)$$
$$= 493.9 \text{ k} \quad > V_u = 315.6 \text{ k}$$

OK for axial resistance of bearing stiffeners in interior and exterior girders.

Example 4.11 Steel Plate Girder Bridge Design (Fatigue Limit State)

□ **Design Requirement**

Design the steel plate girder bridge in Examples 4.9 and 4.10 for the fatigue limit state. The average daily truck traffic for the two lane bridge is estimated as ADTT = 5000 trucks/day. Typically three components in the steel bridge superstructure need to be checked for fatigue limit state: (1) welds of shear connectors to the steel girders, (2) welds connecting the flanges to the web, and (3) welds connecting stiffeners to the web. This example has a focus on the welds between the bottom flange and the web, although they often do not control. In addition, there are no longitudinal or intermediate transverse stiffeners. Fatigue check for stud weld in steel superstructure has been shown in Example 4.8.

Load modifier $\eta = 1.0$

☐ **Dynamic Load Allowance IM**

The fatigue and fracture limit state is given as 15%.

☐ **Fatigue Load Effect**

The maximum moment range is expected at the middle 32-k axle shown in Figure Ex4.11-1:

$$\text{Reaction } R_A = 32\,k\left(\frac{115 - 33.39}{115}\right) + 32\,k\left(\frac{115 - 63.39}{115}\right)$$

$$+ 8\,k\left(\frac{115 - 77.39}{115}\right) = 39.69\,k$$

$$\text{Moment } M_{\text{fatigue}} = 39.69\,k\,(63.39\,\text{ft}) - 32\,k\,(30\,\text{ft}) = 1556\,\text{kft}$$

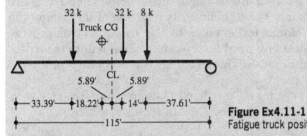

Figure Ex4.11-1
Fatigue truck position for maximum moment range.

Fatigue stress range including dynamic effect IM = 0.15 and load distribution factor: For interior girders

$$\Delta f = \left(\frac{M_{\text{fatigue}}}{S_{\text{bottom girder LL}}}\right)(1+\text{IM})\text{DF}_{\text{interior moment 1}}$$

$$= \frac{1556\,\text{kft}\,(12\,\text{in./ft})}{2081\,\text{in.}^3}(1.15)\frac{0.46}{1.2} = 3.9\,\text{k/in.}^2$$

For exterior girder

$$\Delta f = \left(\frac{M_{\text{fatigue}}}{S_{\text{bottom girder LL}}}\right)(1+\text{IM})\,\text{DF}_{\text{exterior moment 1}}$$

$$= \frac{1556\,\text{kft}\,(12\,\text{in./ft})}{2081\,\text{in.}^3}(1.15)\frac{0.86}{1.2} = 7.4\,\text{k/in.}^2$$

☐ Fatigue Strength

$$\Delta F_n = \left(\frac{A}{N}\right)^{1/3} = \left(\frac{120 \times 10^8 \text{ k/in.}^2}{365\,(75)\,np\,\text{ADTT}}\right)^{1/3}$$

6.6.1.2.5,
3.6.1.4.2

$$= \left(\frac{120 \times 10^8 \text{ k}^3/\text{in.}^6}{365\,(75)1\,(0.85)5000}\right)^{1/3} = 4.7 \text{ k/in.}^2$$

$$< \Delta F_{TH} = 16 \text{ k/in.}^2$$

Then set $\Delta F_n = \Delta F_{TH}$ for infinte life design.
Fatigue category B

☐ Fatigue Limit State Check for Worst Case (Exterior Girders)

$$\gamma\,\Delta f = 1.50\left(7.4 \text{ k/in.}^2\right) = 11.1 \text{ k/in.}^2 < \Delta F_n = 16 \text{ k/in.}^2$$

OK for fatigue in flange–web welds of both interior and exterior girders.

4.7.3 Section Proportioning

As in design of many structural members, steel I-beam design also uses a trial-and-error approach as shown in the flowchart in Figure 4.7-1. Namely a preliminary section is selected first and then the section is checked against a number of requirements. Good preliminary selection of the section can reduce and possibly eliminate iterations of the trial-and-error process. The AASHTO specifications provide several general requirements for the steel beam section to satisfy. The resulting cross section is considered constructable.

For the stability of I-beam web, the depth D should not be too tall compared with the web thickness t_w. They are thus required to satisfy

6.10.2.1.1
$$\frac{D}{t_w} \le 150 \tag{4.7-11}$$

if no longitudinal stiffeners are provided. This is a practical upper limit on the slenderness of the web with respect to the web depth D. This equation allows for easier proportioning of the web in preliminary design. For commercially available rolled I shapes, this requirement is satisfied automatically. Thus this requirement applies to built-up plate girder sections. In this case, the designer is required to select plates of different dimensions to make up the cross section.

With longitudinal stiffeners, this requirement is relaxed to

6.10.2.1.2
$$\frac{D}{t_w} \le 300 \tag{4.7-12}$$

because the stiffeners noticeably stabilize the web.

For the flanges, the following four requirements are given in the AASHTO specifications:

6.10.2.2 $\dfrac{b_f}{2t_f} \leq 12$ $b_f \geq \dfrac{D}{6}$ $t_f \geq 1.1 t_w$ $0.1 \leq \dfrac{I_{yc}}{I_{yt}} \leq 10$

$$(4.7\text{-}13)$$

where b_f = flange width
t_f = flange thickness
I_{yc} = moment of inertia of compression flange about vertical axis in web's plane
I_{yt} = moment of inertia of tension flange about vertical axis in web's plane

As seen, the first requirement is concerned with the slenderness ratio of the flanges for stability (compactness), the second one is for the relation or proportionality between the flange and the web height of the section height, the third is also for the proportionality between the flange and web for their thicknesses, and the last one limits the two flanges to within 10 times each other, or not be too different.

4.7.4 Compactness of Cross Section

Whether the cross section is compact or not, the capacity calculation and strength-checking requirement are different in the AASHTO specifications. For composite sections in noncurved bridges, these are the requirements for compact composite sections:

❏ The specified minimum yield strengths of the flanges do not exceed 70 k/in.[2]

❏ The web satisfies the requirement of Eq. 4.7-11.

❏ The section satisfies the web slenderness limit:

6.10.6.2.2 $\dfrac{2D_{cp}}{t_w} \leq 3.76\sqrt{\dfrac{E}{F_{yc}}}$ $(4.7\text{-}14)$

6.10.2.1.1 $\dfrac{D}{t_w} \leq 150$ $(4.7\text{-}15)$

where D_{cp} = depth of web in compression at plastic moment
t_w = web thickness
E = Young's modulus for steel
F_{yc} = yield strength of compression flange
D = web height

4.7.5 Flexural Strength Design under Strength I Limit State

For compact sections meeting the compactness requirements above, the section's nominal flexural capacity M_n is required to satisfy

6.10.7.1.1 $M_u + \dfrac{1}{3} f_l\, S_{xt} \leq \varphi_f M_n$ $(4.7\text{-}16)$

where ϕ_f = resistance factor for flexure = 1.0
    f_ℓ = flange lateral bending stress
    M_u = factored bending moment about major axis of cross section
    S_{xt} = elastic section modulus about major axis of section to tension flange = M_{yt}/F_{yt}
    M_{yt} = yield moment with respect to tension flange
    F_{yt} = yield stress of tension flange
    M_n = nominal flexural resistance of section determined as discussed below (kin.)

6.10.7.1.2
$$M_n = \begin{cases} M_p & \text{if } D_p \leq 0.1 D_t \\ M_p\left(1.07 - 0.7\dfrac{D_p}{D_t}\right) & \text{otherwise} \end{cases}$$
(4.7-17)

where M_p = plastic moment of composite section
    D_p = distance from top of concrete deck to neutral axis of composite section at plastic moment (in.)
    D_t = total depth of composite section (in.)

For noncompact sections, at the strength limit state, the compression flange shall satisfy

6.10.7.2.1
$$f_{bu} \leq \varphi_f F_{nc}$$
(4.7-18)

and the tension flange shall satisfy for stress control

6.10.7.2.1
$$f_{bu} + \frac{1}{3}f_\ell \leq \varphi_f F_{nt}$$
(4.7-19)

where f_{bu} = flange stress calculated without flange lateral bending
    F_{nc} = nominal flexural resistance of compression flange
    = $R_b R_h F_{yc}$
    F_{nt} = nominal flexural resistance of tension flange = $R_h F_{yt}$
    R_b = web load shedding factor
    R_h = hybrid factor

4.7.6 Shear Strength Design under Strength I Limit State

For steel I beams, it is assumed that only the web carries the shear force. Under the strength limit state of the AASHTO specifications, straight webs need to satisfy the following requirement for shear capacity:

6.10.9.1
$$V_u \leq \varphi_v V_n$$
(4.7-20)

where φ_v = resistance factor for shear = 1.0
    V_n = nominal shear resistance as discussed below for both unstiffened and stiffened webs
    V_u = shear in web due to factored loads of Strength I limit state

For unstiffened webs the nominal shear resistance is given in the AASHTO specifications as

6.10.9.2 $$V_n = V_{cr} = CV_p$$ (4.7-21)

in which V_n is the nominal shear resistance, taken as V_{cr} being the shear-buckling resistance,

6.10.9.2 $$V_p = \text{plastic shear force} = 0.58 F_{yw} D\, t_w$$ (4.7-22)

and C is the ratio of the shear-buckling resistance to the shear yield strength also viewed as a discount factor:

6.10.9.3.2 $$C = \begin{cases} 1 & \text{if } \dfrac{D}{t_w} \le 1.12 \sqrt{\dfrac{Ek}{F_{yw}}} \\[3ex] \dfrac{1.12}{D/t_w} \sqrt{\dfrac{Ek}{F_{yw}}} & 1.12 \sqrt{\dfrac{Ek}{F_{yw}}} \le \dfrac{D}{t_w} \le 1.40 \sqrt{\dfrac{Ek}{F_{yw}}} \\[3ex] \dfrac{1.57}{(D/t_w)^2}\dfrac{Ek}{F_{yw}} & 1.40\sqrt{\dfrac{Ek}{F_{yw}}} \le \dfrac{D}{t_w} \end{cases}$$

(4.7-23)

where k = shear buckling coefficient = 5
          F_{yw} = yield stress of web

The other symbols have been defined earlier in this section.

When unstiffened sections are inadequate, not satisfying the strength requirement Eq. 4.7-20, the web can be stiffened using transverse stiffeners forming panels along the beam length, as seen in Figure 4.7-7. The nominal strength of the stiffened web is given next.

The AASHTO specifications require stiffened webs to have dimensions satisfying

6.10.9.3.2 $$\frac{2 D\, t_w}{b_{fc}\, t_{fc} + b_{ft}\, t_{ft}} \le 2.5$$ (4.7-24)

Figure 4.7-7
(Left) Unstiffened and (right) stiffened webs of steel I beams.

where b_{fc} = width of compression flange
         t_{fc} = thickness of compression flange
         b_{ft} = width of tension flange
         t_{ft} = thickness of tension flange

With Eq. 4.7-24 satisfied, the nominal strength of the web for shear is given in the AASHTO specifications as

6.10.9.3.3

$$V_n = \begin{cases} V_p C & \text{for end panels with} \ \ d_0 \leq 1.5D & (4.7\text{-}25) \\ V_p \left[C + \dfrac{0.87(1-C)}{\sqrt{1+(d_0/D)^2}} \right] & \text{for intermediate panels} & (4.7\text{-}26) \end{cases}$$

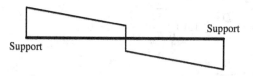

Support

Support

Figure 4.7-8
Typical design shear envelope of simply supported span.

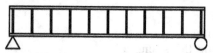

(a) Evenly spaced transverse stiffeners

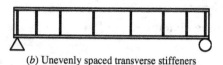

(b) Unevenly spaced transverse stiffeners

Figure 4.7-9
Options of arranging transverse stiffeners in steel I beams.

where d_0 is the space between two transverse stiffeners and C has been defined in Eq. 4.7-23 except the shear buckling coefficient k is not a constant any longer but a function of the stiffeners' spacing versus web height D:

6.10.9.3.2 $$k = 5 + \frac{5}{\left(d_0/D\right)^2} \qquad (4.7\text{-}27)$$

Actually the unstiffened web's shear strength in Eq. 4.7-23 can be viewed as a special case of $d_0 = \infty$ in Eq. 4.7-26 leading to $k = 5$ in Eqs. 4.7-23 and 4.7-27.

Note also that the nominal shear strength formulas in Eqs. 4.7-25 and 4.6-26 refer to a specific cross section along the length of the beam. Since the shear envelope or shear capacity requirement for design is not constant along the beam, the panel size or transverse stiffener spacing d_0 may vary accordingly if the stiffeners are needed at all. Figure 4.7-8 shows a typical design shear envelope for a simply supported beam. Figure 4.7-9 shows two options of constant and variable d_0.

4.7.7 Service Limit State Design   *6.10.4*

As discussed in Chapter 3, Service II limit state load combination intends to control the yield of steel structures and the slip of slip-critical connections due to vehicular live load. It is applicable only to steel bridge structures and components. Such yielding could cause permanent deformation. From the point of view of load level, this combination is approximately halfway between that used for Service I and Strength I limit states as follows:

3.4.1, 6.5.4.2,
6.10.4 $$1.0\,\text{DC} + 1.0\,\text{DW} + 1.3\,\text{LL}(1 + \text{IM}) \leq \varphi_y F_y \qquad (4.7\text{-}28)$$

where $\phi_y = 0.95$ according to the AASHTO specifications for yield stress F_y.

For composite members with shear connectors, flexural stresses in the structural steel caused by service II loads applied to the composite section may be computed using the short- or long-term composite section as appropriate. This refers to the effect of concrete creep with time. For short-term stresses, such as live-load stresses, the Young's modulus ratio between the steel and concrete equal to 8 to 10 is typically used. For long-term stresses, such as those due to wearing surface and railings, a smaller ratio of 3 is commonly used to account for sustained dead-load effect. The concrete deck may be assumed effective subject to both positive and negative moments provided that the maximum longitudinal tensile stresses in the concrete deck at the section under consideration caused by the service II loads are smaller than $2f_r$, where f_r is the concrete modulus of rupture.

For sections that are composite for negative flexure with maximum longitudinal tensile stresses in the concrete deck greater than or equal to $2f_r$, the flexural stresses in the structural steel caused by service II loads shall be computed ignoring the concrete's contribution.

When deflection is concerned, the service I load combination is prescribed in the AASHTO specifications as the load:

2.5.2.6, 3.4.1 $$1.0 \, LL(1 + IM) \leq \delta \qquad (4.7\text{-}29)$$

where δ is recommended in the AASHTO specifications, as shown in Table 4.7-1. It should be noted that deflection control is "optional" in the AASHTO specifications. Nevertheless, bridges should not generate undesirable structural or psychological effects due to their deformations. This essentially leaves the decision on quantitative control of deflection to the bridge owner.

On the other hand, dead-load-induced deflection is required to be controlled in the AASHTO specifications. This is usually accomplished using camber. Deflections due to steel weight and concrete weight are required to be reported separately, and so is that due to future loads such as future wearing surface. While estimation of deflection may not be precisely correct with no uncertainty, conservative camber (i.e., overestimated deflection) has been seen used in bridge design.

Table 4.7-1

Recommended and optional live-load deflection limit

Load and Structure	Deflection Limit
Vehicular load, general	Span/800
Vehicular and pedestrian loads	Span/1000
Vehicular load on cantilever arms	Span/300
Vehicular and pedestrian loads on cantilever arms	Span/375

4.7.8 Load-Induced Steel Fatigue and Fatigue Limit State

As discussed in Chapter 3, the AASHTO specifications include provisions for dealing with load-induced fatigue cracking in both design and evaluation. Under cyclic load such as truck load to metal components, cracking as fatigue failure has been observed at welds. Metal fatigue failure, particularly the steel fatigue failure discussed in this section, as a failure mode is different from strength failure modes. The former is believed due to a number of cycles of stress range applied at a level lower than that required to fail the material applied only once. Therefore, not only the magnitude of the stress range but also the number of times applied needs to be considered when fatigue failure is of concern, such as in a design process.

Experimental results have shown that the relationship between required number of cycles, N, under a given stress range S is approximately linear in the logarithm space of the two quantities:

$$3 \log S = -\log N + \log A \qquad (4.7\text{-}30)$$

where
S = nominal stress range applied to detail
N = number of cycles of S applied
A = model constant depending on quality of weld or fatigue strength category

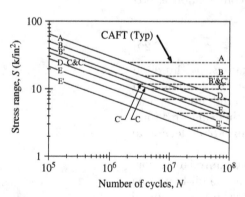

Figure 4.7-10
S–N curves in AASHTO specifications. *Source: AASHTO LRFD Bridge Design Specifications*, 2012. Used by permission.

The AASHTO specifications provide this relation in a graphic way, as shown in Figure 4.7-10. These straight lines in the log–log space are conventionally referred to as S–N curves. They form the basis for designing or checking steel weld details to prevent or control fatigue failure.

As seen, eight fatigue detail categories are included in Figure 4.7-10, represented by the seven straight lines. Each category or straight line represents a value of A in Eq. 4.7-30 for a group of details with similar welding quality, except categories C and C′ represented by one curve. Welding quality here refers to the possibility of leaving microscopic discontinuity in the material for cracking initiation. It is considered to be very much correlated with how easy or difficult the details can be made. For example, welding done in the shop is believed to be of higher quality in general than that in the field. In the shop, welding may be done using a robot with a controlled speed along a guided track. In addition, after-welding inspection for weld quality (to tell whether discontinuity has been left in the material) can be more easily done in the shop than in the field. In the field, most welding has to be done manually and quality control is much more difficult to exercise or the result is more difficult to be of high quality. When low-quality welds result more frequently or likely, lower values of A in Eq. 4.7-30 are assigned for them in the specifications, as shown in Figure 4.7-10 and quantitatively in Table 4.7-4. Table 4.7-2 gives several

Table 4.7-2
Definitions of detail categories for fatigue design and evaluation

Description	Category	Potential Crack Initiation Point	Illustrative Examples (Arrow for stress direction)
Base metal, except noncoated weathering steel, with rolled or cleaned surfaces. Flame-cut edges with surface roughness value of 1000 μin. or less, but without reentrant corners.	A	Away from all welds or structural connections	
Noncoated weathering steel base metal with rolled or cleaned surfaces. Flame-cut edges with surface roughness value of 1000 μin. or less, but without reentrant corners.	B		
Base metal at the net section of eyebar heads or pin plates.	E	In the net section originating at the side of the	
Base metal at the termination of partial-length welded cover plates having square or tapered ends that are narrower than the flange, with or without welds across the ends, or cover plates that are wider than the flange with welds across the ends: Flange thickness ≤ 0.8 in. Flange thickness > 0.8 in.	E E'	In the flange at the toe of the end weld or in the flange at the termination of the longitudinal weld or in the edge of the flange with wide cover	
Base metal at the termination of partial-length welded cover plates that are wider than the flange and without welds across the ends.	E'	In the edge of the flange at the end of the cover plate weld	No end weld
Base metal at the toe of transverse stiffener-to-flange fillet welds and transverse stiffener-to-web fillet welds. (Note: includes similar welds on bearing stiffeners and connection plates.)	C'	Initiating from the geometrical discontinuity at the toe of the fillet weld extending into the base metal	
Base metal at the termination of longitudinal stiffener-to-web or longitudinal stiffener-to-box flange welds: With the stiffener attached by fillet welds and with no transition radius provided at the termination: Stiffener thickness: <1.0 in. Stiffener thickness: ≥1.0 in.	E E'	In the primary member at the end of the weld at the weld toe	Stiffener Fillet, CJP or PJP Web or flange w/o Transition radius

Source: Adapted from *AASHTO LRFD Bridge Design Specifications*, 2012. Used by permission.

examples of fatigue strength category commonly seen in modern US highway bridges.

It should be added that physical testing has also found that when the stress range is kept very low, the weld will never fail or develop fatigue cracking. These stress thresholds are indicated in Figure 4.7-10 using horizontal lines near the right lower corner, depending on detail category (of fatigue strength). These thresholds are referred to as constant-amplitude fatigue threshold (CAFT) $(\Delta F)_{TH}$ in the AASHTO specifications, as indicated in Figure 4.7-10.

Accordingly, there are two design philosophies offered in the AASHTO specifications: (1) design for an infinite life or (2) design for a finite life of 75 years as the target service life required in the AASHTO specifications. These two approaches are presented next with more detail.

The infinite-life design concept is implemented in the specifications quite straightforwardly since the stress level is the only thing that needs to be controlled in design to accomplish the goal:

$$3.4.1, 6.6.1.2.2, 6.6.1.2.5 \qquad \gamma(\Delta f) \leq (\Delta F)_n = (\Delta F)_{TH} \qquad (4.7\text{-}31)$$

where $\gamma = 1.50$ according to the fatigue I limit state presented in Section 3.4, Load Combinations. Δf is the real stress range modeled using the AASHTO fatigue truck, also presented in Chapter 3. The code-specified CAFT values $(\Delta F)_{TH}$ for design are shown in Table 4.7-3. Here $(\Delta F)_{TH}$ is treated as a "nominal fatigue resistance $(\Delta F)_n$" according to the specifications. In other words, if the real stress range at the detail is designed to be smaller than this "resistance," the detail is considered to be safe for a life of infinite length.

The finite-life design concept and procedure involve estimation of the number of cycles of stress expected over the targeted life span, specified as 75 years in the ASHTO specifications. This number is apparently a function of expected truck traffic volume as discussed next. The $S-N$ curve in

Table 4.7-3

Constant-amplitude fatigue thresholds for infinite life

Detail Category	Threshold (k/in.^{2})
A	24
B	16
B′	12
C	10
C′	12
D	7
E	4.5
E′	2.6

Eq. 4.7-30 is also seen in the literature in another form:

$$S^3 N = A \qquad (4.7\text{-}32)$$

Thus, the allowed stress range or "nominal fatigue resistance $(\Delta f)_n^"$ as defined in the AASHTO specifications is then

6.6.1.2.5
$$(\Delta f)_n = S = \left(\frac{A}{N}\right)^{1/3} \qquad (4.7\text{-}33)$$

Namely, if the real stress range is kept below this nominal fatigue resistance $(\Delta f)_n$ for the expected use of the bridge (weld detail) of N cycles, the detail will be considered safe from fatigue cracking over its 75-year life. The A values for using Eq. 4.7-30 are listed in Table 4.7-4. This design procedure is also referred to as fatigue II limit state design in the AASHTO specifications. Its live load factor is 0.75 as shown in Table 3.4-8.

In Eq. 4.7-33, N is given in the specifications as

6.6.1.2.5
$$N = 365\,(75)\,n\,\text{ADTT}_{\text{SL}} \qquad (4.7\text{-}34)$$

where 365 is the number of days per year, 75 is the target number of years for new bridges to serve as specified by the AASHTO, n is the number of stress cycles per truck crossing the bridge span given in Table 4.7-5 taken

Table 4.7-4
Constant A for detail categories for finite-life design

Detail Category	Constant A $(\text{k/in.}^2)^3$
A	250×10^8
B	120×10^8
B'	61×10^8
C	44×10^8
C'	44×10^8
D	22×10^8
E	11×10^8
E'	3.9×10^8

Table 4.7-5
Number of cycles per truck crossing, n

Longitudinal Members	Span Length (ft)	
	>40	≤40
Simple-span girders	1.0	2.0
Continuous girders		
1. Near interior suport	1.5	2.0
2. Elsewhere	1.0	2.0
Cantilever girders	5.0	
Trusses	1.0	

from the specifications, and $ADTT_{SL}$ stands for average daily truck traffic in one single lane (the driving lane).

The single lane specified here is due to the fact that the likelihood or probability is low to have other heavy (fully loaded rather than empty) trucks in other lanes simultaneously on the span. In addition, even if there are such heavy trucks simultaneously on the span, their contribution to the total stress at the detail is also low because members directly under those lanes would carry a more significant share of the total load. It will then distribute very little to the detail being designed.

Example 4.11 shows the design for the fatigue limit state of a plate girder span of highway bridge.

4.7.9 Shear Stud Design
6.10.10

Shear studs are commonly used as shear connectors in steel beam bridges. This section has a focus on shear stud design. Typical shear connector studs are made from cold-drawn bars, Grades 1015, 1018, or 1020 conforming to the AASHTO material specifications. They are required to have a specified minimum yield and tensile strength of 50 and 60 k/in.^{2}, respectively. There are three diameters of shear studs commonly available: $\frac{5}{8}$, $\frac{3}{4}$, and $\frac{7}{8}$ in. Two possible failure modes are usually of concern when designing such shear studs: the fatigue limit state and strength limit state. Given a stud diameter, the focus of shear stud design is to decide on the number of studs needed and their arrangement on the beam in both the longitudinal and transverse directions. Figure 4.7-11 shows a typical arrangement on a steel beam before the steel reinforcement and concrete are placed for the reinforced concrete deck slab.

The AASHTO specifications include general geometric requirements for shear stud placement to ensure constructability. Accordingly, shear studs are

Figure 4.7-11
Shear stud arrangement on steel beam.

to be placed transversely across the top flange of the steel section and may be spaced at regular or variable intervals. They shall not be closer than 4 stud diameters center to center in the transverse direction. The clear distance between the edge of the top flange and the edge of the nearest shear connector shall not be less than 1 in. The concrete cover over the tops of the shear connectors should not be less than 2 in. and the studs should be at least 2 in. into the concrete deck. *6.10.10.1*

Fatigue Limit State
One can start from understanding a typical stud's fatigue strength to learn how to design them for a steel beam. As discussed above on steel fatigue, two philosophies of design may be used in steel shear connector design considering fatigue: infinite life and finite life. Accordingly, the equivalent fatigue strength of shear stud is given in the AASHTO specifications as follows:

6.10.10.2
$$Z_r = \begin{cases} 5.5d^2 & \text{for infinite life} & (4.7\text{-}35) \\ (34.5 - 4.28 \log N)d^2 & \text{for finite life} & (4.7\text{-}36) \end{cases}$$

where d is the diameter of the stud and the coefficient multiplied by d^2 (either 5.5 or $34.5 - 4.28 \log N$) is equivalent to $(\Delta F)_n$ in Eqs. 4.7-31 and 4.7-33. Note that the coefficient in Eq. 4.7-35 or 4.7-36 has a dimension of stress (k/in.^{2}) to make Z_r a force strength for fatigue. As seen in application of $(\Delta F)_n$ in Eq. 4.7-33 for finite life, the strength $34.5 - 4.28 \log N$ is a function of expected number of cycles, N, over the 75-year life span. Namely, a large number of cycles, N, means a smaller strength because $4.28 \log N$ has a negative sign, reducing the total when summed with the constant term 34.5.

With Z_r available, the number of studs per unit length p needed can be readily found as

6.10.10.1.2
$$p = \frac{nZ_r}{V_{sr}} \qquad (4.7\text{-}37)$$

where n is the number of studs per row in the transverse direction to the beam and V_{sr} is the load effect in horizontal shear range per length. As seen, the pitch p is inversely proportional to the load effect V_{sr}, shear range per unit length along the beam. Namely, the larger the shear range, the smaller the pitch or the more studs will be needed.

The shear range per unit length is proportional to shear range as follows, according to shear stress distribution based on mechanics of material:

6.10.10.1.2
$$V_{sr} = \frac{V_f Q}{I} \qquad (4.7\text{-}38)$$

where V_f is the shear range as a function of the cross section's location along the beam length; Q is the first moment of the transformed short-term area of the concrete deck about the neutral axis of the short-term composite section (in.^{3}), as shown in Figure 4.7-12; and I is the

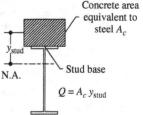

Figure 4.7-12
Computation of first moment Q for shear stress at stud base.

moment of inertia of the short-term composite section including both the concrete and the steel.

As noted above, the shear range V_f is a function of the cross section's location. In other words, V_f varies over the beam length. The following procedure may be helpful in determining this function over the entire beam length. Example 4.8 illustrates the application of this procedure.

1. Divide the beam length into Y sectors: $Y = 10$ is often used. Use $x = 1$, $x = 2$, $x = 3, \ldots, x = Y$, $x = Y + 1$ to name each endpoint of the Y sectors from left to right.
2. For $x = 1$:
3. Find the maximum (positive) shear induced by the AASHTO fatigue truck presented in Chapter 3 (i.e., the HL93 design truck but with the second axle spacing fixed at 30 ft) by running it through the span in the intended direction of traffic.
4. At the same location x, find the minimum shear (maximum negative shear) induced by the same truck in the same direction.
5. Compute $V_f(x) = $ maximum shear − minimum shear.
6. Set $x = x + 1$ and go to step 3 until $x = Y + 1$. If the beam is symmetrically designed, only half of the beam needs to be covered.
7. Multiply $V_f(x)$ by the distribution factor for one lane loaded without the multiple presence factor in Table 3.3-1, $1 + $ IM for dynamic effect, and the load factor 1.5 for infinite life or 0.75 for finite life.

The resulting $V_f(x)$ is then used in Eq. 4.7-37 to find the pitch p as a function of x, which will be used as the basis for arranging shear studs. When the span is not very long, a constant pitch, namely the smallest for $x = 1, 2, \ldots, Y + 1$, may be used. In this option, the $V_f(x)$ calculation may need to be done only for the controlling point. This controlling point is usually at a support section, because the shear range at a support is the maximum value of all $V_f(x)$ values. Otherwise, varying pitch p along the beam length can be used to be more economical. However, not too many different pitches should be used along the beam length to avoid too much additional site work.

Note that the concepts and procedures presented here are applicable only to those bridge spans that are not subjected to radial fatigue forces such as horizontally curved spans. For the latter, $V_f(x)$ will need to include another component perpendicular to the longitudinal one considered here using the vector sum, as specified in the AASHTO specifications.

Strength Limit State

The strength limit state design for shear stud connectors is to address the possible failure of continuity between the connector and other structural components. Failure may occur to the shear stud or the material connected to it, such as the steel beam to which the shear connector is welded or the concrete in which the shear stud is embedded. In general, the design or strength requirement is given in the AASHTO specifications as

$$6.5.4.2, 6.10.10.4.1, \qquad n_{\text{required}} = \frac{P}{\varphi_{sc} Q_n} \qquad (4.7\text{-}39)$$
$$6.10.10.4.2$$

where n_{required} is the required number of stud connectors, P is the design load, $\phi_{sc} = 0.85$ is the resistance factor for the shear stud, and Q_n is the nominal strength of one stud. Note that the stud strength can be the strength of the steel itself and also the concrete around it. More details are given next.

The AASHTO specifications further define the design load P and one stud strength Q_n as

$$P = \min \left(P_{1p}, P_{2p} \right)$$

where

$$P_{1p} = 0.85 f_c' \, b_s t$$

and

6.10.10.4.2 $\qquad P_{2p} = F_{yw} D t_w + F_{xt} b_{ft} t_{ft} + F_{yc} b_{fc} t_{fc} \qquad (4.7\text{-}40)$

or where the flanges and web use the same steel

$$P_{2p} = F_y A_{\text{NC}}$$

6.10.10.4.3 $\qquad Q_n = \min \left(0.5 A_{sc} \sqrt{f_c' E_c}, A_{sc} F_u \right) \qquad (4.7\text{-}41)$

where
- $f_c' = $ 28-day concrete compressive strength
- $b_s = $ effective width of concrete of composite section, equal to beam spacing for interior beams and edge to a half of beam spacing for exterior beams 4.6.2.6
- $t = $ thickness of concrete deck slab
- $F_{yw} = $ yield strength of web
- $F_{yt} = $ yield strength of tension flange
- $F_{yc} = $ yield strength of compression flange
- $F_y = $ yield strength of steel if $F_{yw} = F_{yt} = F_{yc}$
- $D = $ height of web
- $t_w = $ thickness of web
- $b_{ft} = $ width of tension flange
- $b_{fc} = $ width of compression flange
- $t_{ft} = $ thickness of tension flange
- $t_{fc} = $ thickness of compression flange
- $A_{\text{NC}} = $ cross-sectional area of noncomposite beam
- $A_{sc} = $ cross section area of one stud
- $E_c = $ Young's modulus of concrete
- $= 1820\sqrt{f_c'}$ (k/in.²) with f_c' given in k/in.²
- $F_u = $ ultimate strength of stud

Equation 4.7-40 shows that the design load is taken as the minimum of the capacities of the concrete and the steel beam. This approach is commonly used in structural design, which uses the member's strength/capacity as the load or demand to design the connection. This approach ensures that the connection is designed stronger than the members connected by the connection. For this case, Eq. 4.7-40 requires the stud connection strength to be greater than that of the members (the steel beam or concrete capacity), so that the connection would not fail before the members.

Example 4.8 illustrates shear stud design for a typical highway bridge span.

4.7.10 Constructability Check
6.10.3

Bridges are also required to be designed to introduce no undue difficulty or distress in fabrication and erection and to maintain locked-in construction force effects within tolerable limits. When the designer has assumed a particular sequence of construction, that sequence shall be defined in the contract documents and then relevant critical stages need to be investigated in design. Constructability issues should include, but certainly are not be limited to, deflection, overstress in steel and concrete, and stability. Several specific situations or stages are noted next.

Uplifting of Superstructure
Prior to or after completion of construction, certain parts of the superstructure may rise off the supporting member, such as steel beams of continuous spans due to construction stress without the full dead load applied yet in construction or a corner of continuous superstructure after construction completion due to thermal deformation. If uplift is expected to occur at a stage of construction, a temporary load may be placed to prevent liftoff. The magnitude and position of any required temporary load should be provided in the contract documents. Factored forces at high-strength bolted joints of load-carrying members are limited to the slip resistance of the connection during each critical construction state to ensure that the correct or appropriate geometry of the structure is maintained. Potential uplift at bearings shall be investigated at each critical construction stage.

Dead-Load Camber
Steel structures should be cambered during fabrication to compensate for at least dead-load deflection and vertical alignment. As commented on earlier, deflection due to steel member self-weight and concrete member self-weight shall be reported separately, and so shall be those due to future wearing surfaces or other loads. Vertical camber shall be specified to account for the computed dead-load deflection. When staged construction is specified, the sequence of load application should be considered when determining the camber.

Concrete Deck Placement
During deck concrete placement, a composite cross section's components in the final condition can be subjected to construction loads at the time they are noncomposite during construction. These cross sections are required to be investigated for flexure by the AASHTO specifications during

various stages of deck concrete placement. Geometric properties and bracing lengths and stresses used in calculating the nominal flexural resistance shall be for steel section only. Changes in load, stiffness, and bracing during various stages of deck placement shall be considered. The effects of forces from deck overhang brackets acting on the fascia girders during construction should also be considered.

4.8 Design of Prestressed Concrete I Beams

Prestressed concrete beams have found wide use in highway bridges in the past three or four decades in the United States. In many states, they have largely replaced the reinforced concrete beam bridge spans that were popular for some time. Figures 4.8-1 through 4.8-3 show three of the most

Spread
box beam

Figure 4.8-1
Highway bridge with prestressed concrete spread box beams.

Adjacent box
beams next to
each other

Figure 4.8-2
Highway bridge with prestressed concrete adjacent box beams.

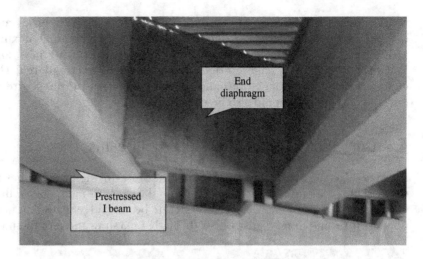

Figure 4.8-3
Highway bridge with
prestressed concrete I beams.

common types of prestressed concrete beam bridges currently in the
United States for short and medium spans spread box beams, adjacent box
beams, and I beams, without significance of the order. While these beams
have been standardized, their application has been widely spread in the
United States. Another group of popular prestressed concrete beams are
the so-called T beams or bulk T beams, although less popular than the
above three types.

Prestressed concrete I beams are used in highway bridges similarly to
steel beams, also with intermediate diaphragms (Figure 4.8-4) to form a

Figure 4.8-4
Prestressed concrete I-beam
superstructure with
diaphragms.

superstructure frame. They are as accommodating as steel beams for both straight and skewed structures. On the other hand, prestressed concrete box beams may be used only in structures with small skew angles. For example, some bridge owners limit prestressed concrete box beams to those spans with skew angle below 30°. Except this limit, spread-box-beam superstructures work similarly to their I-beam counterpart, with respect to beam spacing, associated deck thickness, and so on.

Adjacent-box-beam superstructures save forming for the concrete deck. They also reduce the deck thickness requirement by several inches, while the box beam top flanges are considered part of the concrete deck. However, they require longitudinal shear keys between two box beams formed using grout material after the beam placement. These shear keys also require "tightening" in the transverse direction, often realized using posttentioning. Nevertheless, there have been reports of premature failure of the shear keys, made very visible by cracking on the bridge deck surface right above the shear keys or between the box beams.

Figure 4.8-5 displays a typical and brief procedure for the design of prestressed I beams in beam bridges of short- to medium-span lengths. This procedure is also general and conceptually applicable to prestressed concrete beams of other cross-sectional shapes, such as box and bulk T. This section discusses the details of each step in the flowchart. Before doing that, the basic concept of prestressed concrete beams is presented as a foundation for understanding the AASHTO design requirements.

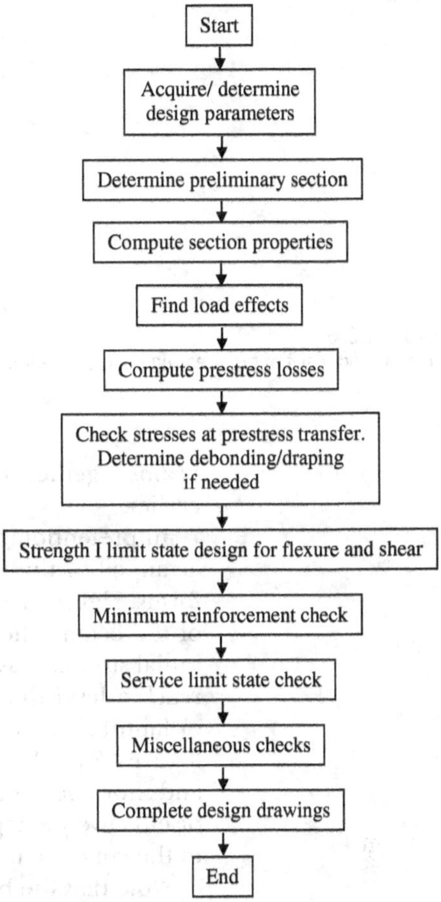

Figure 4.8-5
Typical design procedure for prestressed concrete beams.

PRESTRESSING CONCRETE BEAMS

4.8.1 Concepts of Prestressed Concrete Beams for Bridge Construction

Bridge beams with prestress are usually made using the technique of pretentioning. The prestress is applied through prestressing steel strands prior to hardening of concrete, hence the prefix "pre" in pretentioning. Posttentioning is another technique widely used in prestressed concrete member fabrication but rarely for bridge beams. This is mainly because posttentioning is performed after hardening of concrete and thus can be more expensive, especially when it is done at the site, and it almost always is. In contrast, pretentioning is always done at the fabrication facility if possible. On the other hand, posttentioning is often used in the transverse direction of a bridge, for example, for "tightening" several adjacent box

(a) Forming

(b) Prestressing tendons

(c) Concrete placing

(d) Cutting tendons to
apply prestress,
resulting in
beam camber

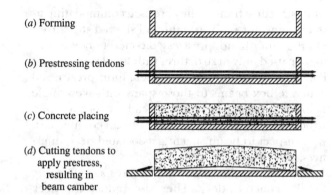

Figure 4.8-6
Process and result of prestressing concrete beam.

beams together to form a superstructure frame. This has been mentioned earlier.

In pretentioning a beam in fabrication, the tendons or strands are first arranged to the locations required and then prestressed in the concrete forms. Figures 4.8-6 *a* and *b* illustrate the operation, showing the elevation of the beam. Then concrete is placed in the forms and cured to develop initial strength, as sketched in Figure 4.8-6*c*. This initial strength needs to reach a level that can withstand the expected prestress without cracking or failure. The strands are then released to apply prestress to the concrete. Figure 4.8-6*d* shows the resulting condition of the concrete beam: cambered and shortened as a result of prestressing due to eccentric prestressing force or effectively a negative bending moment plus a compressive force applied to the concrete beam by the prestressing tendons.

Note that the beam's shortening also causes a partial loss of prestress in the strands and in turn in the concrete, because the strands are also shortened and thus partially unstressed. Quantification of this prestress loss is an important step in prestress concrete beam design and will be discussed in the next section. Since the amount of prestress in the concrete is treated as part of the beam's strength to offset load induced stress, overestimating the prestress is unconservative. In other words, prestress loss or reduction in prestress is required to be adequately estimated to ensure the beam's safety.

The location on the concrete cross section to apply prestress is also an important factor to determine in design because it dictates the magnitude of the moment applied by the prestress. This moment can increase the prestress in the outmost fiber and the capacity to resist the load-induced moment. On the other hand, this moment can introduce a large tensile stress in the outmost fiber at the other side of the beam (usually top), possibly causing cracking in the concrete beam and increasing concern with the beam's durability. Therefore the AASHTO specifications prescribe requirements for controlling stresses at these different stages of the concrete beam, referred to as the initial stage (or prestress transfer stage or prestress release stage).

FUNCTIONS OF PRESTRESSING AND STRESS ANALYSIS

This section quantitatively discusses the functions of prestress and derives the needed formulas for stress calculation and member design.

For structural analysis and design of prestressed concrete members, the stage of prestress release illustrated in Figure 4.8-6*d* marks the first significant step of stressing. This is the first time the concrete is subjected to significant stress. In the literature, this stage is also referred to as the initial stage, prestress release, or prestress application. Apparently, the name "initial stage" alludes to the fact that there are other stages of stressing in the expected life of the beam, such as service or operation stage and possibly the ultimate stage reaching the strength capacity. The name "prestress release" refers to the fact that the prestress applied to the strands is released from the equipment applying the prestress. As a result, this prestress release is constrained by the hardened concrete surrounding the strands through a chemical bond. It then applies the prestress to the concrete through the bond, hence the name "prestress application."

According to stress analysis theory, the stress state for a typical cross section of a prestressed concrete beam (without the composite deck) can be readily written as follows. It is for the initial stage when prestress is applied to the concrete:

$$f(y) = -\frac{F_{pe}}{A_{NC}} \pm \frac{F_{pe} e_{NC} y}{I_{NC}} \mp \frac{M_{beam} y}{I_{NC}} \qquad (4.8\text{-}1)$$

where
    $-$ = compressive stress
    $+$ = tensile stress
    F_{pe} = prestress force $= f_{pe} A_{ps}$
    f_{pe} = effective prestress, which is applied prestress (initial stress) less prestress loss
    A_{ps} = area of prestress strands
    A_{NC} = noncomposite cross-sectional area of beam
    e_{NC} = eccentricity of prestress strands' center of gravity from neutral axis
    y = distance of fiber of interest from neutral axis
    I_{NC} = moment of inertia of noncomposite cross section
    M_{beam} = bending moment due to beam self-weight

Equation 4.8-1 is based on an assumption of the linear distribution of flexural stress, illustrated in Figure 4.8-7. The total stress at distance y from the neutral axis is the sum of a uniformly distributed compressive stress due to the prestress force, a bending stress due to the prestress-induced moment (prestress force multiplied by its eccentricity), and another bending moment due to the beam's self-weight.

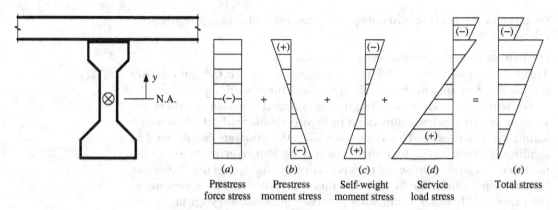

(a)	(b)	(c)	(d)	(e)
Prestress force stress	Prestress moment stress	Self-weight moment stress	Service load stress	Total stress

Figure 4.8-7
Stress distribution and superposition in prestressed concrete beam.

In general, Figure 4.8-7 shows that elastic stresses are superimposed on the total stress from five different loads on the cross section:

Total stress = prestress force stress

+ prestress moment stress

+ self-weight moment stress

+ service dead-load moment stress

+ service live-load moment stress   (4.8-2)

Depending on the construction stage or service stage being focused on, some of these stresses may not be present. For example, the service dead load (e.g., deck self-weight on the beam) and the service live load (truck load) are not present when the prestress is transferred.

At the initial stage, concrete cracking is of concern while full concrete strength may not have been reached yet. Excessive prestress may cause that. Therefore, tensile stresses in the concrete need to be checked so as not to exceed the concrete strength at the time. For example, for simply supported beams, tensile stresses may occur at the top fiber at midspan. That stress can be found using the following formulas:

$$f\left(y_{\text{top}}\right) = -\frac{F_{pe}}{A_{\text{NC}}} + \frac{F_{pe}\,e_{\text{NC}}y_{\text{top}}}{I_{\text{NC}}} - \frac{M_{\text{beam}}y_{\text{top}}}{I_{\text{NC}}}$$

$$= -\frac{F_{pe}}{A_{\text{NC}}} + \frac{F_{pe}\,e_{\text{NC}}}{S_{\text{NC top}}} - \frac{M_{\text{beam}}}{S_{\text{NC top}}} \qquad (4.8\text{-}3)$$

where $S_{\text{NC top}}$ is the section modulus of the noncomposite cross section with respect to the top fiber at a distance y_{top} from the neutral axis, as shown in Figure 4.8-8:

$$S_{\text{NC top}} = \frac{I_{\text{NC}}}{y_{\text{top}}} \qquad (4.8\text{-}4)$$

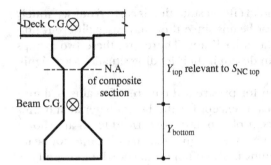

Figure 4.8-8
Typical AASHTO I-beam cross section.

Other stresses at different distances from the neutral axis can also be readily computed using the corresponding distance y in Eq. 4.8-1.

Figure 4.8-7 also includes stresses in the service condition after the initial stage of prestress transfer. It is seen that the tensile stress in the concrete beam caused by the service load (mainly truck live load) is mitigated or even canceled out by the prestress-caused compressive stress, resulting in a very small total stress. This is indeed the intention or advantage of prestressing. This needs to be accomplished by careful design.

QUANTIFYING PRESTRESS INCLUDING PRESTRESS LOSS
The effective prestress f_{pe} in Eq. 4.8-1 is the difference between the initial prestress f_{pi} when first applied and the prestress loss $\Delta f_{p,\,ES}$:

$$f_{pe} = f_{pi} - \Delta f_{p,\,ES} \tag{4.8-5}$$

where $\Delta f_{p,\,ES}$ is the prestress loss due to elastic shortening of concrete as a result of prestress transfer:

$$\Delta f_{p,\,ES} = \frac{E_{tendon}}{E_{ci}} f_{CGp} \tag{4.8-6}$$

where E_{strand} = Young's modulus of prestress strands
E_{ci} = Young's modulus of concrete at initial stage or prestress transfer
f_{CGp} = concrete stress at strand center of gravity due to prestress

Equation 4.8-6 is derived from the principle of strain compatibility, that is, the concrete strain is equal to the strand strain at the center of gravity of the strands:

$$\frac{\Delta f_{p,\,ES}}{E_{tendon}} = \frac{f_{CGp}}{E_{ci}} \tag{4.8-7}$$

Example 4.13 includes illustration of prestress loss estimation applied to prestressed concrete I beams.

4.8.2 Dead-Load Effects

As in steel beam design, dead-load effect estimation is carried out separately for exterior beams and interior beams since they may carry different portions of the deck and other facilities if any. Therefore, these two groups should be treated differently in design if their loading situations are significantly different.

Dead-load effect calculation for prestressed concrete I beams is almost identical with that for steel beams except for the beam material and thus the self-weight using the concept of tributary area. Dead loads are shown in Figure 4.8-9 on an exterior beam and Figure 4.8-10 on an interior beam based on this concept. The railing is also often evenly distributed to all the beams, as seen in Figures 4.8-9 and 4.8-10. Depending on the stiffness of the deck, this uniform distribution may or may not be realistic. However, since the deck usually occupies the majority of the dead load on a beam,

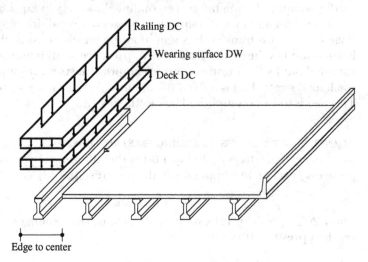

Figure 4.8-9
Dead load on exterior prestressed concrete I beam.

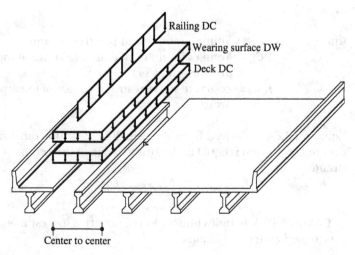

Figure 4.8-10
Dead load on interior prestressed concrete I beam.

inaccuracy in estimating the portion of the railing carried by a particular beam may not be that critical. Example 4.12 includes calculations for the dead-load effect on a prestressed concrete I-beam bridge.

The AASHTO specification beamline analysis method for finding the live-load shear in a beam is still applicable for prestressed concrete beams as long as the following requirements are met:

- ❑ The deck width is constant.
- ❑ The number of beams is not less than 4.
- ❑ The beams are parallel and have approximately the same stiffness.
- ❑ The overhang width minus the barrier width is less than 3.0 ft.
- ❑ Curvature in the plan is zero.
- ❑ The following applicability requirements are met: (i) $3.5 \text{ ft} \leq S \leq 16 \text{ ft}$, where S is the beam spacing; (ii) $4.5 \text{ in. } \leq t \leq 12 \text{ in.}$, where t is the deck slab thickness; (iii) $20 \text{ ft} \leq L \leq 240 \text{ ft}$, where L is the beam span length; (iv) $10{,}000 \text{ in.}^4 \leq K_g \leq 17{,}000{,}000 \text{ in.}^4$, where K_g is a stiffness factor to de defined below; (v) $-1 \text{ ft} \leq d_e \leq 5.5 \text{ ft}$, where d_e is the transverse distance between the inside web face of the exterior girder and the toe of the curb in feet. When the former is inside the latter d_e is positive, otherwise it is negative.

The live-load distribution factor for prestressed concrete I beams supporting a reinforced concrete deck slab is actually identical with that for steel I beams supporting the same deck. For convenience, it is given here with a few comments on the specific beam type. Example 4.12 illustrates application to a simple span of a two-span bridge.

Distribution Factors DF for Interior Beams

- ❑ DF for moment for one lane loaded: *4.6.2.2*

$$
4.6.2.2 \quad DF_{\text{moment interiior 1}} = 0.06 + \left(\frac{S}{14}\right)^{0.4} \left(\frac{S}{L}\right)^{0.3} \left[\frac{K_g}{12\, L\, t^3}\right]^{0.1} \quad (4.8\text{-}8)
$$

where　　S = beam spacing (ft)
　　　　　L = beam span length (ft)
　　　　　t = concrete deck slab thickness (in.)
　　　　　K_g = longitudinal stiffness parameter (in.^{4}) given as

$$
4.6.2.2 \qquad\qquad K_g = n\left(I + A e_g^2\right)
$$

where　　$n = E_B/E_D$, the ratio of Young's modulus of beam and deck slab *4.6.2.2*
　　　　　I = moment of inertia of noncomposite beam (in.^{4})

$$A = \text{cross-sectional area of noncomposite beam (in.}^2)$$
$$e_g = \text{distance between centers of gravity of noncomposite}$$
$$\text{beam and deck (in.)}$$

Note that, with the bridge owner's concurrence, $\left[K_g / \left(12\, L\, t_s^3 \right) \right]^{0.1} = 1.09$ is allowed in the specifications for prestressed concrete I-beam cross sections *(4.6.2.2.1)*. It can be conveniently used particularly when the cross section has not been determined. It also should be emphasized that the multiple presence factor of 1.2 in Table 3.3-1 for one lane has already been included in Eq. 4.8-8 and similar formulas in the specifications but not those referring to the lever rule.

❑ DF for moment for two or more lanes loaded:

$$\text{4.6.2.2} \qquad \text{DF}_{\text{moment interior 2}} = 0.075 + \left(\frac{S}{9.5} \right)^{0.6} \left(\frac{S}{L} \right)^{0.2} \left[\frac{K_g}{12\, L\, t^3} \right]^{0.1}$$

$$(4.8\text{-}9)$$

Between the DF values for one lane and multiple lanes loaded with the multiple presence factor included, whichever is larger will control the capacity of the beam. This concept is applicable in designing shear and other beams. Again note that, with the bridge owner's concurrence, $\left[K_g / \left(12\, L\, t_s^3 \right) \right]^{0.1} = 1.09$ is allowed in the specifications for prestressed concrete I-beam cross sections. *(4.6.2.2.1)*

❑ DF for shear for one lane loaded:

$$\text{4.6.2.2} \qquad\qquad \text{DF}_{\text{shear interior 1}} = 0.36 + \frac{S}{25} \qquad\qquad (4.8\text{-}10)$$

❑ DF for shear for two or more lanes loaded:

$$\text{4.6.2.2} \qquad \text{DF}_{\text{shear interior 2}} = 0.2 + \left(\frac{S}{12} \right) - \left(\frac{S}{35} \right)^{2} \qquad (4.8\text{-}11)$$

Again, whichever is larger between DFs for one lane and multiple lanes loaded shall be used to distribute the load to the beam in design.

Distribution Factors DF for Exterior Beams

❑ DF for moment for one lane loaded: Use the lever rule along with the multiple presence factor of 1.2 in Table 3.3-1. *3.6.1.1.2*

The so-called lever rule is referred to many times in the AASHTO specifications for structural analysis. It is a simple mechanics model thought to be analogical to load distribution here. It can be understood using the simple graph in Figure 4.8-11. Under a unit load 1 consisting of two wheel loads

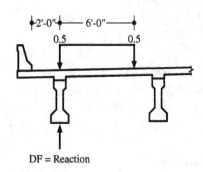

Figure 4.8-11
Lever rule.

each at 0.5 as shown, the reaction at the support is taken as the distribution factor of the support beam, an exterior beam in this case. Apparently, this approach does not include consideration to multiple lanes simultaneously occupied by trucks. Thus, the 1.2 multiple factor in Table 3.3-1 needs to be explicitly applied to the resulting distribution factor.

❑ DF for moment for two or more lanes loaded:

4.6.2.2
$$\mathrm{DF}_{\text{moment exterior 2}} = e\mathrm{DF}_{\text{moment interior 2}} \qquad e = 0.77 + \frac{d_e}{9.1}$$
$$(4.8\text{-}12)$$

As shown in Eq. 4.8-12, the exterior beam's distribution factor is given in the specifications as the product of the interior beam's distribution factor and a correction factor e. In that correction factor, distance d_e is between the web of the exterior girder and the interior edge of the curb in feet. If the former is inside the interior edge, d_e is positive; otherwise it is negative. Assume d_e also is limited in the range $-1 \le d_e \le 5.5$.

❑ DF for shear for one lane loaded: Use the lever rule, that is,

4.6.2.2
$$\mathrm{DF}_{\text{shear exterior 1}} = \mathrm{DF}_{\text{moment exterior 1}} \qquad (4.8\text{-}13)$$

Note that the multiple presence factor 1.2 needs to be applied, since the lever rule method is used.

❑ DF for shear for two or more lanes loaded:

4.6.2.2
$$\mathrm{DF}_{\text{shear exterior 2}} = e\mathrm{DF}_{\text{shear interior 2}} \qquad e = 0.6 + \frac{d_e}{10} \qquad (4.8\text{-}14)$$

Note that, in Eq. 4.8-11, $\mathrm{DF}_{\text{shear interior 2}}$ has the multiple presence factor imbedded.

Additional Investigation for Distribution Factor

In beam bridges, the cross section is often braced using diaphragms or cross frames, as seen earlier in this chapter. The distribution factor for the exterior beam shall not be taken to be less than that which would be obtained by assuming that the cross section deflects and rotates as a rigid cross section. To satisfy this code requirement, the following distribution factor for shear along with the multiple presence factor in Table 3.3-1 needs to be computed to be compared with the corresponding DF values discussed above. The larger DF shall be used in beam design:

C4.6.2.2.2d
$$\mathrm{DF}_{\text{shear exterior}, N_L} = m_{N_L}\left(\frac{N_L}{N_b} + \frac{x_{\text{ext}}\displaystyle\sum_{i=1}^{N_L} e_i}{\displaystyle\sum_{j=1}^{N_b} x_j^2}\right) \qquad (4.8\text{-}15)$$

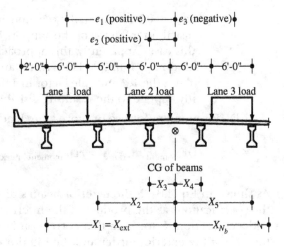

Figure 4.8-12
Truck placement in additional investigation for exterior beam distribution factor.

where (see Figure 4.8-12)   DF = reaction on exterior beam in terms of number of lanes of load, N_L
N_L = number of loaded lanes under consideration
e_i = eccentricity of design truck or lane load in lane i from center of gravity of beams
x_j = horizontal distance from center of gravity of beams to beam j
x_{ext} = horizontal distance from center of gravity of beams to exterior beam
N_b = number of beams
M_{N_L} = multiple presence factor in Table 3.3-1 for N_L lanes

This additional investigation is required because the distribution factor for beams in a multibeam cross section was determined without consideration of diaphragms or cross frames. This check is an interim provision until research provides a better solution.

Correction for Effect of Skew
Skewed support is commonly used in bridges, particularly beam bridges. Apparently it is relatively simpler to include skew in beam bridges than in truss, arch, cable-stayed, and suspension bridges. As discussed earlier in this chapter, skewed supports may save the cost of additional acquisition of right of way when the road carried by the bridge intersects at a nonstraight angle with a roadway, waterway, railway, and so on.

The correction factor to DF given in the AASHTO specifications is as follows:

❑ For moment:

4.6.2.2

$$\text{Correction factor} = 1 - 0.25 \left(\frac{K_g}{12\,Lt^3}\right)^{0.25} \left(\frac{S}{L}\right)^{0.5} \tan^{1.5}\theta$$

$$30° \le \theta \le 60° \quad \text{for } \theta > 60° \quad \text{use } \theta = 60° \qquad (4.8\text{-}16)$$

The skew is quantified using angle θ between the longitudinal axis of the bridge and the normal of the support, as shown in Figure 4.7-6. According to Eq. 4.8-16, the effect of skew is expected to reduce the design moment because skew reduces the effective span length. With the bridge owner's concurrence $\left[K_g/\left(12\,Lt^3\right)\right]^{0.25} = 1.15$ can be used for simplification. *(4.6.2.2.1)*

❑ For obtuse corner shear:

4.6.2.2

$$\text{Correction factor} = 1 + 0.20 \left(\frac{12Lt^3}{K_g}\right)^{0.3} \tan\theta$$

$$0° \le \theta \le 60° \qquad (4.8\text{-}17)$$

As seen, the effect of skew is to increase the obtuse shear. This effect is induced because skew makes adjacent beams behave differently since their respective effective span lengths become different from each other under the load, which induces additional torsion requiring extra shear forces to maintain equilibrium. With the bridge owner's concurrence, $\left(12\,L\,t_s^3/K_g\right)^{0.3} = .85$ is allowed in the specifications for prestressed concrete I-beam cross sections. *4.6.2.2.1*

Example 4.12 illustrates how load effect analysis is performed for a prestressed concrete beam superstructure that is continuous over two symmetric spans. The strength and service limit design of the beams in the same superstructures is covered in Examples 4.13 and 4.14.

Example 4.12 Prestressed Concrete I-Beam Bridge Design (Load Effects)

❑ **Design Requirement**

For the purpose of design, analyze a 35° skewed precast prestressed beam bridge with six Type III prestressed I girders spaced at 7 ft 6 in. and an overhang 3 ft 6 in. wide. The bridge

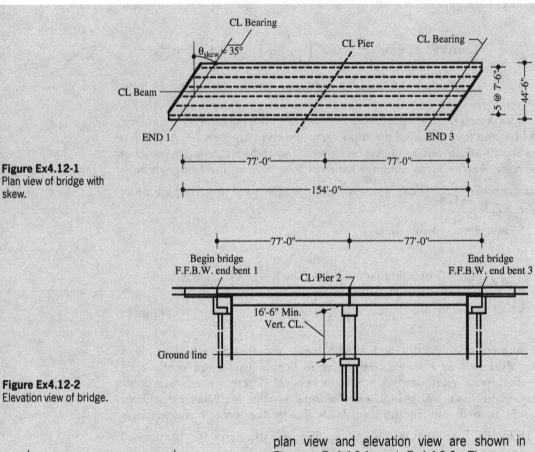

Figure Ex4.12-1
Plan view of bridge with skew.

Figure Ex4.12-2
Elevation view of bridge.

Figure Ex4.12-3
Cross section of bridge.

plan view and elevation view are shown in Figures Ex4.12-1 and Ex4.12-2. The cross section is shown in Figure Ex4.12-3.
Load modifier $\eta = 1.0$

❑ **Design Parameters**
Overall bridge length $L_{bridge} = 154$ ft
Span length $L_{span} = 77$ ft
Beam number $N_{beams} = 6$
$\theta_{skew} = 35°$
Beam spacing $S = 7$ ft 6 in.
Overhang width $S_{overhang} = 3$ ft 6 in.
Average hunch depth $h_{hunch} = 1$ in.
Diaphragm thickness $t_{diaphragm} = 8.5$ in.
Roadway clear width $w_{roadway} = 39$ ft 6 in.
Number of design traffic lanes per roadway $= 3$
Top cover $C_{top} = 2$ in.

Compressive strength of concrete for slab, $f'_{c,\,slab} = 4.5\ k/in.^2$
Compressive strength of concrete for prestressed beams,
$f'_{c,\,beam} = 6.5\ k/in.^2$
Concrete density $W_{concrete} = 0.145\ k/ft^3$ (normal weight
$f'_c \leq 5\ k/in.^2$) 3.5.1
Modulus of elasticity for slab:

$$C\ 5.4.2.4 \qquad E_{slab} = 1820\sqrt{f'_{c,\,slab}} = 1820 \times \sqrt{4.5\ k/in.^2} = 3861\ k/in.^2$$

Modulus of elasticity for prestressed concrete:

$$C\ 5.4.2.4 \qquad E_{beam} = 1820\sqrt{f'_{c,\,beam}} = 1820 \times \sqrt{6.5\ k/in.^2} = 4640\ k/in.^2$$

Wearing surface (WS) weight $W_{WS} = 0.029\ k/ft^2$
Slab thickness $t_{slab} = 8.5$ in. (minimum 7 in.) 9.7.1.1
Deck overhang thickness $t_{overhang} = 8.5$ in. (minimum 8 in.)
    13.7.3.1.2
Modulus of elasticity for reinforcing steel, $E_{steel} = 29,000\ k/in.^2$
Ultimate tensile strength for prestressing tendon, $f_{tendon} = 270\ k/in.^2$
Modulus of elasticity for prestressing tendon, $E_{tendon} = 28,500\ k/in.^2$
Traffic barrier weight $W_{barrier} = 0.54\ k/ft$
Weight of stay-in-place metal forms, $W_{forms} = 0.015\ k/ft^2$
Distance from centerline pier to centerline bearing, $K = 11$ in.
Distance from end of beam to centerline of bearing, $J = 6$ in.
Beam length $L_{beam} = L_{span} - 2(K - J) = 77\ ft - 2(11\ in. - 6\ in.) = 76\ ft\ 2$ in.
Beam design length $L_{design} = L_{span} - 2K = 77\ ft - 2(11\ in.) = 75\ ft\ 2$ in.

❑ Noncomposite (NC) Cross-Sectional Properties for Type III I Beams (Figure Ex4.12-4)

Moment of inertia $I_{NC} = 125,390$ in.^{4}
Section area $A_{NC} = 560$ in.^{2} $y_{t,\,NC} = 24.73$ in., $y_{b,\,NC} = 20.27$ in.

Section modulus top $S_{t,\,NC} = \dfrac{I_{NC}}{y_{t,\,NC}} = \dfrac{125,390\ in.^4}{24.73\ in.}$

$= 5070$ in.^{3}

Section modulus bottom $S_{b,\,NC} = \dfrac{I_{NC}}{y_{b,\,NC}} = \dfrac{125,390\ in.^4}{20.27\ in.}$

$= 6189$ in.^{3}

❑ Design Check Sections

Support section: $x_{support} = 0$ ft
Shear check section: $x_{shear\ check} = 0.72(h_{NC} + h_{hunch} + t_{slab})$ in. $= 39.24$ in. $= 3.27$ ft

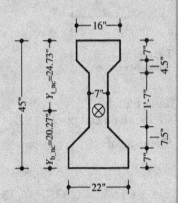

Figure Ex4.12-4
AASHTO Type III prestressed I-beam section.

Debond section: $x_{debond} = 16$ ft

Midspan section: $x_{midspan} = 0.5L_{design} = 0.5 \times 75.17$ ft $= 37.59$ ft

Effective Flange Width

Interior Beams:

4.6.2.6.1 $b_{effective\ interior} = 7.5$ ft $(12$ in./ft$) = 90$ in.

Exterior Beams:

$$b_{effective\ exterior} = \frac{b_{effective\ interior}}{2} + w_{overhang}$$

$$= \frac{90\ \text{in.}}{2} + 3.5\ \text{ft}\ (12\ \text{in./ft}) = 87\ \text{in.}$$

Transformed Properties

To find composite section properties, the effective flange width of the slab is transformed equivalent to the beam properties.

Modular ratio between the deck and beam:

$$n = \frac{E_{slab}}{E_{beam}} = \frac{3861\ \text{k/in.}^2}{4640\ \text{k/in.}^2} = 0.832$$

Transformed slab width for interior beams:

$$b_{transformed\ interior} = n\ b_{effective\ interior}$$

$$= 0.832\ (90\ \text{in.}) = 74.88\ \text{in.}$$

Transformed slab width for exterior beams:

$$b_{transformed\ exterior} = n\ b_{effective\ exterior}$$

$$= 0.832\ (87\ \text{in.}) = 72.38\ \text{in.}$$

Composite Section Properties

Interior Beams:

The section properties of the interior beams are calculated in Table Ex4.12-1 and subsequent computations, according to the cross section in Figure Ex4.12-5.

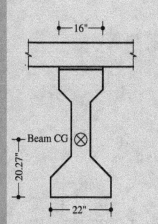

Figure Ex4.12-5
Composite section.

Table Ex4.12-1
Interior beam composite section properties

Component	A (in.)	Y (in.)	AY (in.³)	AY² (in.⁴)	I_0(in.⁴)
Type III	560	20.27	11,351	230,098	125,390
Hunch (n)	0.832(16) = 13.31	45.50	606	27,555	$\frac{(13.31)\,1^3}{12} = 1$
Slab (n)	636.5	50.25	31,983	1,607,202	$\frac{74.88\,(8.5)^3}{12} = 3832$
Sum	1210		43,941	1,864,846	129,222

$$I_z = \sum I_0 + \sum Ay^2 = 129{,}222 \text{ in.}^4 + 1{,}864{,}846 \text{ in.}^4 = 1{,}994{,}068 \text{ in.}^4$$

$$Y' = \frac{\sum Ay}{\sum A} = \frac{43{,}941 \text{ in.}^3}{1210 \text{ in.}^2} = 36.31 \text{ in.}$$

$$Y_t = 45 \text{ in.} + 1 \text{ in.} + 8.5 \text{ in.} - 36.31 \text{ in.}$$
$$= 18.19 \text{ in.}$$

$$I_x = I_z - \left(\sum A\right)\left(Y'\right)^2 = 1{,}994{,}068 \text{ in.}^4 - 1210 \text{ in.}^2 \, (36.31 \text{ in.})^2$$
$$= 398{,}785 \text{ in.}^4$$

$$S_t = \frac{398{,}785}{18.19} = 21{,}923 \text{ in.}^3 \qquad S_{\text{beam top}} = \frac{398{,}785}{45 - 36.31} = 45{,}890 \text{ in.}^3$$

$$S_b = \frac{398{,}785}{36.31} = 10{,}983 \text{ in.}^3$$

Exterior Beams:

Table Ex4.12-2
Exterior beam composite section properties

Component	A (in.^{2})	Y (in.)	AY (in.^{3})	AY^2 (in.^{4})	I_0 (in.^{4})
Type III	560	20.27	11,351	230,098	125,390
Hunch (n)	0.832(16)	45.50	606	27,555	$\dfrac{(13.31)\,1^3}{12} = 1$
Slab (n)	72.38(8.5)= 615.2	50.25	30,914	1,553,418	$\dfrac{72.38\,(8.5)^3}{12} = 3704$
Sum	1189		42,872	1,811,062	129,095

$$I_z = \sum I_0 + \sum Ay^2 = 129{,}095 \text{ in.}^4 + 1{,}811{,}062 \text{ in.}^4 = 1{,}940{,}157 \text{ in.}^4$$

$$Y' = \frac{\sum Ay}{\sum A} = \frac{42{,}872 \text{ in.}^3}{1189 \text{ in.}^2} = 36.06 \text{ in.}$$

$$Y_t = 45 \text{ in.} + 1 \text{ in.} + 8.5 \text{ in.} - 36.06 \text{ in.} = 18.44 \text{ in.}$$

$$I_x = I_z - \left(\sum A\right)\left(Y'\right)^2 = 1{,}940{,}157 \text{ in.}^4 - 1189 \text{ in.}^2 \, (36.06 \text{ in.})^2$$
$$= 394{,}072 \text{ in.}^4$$

$$S_t = \frac{394{,}072}{18.44} = 21{,}370 \text{ in.}^3 \qquad S_{\text{beam top}} = \frac{394{,}072}{45 - 36.06} = 44{,}080 \text{ in.}^3$$

$$S_b = \frac{394{,}072}{36.06} = 10{,}928 \text{ in.}^3$$

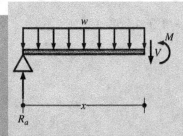

Figure Ex4.12-6
Dead load moment M and shear V.

☐ **Dead-Load Effects**

Calculate the moments and shears as functions of x, where x represents a point along the beam from zero to L_{design} in feet.

According to Figure Ex4.12-6, moment M and shear V at location x due to dead load w are as follows, with w given in Tables Ex4.12-3 and Ex4.12-4, respectively for the interior and exterior beams:

Table Ex4.12-3
Uniformly distributed dead load w for interior beams

		w
DC	Beam	$\dfrac{560 \, in.^3}{12 \, in./ft \, (12 \, in./ft)} 0.245 \, k/ft^3 = 0.5639 \, k/ft$
	Steel forms	$\dfrac{90 \, in. - 16 \, in.}{12 \, in./ft} 0.015 \, k/ft^2 = 0.0925 \, k/ft$
	Hunch + slab	$\dfrac{8.5 \, in.(90 \, in.) + 1 \, in. \, (16 \, in.)}{12 \, in./ft \, (12 \, in./ft)} 0.145 \, k/ft^3 = 0.7864 \, k/ft$
	Barriers	$\dfrac{2 \, (0.54 \, k/ft)}{6 \, beams} = 0.18 \, k/ft$
DW:	WS	$7.5 \, ft \, (0.029 \, k/ft^2) = 0.2175 \, k/ft$

Table Ex4.12-4
Uniformly distributed dead load w for exterior beams

		w
DC	Beam	$\dfrac{560 \, in.^3}{12 \, in./ft \, (12 \, in./ft)} 0.245 \, k/ft^3 = 0.5639 \, k/ft$
	Steel forms	$\dfrac{86 \, in. - 16 \, in.}{12 \, in./ft} 0.015 \, k/ft^2 = 0.0875 \, k/ft$
	Hunch + slab	$\dfrac{8.5 \, in. \left(\dfrac{90 \, in.}{2}\right) + 9 \, in.(42 \, in.) + 1 \, in.(16 \, in.)}{12 \, in./ft \, (12 \, in./ft)} 0.145 \, k/ft^3 = 0.7819 \, k/ft$
	Barriers	$\dfrac{2 \, (0.54 \, k/ft)}{6 \, beams} = 0.18 \, k/ft$
DW:	FWS	$6 \, ft \, (0.029 \, k/ft^2) = 0.174 \, k/ft$

$$M(x) = \frac{wL}{2}x - \frac{w}{2}x^2 \qquad V(x) = \frac{wL}{2} - wx$$

Figure Ex4.12-7
Placement of HL93 truck for maximum shear at support.

Live-Load Effects

Same for the interior and exterior beams until the live-load distribution factors are applied later.

For Support Section

The following calculations refer to Figure Ex4.12-7 for truck location.

$$V_{truck} = 32\,k + 32\,k \left(\frac{75.17\,ft - 14\,ft}{75.17\,ft}\right)$$

$$+ 8\,k \left(\frac{75.17\,ft - 28\,ft}{75.17\,ft}\right)$$

$$= 32\,k + 32\,k(0.81) + 8\,k(0.63)$$

$$= 63.1\,k$$

$$V_{tandem} = 48.7\,k \quad \text{not controlling} < 63.1\,k$$

$$V_{lane} = 0.64\,k/ft\,\frac{75.17\,ft}{2} = 24.1\,k$$

$$V_{LL,\,support} = 63.1\,k(1 + IM) + 24.1\,k$$

$$= 63.1(1.33) + 24.1 = 108.0\,k$$

Figure Ex4.12-8
Placement of HL93 truck for maximum moment and shear at shear check.

For Shear Check Section

The calculations use the position of load shown in Figure Ex4.12-8.

$$V_{truck} = R_{left}$$

$$= 32\,k \left(\frac{75.17\,ft - 3.27\,ft}{75.17\,ft}\right) + 32\,k \left(\frac{75.17\,ft - 17.27\,ft}{75.17\,ft}\right)$$

$$+ 8\,k \left(\frac{75.17\,ft - 31.27\,ft}{75.17\,ft}\right)$$

$$= 32\,k(0.96) + 32\,k(0.77) + 8\,k(0.58) = 59.9\,k$$

$$V_{tandem} = 46.5\,k \quad \text{not controlling} < 59.9\,k$$

$$M_{truck} = 59.9\,k\,(3.27\,ft) = 195.9\,kft$$

$$M_{tandem} = 152.1\,kft \quad \text{not controlling} < 195.9\,kft$$

$$V_{lane} = 0.64\,k/ft \left(\frac{75.17\,ft}{2}\right) - 0.64\,k/ft\,(3.27\,ft) = 22.0\,k$$

$$M_{lane} = 0.64 \text{ k/ft} \left(\frac{75.17 \text{ ft}}{2}\right) 3.27 \text{ ft} - 0.64 \text{ k/ft} \frac{(3.27 \text{ ft})^2}{2} = 75.2 \text{ kft}$$

$$V_{LL \text{ shear check}} = 59.9 \text{ k} (1.33) + 22.0 \text{ k} = 101.7 \text{ k}$$

$$M_{LL \text{ shear check}} = 195.9 \text{ kft} (1.33) + 75.2 \text{ kft} = 335.7 \text{ kft}$$

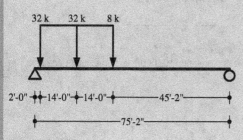

Figure Ex4.12-9
Placement of HL93 truck for maximum moment at debond.

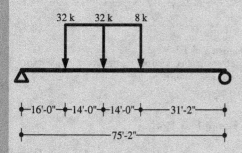

Figure Ex4.12-10
Alterative placement of HL93 truck for possibly higher maximum moment and shear at debond.

For Debond Section

The following calculations refer to the load position in Figure Ex4.12-9 for the section at 2 ft from the support.

$$R_{left} = 32 \text{ k} \left(\frac{75.17 \text{ ft} - 2 \text{ ft}}{75.17 \text{ ft}}\right)$$

$$+ 32 \text{ k} \left(\frac{75.17 \text{ ft} - 16 \text{ ft}}{75.17 \text{ ft}}\right)$$

$$+ 8 \text{ k} \left(\frac{75.17 \text{ ft} - 30 \text{ ft}}{75.17 \text{ ft}}\right)$$

$$= 32 \text{ k}(0.97) + 32 \text{ k}(0.79) + 8 \text{ k}(0.60)$$

$$= 61.6 \text{ k}$$

$$M_{truck} = 61.6 \text{ k}(2 \text{ ft}) = 123.2 \text{ kft}$$

$$V_{truck} = 61.6 \text{ k}$$

The following calculations refer to the load position in Figure Ex4.12-10 for the section at 16 ft from the support.

$$V_{truck} = R_{left} = 32 \text{ k} \left(\frac{75.17 \text{ ft} - 16 \text{ ft}}{75.17 \text{ ft}}\right)$$

$$+ 32 \text{ k} \left(\frac{75.17 \text{ ft} - 30 \text{ ft}}{75.17 \text{ ft}}\right)$$

$$+ 8 \text{ k} \left(\frac{75.17 \text{ ft} - 44 \text{ ft}}{75.17 \text{ ft}}\right)$$

$$= 32 \text{ k}(0.79) + 32 \text{ k}(0.60) + 8 \text{ k}(0.41)$$

$$= 47.8 \text{ k}$$

$$V_{tandem} = 38.0 \text{ k} \quad \text{not controlling} < 47.8 \text{ k}$$

$$M_{truck} = 47.8 \text{ (16 ft)} = 764.2 \text{ kft}$$

$$M_{tandem} = 608 \text{ kft} \quad \text{not controlling} < 764.2 \text{ kft}$$

$$M_{lane} = 0.64 \text{ k/ft} \left(\frac{75.17 \text{ ft}}{2} \right) 16 \text{ ft} - 0.64 \text{ k/ft} \frac{(16 \text{ ft})^2}{2} = 303.0 \text{ kft}$$

$$M_{LL, debond} = 1.33 (764.2 \text{ kft}) + 303.0 \text{ kft} = 1319 \text{ kft}$$

$$V_{lane} = 0.64 \text{ k/ft} \left(\frac{75.17 \text{ ft}}{2} \right) - 0.64 \text{ k/ft} (16 \text{ ft}) = 13.8 \text{ k}$$

$$V_{LL \ debond} = 1.33 (47.8 \text{ k}) + 13.8 \text{ k} = 77.3 \text{ k}$$

For Midspan Section

The design load effect is calculated using the load position in Figure Ex4.12-11.

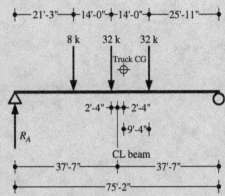

$$R_A = 8 \text{ k} \left(\frac{75.17 \text{ ft} - 21.25 \text{ ft}}{75.17 \text{ ft}} \right)$$
$$+ 32 \text{ k} \left(\frac{75.17 \text{ ft} - 35.25 \text{ ft}}{75.17 \text{ ft}} \right)$$
$$+ 32 \text{ k} \left(\frac{75.17 \text{ ft} - 49.25 \text{ ft}}{75.17 \text{ ft}} \right)$$
$$= 8 \text{ k}(0.72) + 32 \text{ k}(0.53) + 32 \text{ k}(0.34)$$
$$= 33.8 \text{ k}$$

$$M_{midspan \ truck} = 33.8 \text{ k}(35.25 \text{ ft}) - 8 \text{ k}(14 \text{ ft})$$
$$= 1078 \text{ kft}$$

Figure Ex4.12-11
Placement of HL93 truck for maximum moment at midspan.

$$M_{midspan \ lane} = 0.64 \text{ k/ft} \frac{(75.17 \text{ ft})^2}{8} = 452 \text{ kft}$$

$$M_{LL \ midspan} = 1078 \text{ kft}(1 + IM) + 452 \text{ kft}$$
$$= 1078 (1.33) + 452 = 1886 \text{ kft}$$

Live-Load Distribution Factors

The AASHTO distribution factors can be used since the following conditions are met: (1) The deck width is constant. (2) The number of beams is not less than 4. (3) The beams are parallel and have approximately the same stiffness. (4) The overhang width minus the barrier width is less than 3.0 ft. (5) Curvature in the plan is zero. In addition, the following quantitative range applicability requirements are met: (i) $3.5 \text{ ft} \leq S \leq 16 \text{ ft}$; (ii) $4.5 \text{ in.} \leq t_{slab} \leq 12 \text{ in.}$; (iii) $20 \text{ ft} \leq L_{design} \leq 240 \text{ ft}$; (iv) $10,000 \text{ in.}^4 \leq K_g \leq 17,000,000 \text{ in.}^4$; (v) $30° \leq$ skew angle $\theta \leq 60°$; (vi) $-1 \text{ ft} \leq d_e \leq 5.5 \text{ ft}$.

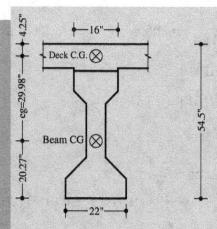

Figure Ex4.12-12
Distance between centers of gravity.

The distance between the centers of gravity of the non-composite beam and the deck (Figure Ex4.12-12) is

$$e_g = h_C - y_{b,\,NC} - \frac{t_{slab}}{2} = 54.5 \text{ in.} - 20.27 \text{ in.}$$

$$- \frac{8.5 \text{ in.}}{2} = 29.98 \text{ in.}$$

The longitudinal stiffness parameter is

$$K_g = \frac{I_{NC} + A_{NC} e_g^2}{n}$$

$$= \frac{125{,}390 \text{ in.}^4 + 560 \text{ in.}^2 \,(29.98 \text{ in.})^2}{0.832}$$

$$= 755{,}569 \text{ in.}^4$$

☐ **Distribution Factors for Interior Beams**
Moment Distribution Factor:
For one lane loaded *4.6.2.2.2b*

$$DF_{moment\ interior\ 1} = 0.06 + \left(\frac{S}{14}\right)^{0.4} \left(\frac{S}{L_{design}}\right)^{0.3} \left(\frac{K_g}{12\,L_{design}\,t_{slab}^3}\right)^{0.1}$$

$$= 0.06 + \left(\frac{7.5 \text{ ft}}{14}\right)^{0.4} \left(\frac{7.5 \text{ ft}}{75.17 \text{ ft}}\right)^{0.3} \left(\frac{755{,}569 \text{ in.}^4}{(12)\,75.17 \text{ ft}(8.5 \text{ in.})^3}\right)^{0.1}$$

$$= 0.46$$

For two or more lanes loaded

$$DF_{moment\ interior\ 2} = 0.075 + \left(\frac{S}{9.5}\right)^{0.6} \left(\frac{S}{L_{design}}\right)^{0.2} \left(\frac{K_g}{12.0\,L_{design}\,t_{slab}^3}\right)^{0.1}$$

$$= 0.075 + \left(\frac{7.5 \text{ ft}}{9.5}\right)^{0.6} \left(\frac{7.5 \text{ ft}}{75.17 \text{ ft}}\right)^{0.2} \left(\frac{755{,}569 \text{ in.}^4}{(12)\,75.17 \text{ ft}(8.5 \text{ in.})^3}\right)^{0.1}$$

$$= 0.64$$

The greater distribution factor is selected for beam moment design.
Shear Distribution Factor:
For one lane loaded *4.6.2.2.3a*

$$DF_{shear\ interior\ 1} = 0.36 + \frac{S}{25} = 0.36 + \frac{7.5 \text{ ft}}{25} = 0.66$$

For two or more design lanes loaded

$$DF_{shear\ interior\ 2} = 0.2 + \frac{S}{12} - \left(\frac{S}{35}\right)^{2.0}$$

$$= 0.2 + \frac{7.5\ ft}{12} - \left(\frac{7.5\ ft}{35}\right)^{2.0}$$

$$= 0.78$$

The greater distribution factor is selected for shear design of the beams.

$$DF_{shear\ interior} = \max\left(DF_{shear\ interior\ 1},\ DF_{shear\ interior\ 2}\right)$$

$$= \max\left(0.66, 0.78\right) = 0.78$$

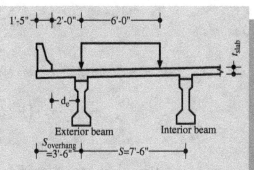

Figure Ex4.12-13
Dimensions for lever rule application.

☐ **Distribution Factors for Exterior Beams**   *4.6.2.2d*
Moment Distribution Factor

For one lane loaded: According to Figure Ex4.12-13 and its lever rule model in Figure Ex4.12-14:

$$DF_{moment\ exterior\ 1} = m_1 \left[0.5 \left(\frac{S + 7.5\ ft - 17\ in. - 2\ ft}{S}\right) \right.$$

$$\left. +0.5 \left(\frac{S + 3.5\ ft - 17\ in. - 2\ ft - 6\ ft}{S}\right) \right]$$

$$= 1.2 \left[0.5 \left(\frac{7.5 + 7.5\ ft - 17\ in. - 2\ ft}{7.5}\right) \right.$$

$$\left. +0.5 \left(\frac{7.5 + 3.5\ ft - 17\ in. - 2\ ft - 6\ ft}{7.5}\right) \right] = 0.73$$

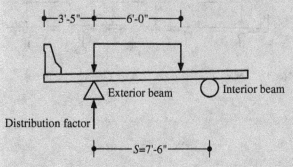

Figure Ex4.12-14
Level rule application model.

For two or more design lanes loaded

$$DF_{moment\ exterior\ 2} = DF_{moment\ interior\ 2} \left(0.77 + \frac{d_e}{9.1} \right)$$

$$= 0.64 \left(0.77 + \frac{3.5\ ft - 17\ in./\left(12\ in./ft \right) - 7\ in./\left[2\left(12\ in./ft \right) \right]}{9.1} \right)$$

$$= 0.62$$

$$DF_{moment\ exterior} = \max \left(DF_{moment\ exterior\ 1}, DF_{moment\ exterior\ 2} \right) = \max \left(0.73, 0.62 \right)$$

$$= 0.73$$

Shear Distribution Factor

For one design lane loaded

The lever rule is applied:

$$DF_{shear\ exterior\ 1} = DF_{moment\ exterior\ 1} = 0.73$$

$$DF_{shear\ exterior\ 2} = DF_{shear\ interior\ 2} \left(0.6 + \frac{d_e}{10} \right)$$

$$= 0.78 \left(0.6 + \frac{3.5 - 1.417 - 3.5/12}{10} \right) = 0.61$$

$$DF_{shear\ exterior} = \max \left(DF_{shear\ exterior\ 1}, DF_{shear\ exterior\ 2} \right)$$

$$= \max \left(0.73, 0.61 \right) = 0.73$$

☐ **Special Analysis for Exterior Beams (Figure Ex4.12-15)**
C4.6.2.2.2d

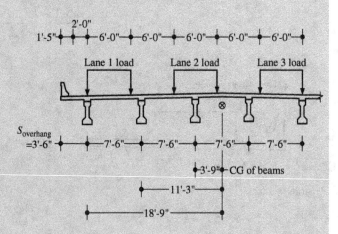

Figure Ex4.12-15
Model for special analysis for exterior beams.

$$DF_{\text{shear exterior 1}} = m_1 \left(\frac{N_L}{N_b} + \frac{x_{\text{ext}} \sum^{N_L} e}{\sum^{N_b} x^2} \right) = 1.2 \left(\frac{1}{6} + \frac{18.75\,(15.83)}{2\left(18.75^2 + 11.25^2 + 3.75^2\right)} \right)$$

$$= 1.2\,(0.47) = 0.56$$

$$DF_{\text{shear exterior 2}} = m_2 \left(\frac{2}{6} + \frac{18.75\,(15.83 + 3.83)}{2\left(18.75^2 + 11.25^2 + 3.75^2\right)} \right) = 1\,(0.71) = 0.71$$

$$DF_{\text{shear exterior 3}} = m_3 \left(\frac{3}{6} + \frac{18.75\,(15.83 + 3.83 - 8.17)}{2\left(18.75^2 + 11.25^2 + 3.75^2\right)} \right) = 0.85\,(0.72) = 0.61$$

Since the above three values are smaller than $DF_{\text{moment exterior}}$ $= DF_{\text{shear exterior}} = 0.73$, they do not control.

☐ **Applying Skew Correction Factor for Moment**   *4.6.2.2e*
A skew modification factor for moments may be used if the supports are skewed and the difference between skew angles of two adjacent supports does not exceed 10°.

$$CR_{\text{moment skew}} = 1 - 0.25 \left(\frac{K_g}{12\,L_{\text{design}}\,t_{\text{slab}}^3} \right)^{0.25} \left(\frac{S}{L_{\text{design}}} \right)^{0.5} \left(\tan \theta_{\text{skew}} \right)^{1.5}$$

$$= 1 - 0.25 \left(\frac{755{,}569 \text{ in.}^4}{(12)\,75.17 \text{ ft}\,(8.5 \text{ in.})^3} \right)^{0.25} \left(\frac{7.5 \text{ ft}}{75.17 \text{ ft}} \right)^{0.5} \left(\tan 35° \right)^{1.5} = 0.95$$

$$DF_{\text{moment interior skew}} = DF_{\text{moment interior}} CR_{\text{moment skew}} = 0.64(0.95) = 0.61$$

$$DF_{\text{moment exterior skew}} = DF_{\text{moment exterior}} CR_{\text{moment skew}} = 0.73(0.95) = 0.69$$

☐ **Applying Skew Correction Factor for Shear**   *4.6.2.2.d*

$$CR_{\text{shear skew}} = 1 + 0.20 \left(\frac{12\,L_{\text{design}}\,t_{\text{slab}}^3}{K_g} \right)^{0.3} \tan \theta_{\text{skew}}$$

$$= 1 + 0.20 \left[\frac{(12)\,75.17 \text{ ft}\,(8.5 \text{ in.})^3}{755{,}569 \text{ in.}^4} \right]^{0.3} \tan 35° = 1.13$$

$$DF_{\text{shear interior skew}} = DF_{\text{shear interior}} CR_{\text{shear skew}} = 0.78\,(1.13) = 0.88$$

$$DF_{\text{shear exterior skew}} = DF_{\text{shear exterior}} CR_{\text{shear skew}} = 0.73\,(1.13) = 0.82$$

❑ **Summary for Live-Load Distribution Factors**

Table Ex4.12-5
Live-load distribution factors

	Moment	Shear
Interior beams	0.61	0.88
Exterior beams	0.69	0.82

❑ **Summary of Dead and Live Loads (without Distribution to Beam) for Cross Sections**

Table Ex4.12-6
Unfactored dead-load and HL93 moment (kft) for interior beams

	Support	Shear Check	Debond	Midspan
x(ft)	0.0	3.3	16.0	37.6
Beam	0.0	66.9	266.9	398.3[a]
Steel forms	0.0	10.9	43.8	65.3
Hunch+slab	0.0	93.3	372.3	555.4
Barrier	0.0	21.3	85.2	127.1
WS	0.0	25.8	103.0	153.6
LL	0.0	204.8	804.4	1,150.5

[a] 409 kft for prestress transfer, considering the beam length of 76 ft 2 in.

Table Ex4.12-7
Unfactored dead-load and HL93 shear (k) for interior beams

	Support	ShearCheck	Debond	Midspan
x(ft)	0.0	3.3	16.0	37.6
Beam	21.2	19.3	12.2	0.0
Steel forms	3.5	3.2	2.0	0.0
Hunch+slab	29.6	27.0	17.0	0.0
Barrier	6.8	6.2	3.9	0.0
WS	8.2	7.5	4.7	0.0
LL	95.0	89.5	68.0	–

Table Ex4.12-8
Unfactored dead-load and HL93 moment (kft) for exterior beams

	Support	ShearCheck	Debond	Midspan
x(ft)	0.0	3.3	16.0	37.6
Beam	0.0	66.9	266.9	398.3[a]
Steel forms	0.0	10.3	41.4	61.8
Hunch+slab	0.0	92.7	370.1	552.3
Barrier	0.0	21.3	85.2	127.1
WS	0.0	20.6	82.4	122.9
LL	0.0	231.8	902.7	1,301.3

[a] 409 kft for prestress transfer, considering the beam length of 76 ft 2 in.

Table Ex4.12-9
Unfactored dead-load shear for exterior beams

	Support	ShearCheck	Debond	Midspan
x(ft)	0	3.27	16	37.6
Beam	21.2	19.4	12.2	0.0
Steel forms	3.3	3.0	1.9	0.0
Hunch+slab	29.4	26.8	16.9	0.0
Barrier	6.8	6.2	3.9	0.0
WS	6.5	6.0	3.8	0.0
LL	88.6	83.3	63.4	–

Example 4.13 Prestressed Concrete I-Beam Bridge Design (Strength I Limit State)

❑ **Design Requirement**

Design the prestressed concrete interior Type III beams for the bridge in Example 4.12 with six beams spaced at 7 ft 6 in. and an overhang 3 ft 6 in. wide. The bridge's cross-sectional view is shown in Figure Ex4.13-1 and the plan view in Figure Ex4.13-2.

Load modifier $\eta = 1.0$

❑ **Design Parameters**

Same as Example 4.12.

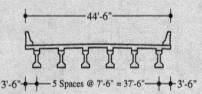

Figure Ex4.13-1
Cross section of bridge.

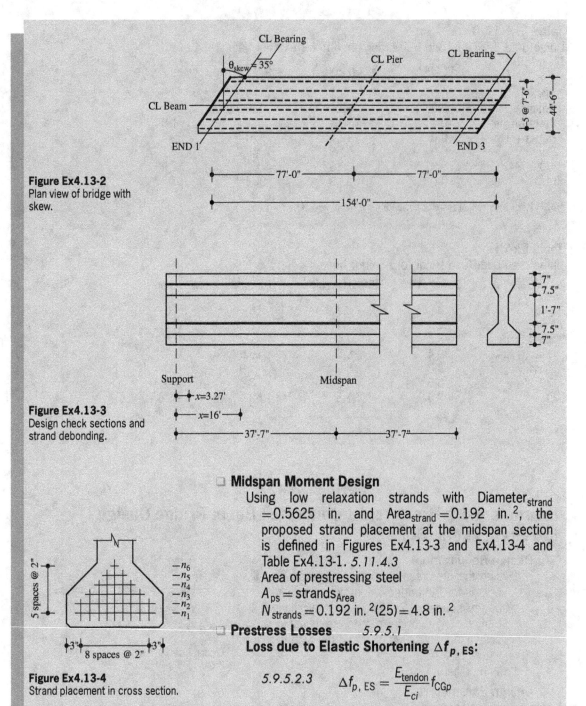

Figure Ex4.13-2
Plan view of bridge with skew.

Figure Ex4.13-3
Design check sections and strand debonding.

Figure Ex4.13-4
Strand placement in cross section.

❑ Midspan Moment Design

Using low relaxation strands with $\text{Diameter}_{strand} = 0.5625$ in. and $\text{Area}_{strand} = 0.192$ in.^{2}, the proposed strand placement at the midspan section is defined in Figures Ex4.13-3 and Ex4.13-4 and Table Ex4.13-1. *5.11.4.3*

Area of prestressing steel

$A_{ps} = \text{strands}_{Area}$

$N_{strands} = 0.192 \text{ in.}^2 (25) = 4.8 \text{ in.}^2$

❑ Prestress Losses　　*5.9.5.1*

Loss due to Elastic Shortening $\Delta f_{p,\,ES}$:

5.9.5.2.3　　$\Delta f_{p,\,ES} = \dfrac{E_{tendon}}{E_{ci}} f_{CGp}$

Table Ex4.13-1
Midspan strand placement data

Distance of Strand from Bottom of Beam (in.)	Number of Strands per Row at Midspan
$y_5 = 11$	$n_5 = 3$
$y_4 = 9$	$n_4 = 5$
$y_3 = 7$	$n_3 = 5$
$y_2 = 5$	$n_2 = 5$
$y_1 = 3$	$n_1 = 7$
Strand$_{CG}$ = 6.36 in.	Total strands, $N_{strands} = 25$

To find concrete stress f_{CGp} at prestress transfer at the midspan, determine the location of strands with reference to the neutral axis for the noncomposite beam section and its section modulus:

$$e_{CG\,NC} = y_{b,\,NC} - strand_{CG} = 20.27 \text{ in.} - 6.36 \text{ in.} = 13.91 \text{ in.}$$

$$S_{strand\,CG\,NC} = \frac{I_{NC}}{e_{CG\,NC}} = \frac{125{,}390 \text{ in.}^4}{13.91 \text{ in.}} = 9014 \text{ in.}^3$$

Initial prestress as 90% of stress in prestressing steel and initial prestress force prior to transfer: *5.9.3.1*

$$f_{pi} = 0.9 \left(0.75 f_{tendon}\right)$$

$$= 0.9 \left(0.75\right) 270 \text{ k/in.}^2$$

$$= 182.3 \text{ k/in.}^2$$

$$F_{ps} = A_{ps} f_{pi} = 4.8 \text{ in.}^2 \left(182.3 \text{ k/in.}^2\right) = 875 \text{ k}$$

Concrete stresses at CG of the prestressing force at transfer and the self-weight of the beam at midspan:

$$f_{CGp} = \frac{-F_{ps}}{A_{NC}} - \frac{F_{ps} e_{CG\,NC}}{S_{strand\,CG\,NC}} + \frac{M_{beam}}{S_{strand\,CG\,NC}}$$

$$= -\frac{875.0 \text{ k}}{560 \text{ in.}^2} - \frac{875.0 \text{ k} \left(13.91 \text{ in.}\right)}{9{,}014 \text{ in.}^3}$$

$$+ \frac{409 \text{ kft} \left(12 \text{ in./ft}\right)}{9{,}014 \text{ in.}^3} = -2.37 \text{ k/in.}^2 \quad \text{(compression)}$$

Losses due to elastic shortening:

5.9.5.2.3

$$\Delta f_{p,ES} = \frac{E_{tendon}}{E_{ci}} f_{CGp} = \frac{28{,}500 \text{ k/in.}^2}{1820 \sqrt{f'_{ci}}} f_{CGp}$$

$$= \frac{28{,}500 \text{ k/in.}^2}{1820 \sqrt{0.75 f'_c}} f_{CGp} = \frac{28{,}500 \text{ k/in.}^2}{1820 \sqrt{0.75 \left(6.5 \text{ k/in.}^2\right)}} 2.37 \text{ k/in.}^2$$

$$= \frac{28{,}500 \text{ k/in.}^2}{4018 \text{ k/in.}^2} 2.37 \text{ k/in.}^2 = 16.81 \text{ k/in.}^2$$

Long-term Prestress Loss $\Delta f_{p,LT}$:

5.9.5.3

$$\Delta f_{p,LT} = 10 \frac{f_{pi} A_{ps}}{A_{NC}} \gamma_h \gamma_{st} + 12 \gamma_h \gamma_{st} + \Delta f_{pR}$$

Correction factor for specified concrete strength at prestress transfer:

5.4.2.3.2

$$\gamma_{st} = \frac{5}{1 + f'_{ci, \text{ beam}}} = \frac{5}{1 + 0.75 f'_{c, \text{ beam}}} = \frac{5}{1 + 0.75 \left(6.5\right)} = 0.85$$

Using an average annual ambient relative humidity $H = 70\%$, the correction factor for humidity is

5.4.2.3.2

$$\gamma_h = 1.7 - 0.01H = 1.7 - 0.01 \times 70 = 1.0$$

Relaxation loss $\Delta f_{pR} = 2.4 \text{ k/in.}^2$ for low relaxation strand 5.9.5.3 Thus,

$$\Delta f_{p,LT} = 10 \frac{f_{pi} A_{ps}}{A_{NC}} \gamma_n \gamma_{st} + 12 \gamma_n \gamma_{st} + \Delta f_{pR}$$

$$= 10 \frac{182.3 \text{ k/in.}^2 \left(4.8 \text{ in.}^2\right)}{560 \text{ in.}^2} 1 \left(0.85\right) + 12(1)0.85 + 2.4 \text{ k/in.}^2$$

$$= 25.86 \text{ k/in.}^2$$

Total Prestress Loss:

5.9.5.1

$$\Delta f_{pT} = \Delta f_{p,ES} + \Delta f_{p,LT}$$

$$= 16.81 \text{ k/in.}^2 + 25.86 \text{ k/in.}^2 = 42.67 \text{ k/in.}^2$$

Prestress in tendon at jacking:

5.9.3 $$f_{pj} = 0.75 f_{pu} = 0.75 \left(270 \text{ k/in.}^2\right) = 202.5 \text{ k/in.}^2$$

Percentage loss of prestress:

$$\text{Prestress loss percentage} = \frac{\Delta f_{pT}}{f_{pj}} = \frac{42.67 \text{ k/in.}^2}{202.5 \text{ k/in.}^2} = 21\%$$

☐ **Beam Concrete Stress Limits**   5.9.4

Table Ex4.13-2
Beam concrete stress limits

	Compression	Tension
Prestress transfer	$0.6 f'_{ci} = 2.93 \text{ k/in.}^2$	$0.24 \sqrt{f'_{ci}} = 0.53 \text{ k/in.}^2$
Service	$0.45 f'_{ci} = 2.93 \text{ k/in.}^2$ (without transient load) $0.6 f'_{ci} = 3.90 \text{ k/in.}^2$ (with transient load)	$0.19 \sqrt{f'_{c}} = 0.48 \text{ k/in.}^2$

☐ **Prestressing Strand Stress Limits**   5.9.3

Table Ex4.13-3
Prestressing strand stress limits

	Tension
Prestress transfer	$0.75 f_{pu} = 202.5 \text{ k/in.}^2$
Service	$0.8 f_{py} = 194.4 \text{ k/in.}^2$

☐ **Concrete Beam Stress Limit Checking at Prestress Transfer for Midspan**

The actual stress and force in strands after losses at transfer:

$$f_{pe} = f_{pj} - \Delta f_{p,ES} = 0.75 \left(270 \text{ k/in.}^2\right) - 16.81 \text{ k/in.}^2 = 185.7 \text{ k/in.}^2$$

$$F_{pe} = f_{pe} A_{ps} = 185.7 \text{ k/in.}^2 \left(4.8 \text{ in.}^2\right) = 891.3 \text{ k}$$

Stress at top of beam at midspan:

$$\sigma_{pi,\, top\; midspan} = -\frac{M_{beam}}{S_{t,\, NC}} - \frac{F_{pe}}{A_{NC}} + \frac{F_{pe}\, e_{CG\, NC}}{S_{t,\, NC}}$$

$$= -\frac{409\; kft\,(12\; in./ft)}{5{,}070\; in.^3} - \frac{891.3\; k}{560\; in.^2} + \frac{891.3\; k\,(13.91\; in.)}{5{,}070\; in.^3}$$

$$= -0.11 \;\; k/in.^2 \;\; (compression) < 2.93\; k/in.^2$$

OK for concrete beam top (compressive) stress at transfer at midspan.

Stress at bottom of beam at midspan:

$$\sigma_{pi,\, bottom\; midspan} = +\frac{M_{beam}}{S_{b,\, NC}} - \frac{F_{pe}}{A_{NC}} - \frac{F_{pe}\, e_{CG\, NC}}{S_{b,\, NC}}$$

$$= +\frac{409\; kft\,(12\; in./ft)}{6189\; in.^3} - \frac{891\,.3\; k}{560\; in.^2} - \frac{891.3\; k\,(13.91\; in.)}{6189\; in.^3}$$

$$= -2.8\; k/in.^2 \;\; (compression) < 2.93\; k/in.^2$$

OK for concrete beam bottom (compressive) stress at transfer for midspan.

☐ **Concrete Beam Stress Limit Check at Prestress Transfer for Support**

Stress at top of beam at support:

$$\sigma_{pi,\, top\; support} = -\frac{F_{pe}}{A_{NC}} + \frac{F_{pe}\, e_{CG\, NC}}{S_{t,\, NC}} = -\frac{891.3\; k}{560\; in.^2} + \frac{891.3\; k\,(13.91\; in.)}{5070\; in.^3}$$

$$= +0.85\; k/in.^2 \;\; (tension) > 0.53\; k/in.^2$$

NG for concrete beam top (tensile) stress, at transfer for support. Debonding is required.

☐ **Strength I Limit State Moment Design for Midspan**   *5.7.3*

3.4.1 $M_u = 1.25DC + 1.5DW + 1.75LL(1 + IM)$

$$= 1.25\,(398.3 + 65.3 + 555.4 + 127.1) + 1.5(153.6) + 1.75(1150.5)$$

$$= 3676\; kft$$

For a rectangular section without compression reinforcement:

5.7.3.2.3 $M_n = A_{ps}\, f_{ps}\left(d_p - \dfrac{a}{2}\right)$ where $a = \beta_1\, c$

With distance between the neutral axis and compressive face $d_p = h - \text{strand}_{CG} = 54.5$ in. - 6.36 in. $= 48.14$ in. and $\beta_1 = 0.85 - 0.05 \left(f'_c - 4 \text{ k/in.}^2\right) = 0.825$:

5.7.3.1.1
$$c = \frac{A_{ps}\, f_{pu}}{0.85 f'_{c,\,slab}\, \beta_1\, b + k A_{ps}\, (f_{pu}/d_p)}$$

$$= \frac{A_{ps}\, f_{pu}}{0.85 f'_{c,\,slab}\, \beta_1\, b + 2\,(1.04 - f_{py}/f_{pu})\, A_{ps}\, (f_{pu}/d_p)}$$

$$= \frac{4.8 \text{ in}^2 (270 \text{ k/in.}^2)}{0.85\,(4.5 \text{ k/in.}^2)\,0.825(90 \text{ in.}) + 2\,(1.04 - 0.9)\,4.8 \text{ in.}^2\,(270 \text{ k/in.}^2/48.14 \text{ in.})}$$

$$= 4.45 \text{ in}$$

Depth of equivalent stress block:

$$a = \beta_1 c = 0.825\,(4.45 \text{ in.})$$

$$= 3.67 \text{ in.} < 8.5 \text{ in.} \quad \text{(confirming rectangular section)}$$

Average stress in prestressing steel:

$$f_{ps} = f_{pu}\left(1 - k\frac{c}{d_p}\right) = 270 \text{ k/in.}^2\left(1 - 0.28\frac{4.45 \text{ in.}}{48.14 \text{ in.}}\right) = 263 \text{ k/in.}^2$$

Moment capacity provided:

5.5.4.2
$$M_{\text{strength 1 midspan}} = \varphi_f A_{ps} f_{ps}\left(d_p - \frac{a}{2}\right)$$

$$= \frac{1\,\left(4.8 \text{ in.}^2\right) 263 \text{ k/in.}^2\,[48.14 \text{ in.} - (3.67 \text{ in.}/2)]}{12 \text{ in./ft}}$$

$$= 4871 \text{ kft} > M_u = 3676 \text{ kft}$$

OK for strength I moment at midspan.

❑ **Minimum Reinforcement Check**   5.7.3.3.2
The modulus of rupture is given as

5.4.2.6; 5.7.3.3.2 $f_r = 0.24\sqrt{f_{c,\,beam}} = 0.24\sqrt{6.5 \text{ k/in.}^2} = 0.61 \text{ k/in.}^2$

Compressive stress in concrete due to effective prestress forces only at extreme fiber of section where tensile stress is caused by externally applied loads.

$$f_{cpe} = \frac{F_{pe}\, e_{CG\,NC}}{S_{b,\,NC}} + \frac{F_{pe}}{A_{NC}}$$

$$= \frac{891.3\,k\,(13.91\,in.)}{6189\,in.^3} + \frac{891.3\,k}{560\,in.^2} = 3.59\,k/in.^2$$

Total unfactored dead-load moment acting on the monolithic or non-composite section:

$$M_{dnc} = M_{beam} + M_{slab} + M_{forms} + M_{barriers} + M_{WS}$$

$$= 12\,in./ft(398.3 + 555.4 + 65.3 + 127.1 + 153.6)\,kft$$

$$= 15{,}596\,kin.$$

Cracking moment:

5.7.3.3.2
$$M_{cr} = 1.0\left[S_b\,(1.6f_r + 1.1f_{cpe}) - M_{dnc}\left(\frac{S_b}{S_{b,\,NC}} - 1 \right) \right]$$

$$= 1.0\left[10{,}983\,in.^3\left((1.6)\,0.61\,k/in.^2 + (1.1)\,3.59\,k/in.^2\right) \right.$$

$$\left. -15{,}596\,kin.\left(\frac{10{,}983\,in.^3}{6189\,in.^3} - 1 \right) \right]$$

$$= 3500\,kft$$

5.7.3.3.2
$$M_{strength\,1\,midspan} = 4871\,kft > \min(M_{cr},\ 1.33\,M_u)$$

$$= \min[3500,\ 1.33(3676)]$$

$$= 3500\,kft$$

OK for minimum reinforcement at midspan.

☐ **Debonding Check**

The number and eccentricity of strands are arranged in Table Ex4.13-4 for the check section other than midspan.

5.11.4.3

Area of prestressing steel:

$$A_{ps,\,support} = Area_{strand}\,N_{strands} = 0.192(19) = 3.65\,in.^2$$

Table Ex4.13-4
Strand placement data for sections of support and debond

Distance of Strand from Bottom of Beam (in.)	Number of Strands per Row at Support
$y_5 = 11$	$n_5 = 3$
$y_4 = 9$	$n_4 = 5$
$y_3 = 7$	$n_3 = 5$
$y_2 = 5$	$n_2 = 5$
$y_1 = 3$	$n_1 = 1$
Strand$_{CG}$ = 7.42 in.	Total strands, $N_{strands}$ = 19

Effective stress and force after elastic losses:

$$f_{pe} = f_{pj} - \Delta f_{p,\,ES} = 202.5 \text{ k/in.}^2 - 16.81 \text{ k/in.}^2 = 185.7 \text{ k/in.}^2$$

$$F_{pe} = f_{pe} \cdot A_{ps} = 185.7 \text{ k/in.}^2 \left(3.65 \text{ in.}^2\right) = 677.8 \text{ k}$$

Eccentricity of strands at support for noncomposite section:

$$e_{CG\,NC} = y_{b,\,NC} - \text{strand}_{CG} = 20.27 \text{ in.} - 7.42 \text{ in.} = 12.85 \text{ in.}$$

Use

$$\sigma = \begin{cases} -\dfrac{M_{beam}}{S} - \dfrac{F_{pe}}{A_{NC}} + \dfrac{F_{pe}\,e_{CG\,NC}}{S} & \text{for top fiber (tension)} \\[3mm] +\dfrac{M_{beam}}{S} - \dfrac{F_{pe}}{A_{NC}} - \dfrac{F_{pe}\,e_{CG\,NC}}{S} & \text{for bottom fiber (compression)} \end{cases}$$

$$A_{NC} = 560 \text{ in.}^2$$

to check stresses at the other check sections at transfer in Table Ex4.13-5.

Table Ex4.13-5
Concrete stress check (T = tension, C = compression)

Location	Fiber	M_{beam} (kft)	S (in.^{3})	σ (k/in.^{2})	Stress Limit (k/in.^{2})	Check
Support	Top	0	5070	0.52(T)	0.53(T)	OK
	Bottom	0	6189	2.62(C)	2.93(C)	OK
Debond	Top	266.9	5070	0.12(C)	2.93(C)	OK
	Bottom	266.9	6189	2.10(C)	2.93(C)	OK

☐ **Find Concrete Shear Strength V_c for Shear Design at Shear Check Section**   *5.8.3.3*

The shear resistance of a concrete member is separated into three components: V_c relying on tensile stresses in the concrete, V_s relying on tensile stresses in the transverse reinforcement, and V_p representing vertical component of the prestressing force.

Nominal shear resistance of concrete section:

$$V_c = 0.0316\beta\sqrt{f_c}\,b_v d_v$$

where effective shear depth d_v is given as

5.8.2.9 $d_v = \max\left(d_e - \dfrac{a}{2},\ 0.9d_e,\ 0.72h_C\right)$

$$= \max\left(54.5 - 7.42 - \frac{3.67}{2}, 0.9(54.5 - 7.42), 0.72\,(54.5)\right)$$

$$= \max(45.2 \text{ in.}, 42.37 \text{ in.}, 39.24 \text{ in.}) = 45.2 \text{ in.}$$

and

$$b_v = 7 \text{ in.}$$

To find β, use an iterative approach as follows. *B5.2*
Find longitudinal strain ε_x for sections with prestressing and transverse reinforcement, assuming transverse reinforcement of two No. 5 bars ($A_s = 0.61$ in.^{2}) and

$$M_u = \max\left(M_{\text{strength 1 shear check}}, V_u d_v\right)$$

with

$$M_{\text{strength 1 shear check}} = 1.25(66.9 + 10.9 + 93.3 + 21.3) + 1.5(25.8) + 1.75(204.8)$$

$$= 637.7 \text{ kft}$$

$$V_u = 1.25(19.3 + 3.2 + 27 + 6.2) + 1.5(7.5) + 1.75(89.5)$$

$$= 237.5 \text{ k}$$

$$d_v = 45.2 \text{ in.}$$

$$M_u = V_u d_v = 237.5 \text{ k}(45.2 \text{ in.}) = 895.6 \text{ kft}$$

$$f_{p0} = 0.70 f_{pu} = 0.7(270 \text{ k/in.}^2) = 189 \text{ k/in.}^2$$

and an initial trial value of $\theta = 22.5°$:

$$\varepsilon_x = \frac{M_u/d_v + 0.5V_u \cot(\theta) - A_{ps}f_{po}}{2(E_sA_s + E_pA_{ps})}$$

$$= \frac{237.5\,k + 0.5\,(237.5\,k)\cot(22.5) - 3.65\,in.^2\left(189\,k/in.^2\right)}{2\,(29,000\,k/in.^2\,(0.61\,in.^2) + 28,500\,k/in.^2\,(3.65\,in.^2))}$$

$$= -0.00068$$

Since the strain is negative, it needs to be recalculated using

B5.2 $A_c = 7\,in.(27.25\,in.) + 7.5\,in.(7.5\,in.) + 2(7\,in.)7.5\,in.$

$$= 352\,in.^2$$

based on Figures Ex4.13-5 and Ex4.13-6.

B5.2

$$\varepsilon_x = \frac{M_u/d_v + 0.5V_u\cot(\theta) - A_{ps}f_{po}}{2(E_cA_c + E_sA_s + E_pA_{ps})}$$

$$= \frac{237.5\,k + 0.5(237.5\,k)\cot(22.5°) - 3.65\,in.^2(189\,k/in.^2)}{2(1820\sqrt{6.5}\,k/in.^2 \times 352\,in.^2 + 29,000\,k/in.^2(0.61\,in.^2) + 28,500\,k/in.^2(3.65\,in.^2))}$$

$$= -0.000047$$

Find

5.8.2.9 $\dfrac{V_u}{f_c'} = \dfrac{V_u - \phi_v V_p}{\phi_v b_v d_v f_c'} = \dfrac{237.5\,k - 0.9\,(0)}{0.9\,(7\,in.)45.2\,in.(6.5\,k/in.^2)}$

$$= 0.13$$

Use

$$\frac{V_u}{f_c'} = 0.13 \quad \text{and} \quad \varepsilon_x \times 1000 = -0.047$$

in Table Ex4.13-6.
Find the angle of inclination of compression stresses,

$$\theta = 22.8°$$

and the factor relating to longitudinal strain on the shear capacity of concrete,

$$\beta = 2.94$$

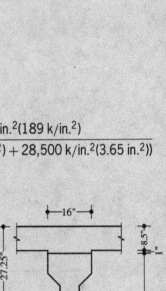

Figure Ex4.13-5
AASHTO Type III prestressed I-beam section.

Figure Ex4.13-6
Area of concrete on the flexural tension side.

Table Ex4.13-6
Values of θ and β for sections with transverse reinforcement   B5.2

$\dfrac{V_u}{f'_c}$	≤ -0.20	≤ -0.10	≤ -0.05	≤ 0	≤ 0.125	≤ 0.25	≤ 0.50	≤ 0.75	≤ 1.00
				$\varepsilon_x \times 1000$					
≤ 0.075	22.3	20.4	21.0	21.8	24.3	26.6	30.5	33.7	36.4
	6.32	4.75	4.10	3.75	3.24	2.94	2.59	2.38	2.23
≤ 0.100	18.1	20.4	21.4	22.5	24.9	27.1	30.8	34.0	36.7
	3.79	3.38	3.24	3.14	2.91	2.75	2.50	2.32	2.18
≤ 0.125	19.9	21.9	22.8	23.7	25.9	27.9	31.4	34.4	37.0
	3.18	2.99	2.94	2.87	2.74	2.62	2.42	2.26	2.13
≤ 0.150	21.6	23.3	24.2	25.0	26.9	28.8	32.1	34.9	37.3
	2.88	2.79	2.78	2.72	2.60	2.52	2.36	2.21	2.08
≤ 0.175	23.2	24.7	25.5	26.2	28.0	29.7	32.7	35.2	36.8
	2.73	2.66	2.65	2.60	2.52	2.44	2.28	2.14	1.96
≤ 0.200	24.7	26.1	26.7	27.4	29.0	30.6	32.8	34.5	36.1
	2.63	2.59	2.52	2.51	2.43	2.37	2.14	1.94	1.79
≤ 0.225	26.1	27.3	27.9	28.5	30.0	30.8	32.3	34.0	35.7
	2.53	2.45	2.42	2.40	2.34	2.14	1.86	1.73	1.64
≤ 0.250	27.5	28.6	29.1	29.7	30.6	31.3	32.8	34.3	35.8
	2.39	2.39	2.33	2.33	2.12	1.93	1.70	1.58	1.50

Source: AASHTO LRFD Bridge Design Specifications, 2012. Used by permission.

Use the updated θ to reiterate for ε_x:

$$\varepsilon_x = \frac{M_u/d_v + 0.5V_u \cot(\theta) - A_{ps}f_{po}}{2\left(E_c A_c + E_s A_s + E_p A_{ps}\right)}$$

$$= \frac{237.5\ \text{k} + 0.5\,(237.5\ \text{k})\ \cot(22.8°) - 3.65\ \text{in.}^2\left(189\ \text{k/in.}^2\right)}{2\left(1820\sqrt{6.5}\ \text{k/in.}^2 \times 352\ \text{in.}^2 + 29{,}000\ \text{k/in.}^2\,(0.61\ \text{in.}^2) + 28{,}500\ \text{k/in.}^2\,(3.65\ \text{in.}^2)\right)}$$

$$= -0.000048$$

Accept $\theta = 22.8°$ and $\beta = 2.94$ using Table Ex4.13-6. *B5.2*
Nominal shear resistance of concrete section:

$$V_c = 0.0316\ \beta\sqrt{f_{c,\,beam}}\,b_v d_v$$

$$= 0.0316\,(2.94)\ 2.55\ \text{k/in.}^2\,(7\ \text{in.})\ 45.2\ \text{in.}$$

$$= 75.0\ \text{k}$$

❑ Transfer Length Check

The prestressing strand force becomes effective with the transfer length:

5.11.4.1 $\qquad L_{transfer} = 60 \text{Diameter}_{strand} = 60\,(0.5625) = 2.8 \text{ ft}$

Since the transfer length $L_{transfer} = 2.8$ ft is shorter than the shear check location's distance to the CL bearing, $x_{shear\ check} = 3.27$ ft, the full force of the strands is developed.

❑ Design Transverse Reinforcement (Stirrups) Spacing s for Shear Strength

Find strength requirement for transverse steel:

$$V_n = \min\left(\frac{V_u}{\phi_v},\ 0.25 f_{c,\,beam} b_v d_v + V_p\right)$$

$$= \min\left(\frac{237.5 \text{ k}}{0.9},\ 0.25\left(6.5 \text{ k/in.}^2\right) 7 \text{ in.}(45.2 \text{ in.}) + 0\right)$$

$$= \min\,(264.0 \text{ k},\ 514.8 \text{ k}) = 264.0 \text{ k}$$

$$V_s = V_n - V_c - V_p = 264.0 \text{ k} - 75.0 \text{ k} - 0 \text{ k} = 189.0 \text{ k}$$

Spacing of transverse reinforcement of two No. 5 bars in a stirrup needs to meet the following three conditions:

5.8.2.5 $\qquad s_{minimum\ transverse\ reinforcement} = \dfrac{A_v f_y}{0.0316 b_v \sqrt{f'_{c,\,beam}}}$

$$= \frac{0.61 \text{ in.}^2\left(60 \text{ k/in.}^2\right)}{0.0316\,(7 \text{ in.})\left(\sqrt{6.5 \text{ k/in.}^2}\right)}$$

$$= 64.9 \text{ in.}$$

5.8.3.3 $\qquad s_{strength\ required} = \dfrac{A_v f_y d_v \cot(\theta)}{V_s}$

$$= \frac{\left(0.61 \text{ in.}^2\right) 60 \text{ k/in.}^2\,(45.2 \text{ in.})\,\cot 22.8}{189.0 \text{ k}}$$

$$= 20.9 \text{ in.}$$

$$s_{maximimum\ spacing} = \min\,(0.4 d_v,\ 12 \text{ in.})$$

$$= \min\,(18.1 \text{ in.},\ 12 \text{ in.}) = 12 \text{ in.}$$

5.8.2.7 for $v_u = \dfrac{V_u - \phi_v V_p}{\phi_v b_v d_v} = \dfrac{237.5\,k}{0.9(7\,in.)45.2\,in.} = 0.83\,k/in.^2$

$> 0.125 f'_{c,\,beam} = 0.125\left(6.5\,k/in.^2\right) = 0.81\,k/in.^2$

Thus, $s = \min(s_{minimum\,transverse\,reinforcement}, \; s_{strength\,required},$
$s_{maximum\,spacing}) = 12$ in.

☒ for transverse reinforcement spacing.

☒ for minimum transverse reinforcement.

☐ Design Longitudinal Reinforcement

For sections not subjected to torsion, the longitudinal reinforcement shall be proportioned so that at each section the tensile capacity of the reinforcement on the flexural tension side of the member, takes into account any lack of full development of that reinforcement.

Force in longitudinal reinforcement:

$$T = \frac{M_u}{d_v \varphi_f} + \left(\frac{V_u}{\varphi_v} - 0.5 V_s - V_p\right) \cot(\theta)$$

where

5.8.3.5

$$V_s = \min\left[\frac{\left(0.61\,in.^2\right)60\,k/in.^2\,(45.2\,in.)\,2.38}{12\,in.}, \frac{237.5\,k}{0.9}\right]$$

$= \min(328.3\,k,\,264.0\,k) = 264.0\,k$

$T = \dfrac{637.7\,kft}{(45.2\,in./12\,in./ft)\,0.9} + \left[\dfrac{237.5\,k}{0.9} - 0.5\,(264\,k) - 0\right]\cot(22.8°)$

$= 501.9\,k$

$< F_{ps} = f_{ps} A_{ps} = 185.7\,k/in.^2(3.65\,in.^2) = 677.8\,k$

☒ for longitudinal reinforcement.

☐ Summary

The arrangement of the prestress strands is shown in Figure Ex4.13-7.

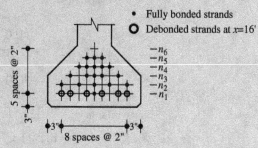

Figure Ex4.13-7
Arrangement of prestress strands in cross sections.

One of the advantages of using prestressed concrete beams is that tensile stresses in concrete can be eliminated, particularly in the service stage. For constructing prestressed concrete beams, the AASHTO specifications prescribe stress limits to control tensile and compressive stresses in both the concrete and prestressing tendons, as well as the strength limit state requirements to be discussed below.

Table 4.8-1 gives the AASHTO stress limits for concrete at the prestress transfer and service stages, and Table 4.8-2 gives those for the prestressing tendons.

It is advised that these stress limits be checked when laying out the prestressing tendons as to where to drape or debond them if needed. For example, for simply supported beams expected to be subjected to positive moments in service, prestressing tendons are placed near the beam bottom to maximize the counter moment induced by the prestress force. At another cross section, however, where the service-load-induced moment becomes smaller or even negative, such as a section at a support for a continuous span or a section at the end support of a simply supported span, this prestress-induced negative moment can become excessive and possibly cause concrete cracking. This cracking may occur as early as prestress transfer before the service load is applied because the positive moment is small then.

Accordingly, reducing the negative moment is desired at those sections where a positive service load moment is small. This may be accomplished by changing the vertical location of the tendons at these sections or reducing the prestress force by debonding some of the tendons at these sections. To understand where this may be needed, stress limit checking at

4.8.4 Stress Limit Design

Table 4.8-1
Beam concrete stress limits   *5.9.4*

	Compression	**Tension**
Prestress transfer	$0.6f'_{ci}$	$0.24\sqrt{f'_{ci}}$
Service	Use Service I Limit State: $0.45f'_c$ (without transient load), $0.6f'_c$ (with transient load)	Use Service III Limit State: $0.19\sqrt{f'_c}$

Table 4.8-2
Prestressing steel stress limits   *5.9.3*

	Tension
Prestress transfer	$0.75f_{pu}$
Service	$0.8f_{py}$

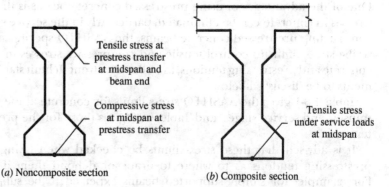

Figure 4.8-13
Locations of possible critical stresses.

these sections will need to be performed. Therefore, it is advantageous to perform this stress limit checking before the strength limit state design and design the prestressing tendons' arrangement. Then the limit state design can be performed.

Several formulas for stress limit checking are given below to illustrate application. They are particularly relevant to simply supported beams. For other span types and support arrangements, the concept of stress limit checking can be applied similarly using the concept of elastic stress analysis based on the theory of the strength of material.

❏ Tensile stress at prestress transfer at top fiber of midspan (see Figure 4.8-13 for location)

$$f_{i, \text{top midspan}} = -\frac{F_{pi}}{A_{NC}} + \frac{F_{pi}e_{NC}y_{top}}{I_{NC}} - \frac{M_{beam}y_{top}}{I_{NC}} \leq 0.24\sqrt{f'_{ci}} \quad (4.8\text{-}18)$$

where i = initial stage or stage of prestress transfer prior to prestress loss

F_{pi} = prestress force at prestress transfer at centroid of tendons

A_{NC} = cross-sectional area of noncomposite beam

e_{NC} = eccentricity of prestress tendon centroid to neutral axis of noncomposite beam

I_{NC} = moment of inertia of noncomposite beam

y_{top} = distance of top fiber to neutral axis

M_{beam} = moment at midspan due to beam self-weight

f'_{ci} = concrete compressive strength at prestress transfer

❏ Compressive stress at prestress transfer at bottom fiber of midspan (see Figure 4.8-13 for location):

$$f_{i, \text{bottom midspan}} = \left| -\frac{F_{pi}}{A_{NC}} - \frac{F_{pi}e_{NC}y_{bottom}}{I_{NC}} + \frac{M_{beam}y_{bottom}}{I_{NC}} \right| \leq 0.6f'_{ci}$$

$$(4.8\text{-}19)$$

Most of the symbols have been defined in Eq. 4.8-18, except for y_{bottom} the distance of the bottom fiber to the neutral axis and $|\cdot|$, which stands for absolute value.

Note that in rare situations the quantity within the absolute value sign in the above formula becomes positive, indicating tensile stress; then the corresponding tensile stress limit $0.24\sqrt{f'_{ci}}$ should replace the compressive stress limit $0.6f'_{ci}$. Similarly, if stress $f_{i,\text{ top midspan}}$ in Eq. 4.8-18 turns out to be negative, its absolute value should be used to be compared with the compressive stress limit $0.6f'_{ci}$ replacing the tensile stress limit $0.24\sqrt{f'_{ci}}$.

❑ Tensile stress at presrestress transfer at top of beam end (see Figure 4.8-13):

$$f_{i,\text{ top beam end}} = -\frac{F_{pi}}{A_{\text{NC}}} + \frac{F_{pi}\,e_{\text{NC}}y_{\text{top}}}{I_{\text{NC}}} \leq 0.24\sqrt{f'_{ci}} \qquad (4.8\text{-}20)$$

In the area near the simply supported beam end, the moment due to the beam self-weight becomes zero or very insignificant depending on location of the cross section. Thus the stress that results from the self-weight has been deleted in the stress calculation formula above. As a result, the tensile stress in Eq. 4.8-20 likely can become excessive since there is no or almost no stress due to self-weight moment to cancel some of the tensile stress due to prestressing. This often is the cause of cracking in fabrication or construction. It is also the reason draping or debonding the prestressing tendons is required in the design to reduce the prestressing moment and thus avoid cracking in fabrication or construction.

❑ Compressive stress at prestress transfer at bottom of beam end (see Figure 4.8-13):

$$f_{i,\text{ bottom beam end}} = \left|-\frac{F_{pi}}{A_{\text{NC}}} - \frac{F_{pi}\,e_{\text{NC}}y_{\text{bottom}}}{I_{\text{NC}}}\right| \leq 0.6f'_{ci} \qquad (4.8\text{-}21)$$

This formula is similar to Eq. 4.8-20 in that the moment due to the beam self-weight has been eliminated since it is either zero or negligible depending on the location of the cross section. It is also similar to Eq. 4.8-19 in that negative or compressive stress is expected – hence the absolute value sign.

❑ Tensile stress in service at bottom fiber of midspan (see Figure 4.8-13 for location):

$$f_{i,\text{ bottom midspan}} = -\frac{F_{pe}}{A_{\text{NC}}} - \frac{F_{pe}\,e_{\text{NC}}y_{\text{bottom}}}{I_{\text{NC}}} + \frac{M_{\text{beam}}y_{\text{bottom}}}{I_{\text{NC}}}$$

$$+ \frac{M_{\text{DC+DW}}y_{\text{bottom long term}}}{I_{C,\text{ long term}}} + \frac{M_{\text{LL}}y_{\text{bottom short term}}}{I_{C,\text{short term}}} \leq 0.19f'_{c}$$

$$(4.8\text{-}22)$$

where F_{pe} = prestress force in service (after prestress loss)

M_{DC+DW} = midspan moment due to superimposed wearing surface DW and dead load DC such as railings

$I_{c,\,\text{long term}}$ = moment of inertia of composite section including long-term effect such as concrete creep

$y_{\text{bottom long term}}$ = distance of bottom fiber to neutral axis of composite section, including long-term effect

M_{LL} = midspan moment due to live- (truck) load effect, including impact IM

$I_{c,\,\text{short term}}$ = moment of inertia of composite section without long-term effect

$y_{\text{bottom short term}}$ = distance of bottom fiber to neutral axis of composite section without long-term effect

and other symbols are as defined earlier.

These five stresses are superimposed according to Figure 4.8-7. Since $I_{c,\,\text{long term}}$ is not equal to $I_{c,\,\text{short term}}$, their corresponding neutral axes are not at the same location in the composite cross section, and neither are the relative distances of the bottom fibers to the respective axes. Example 4.14 illustrates the application of a service limit check applied to a two-span prestressed concrete I-beam bridge.

Example 4.14 Prestressed Concrete I-Beam Bridge Design (Service Limit States)

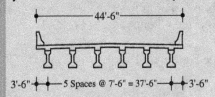

Figure Ex4.14-1
Cross section of bridge.

☐ **Design Requirement**
Check the design of the prestressed concrete interior Type III beams in Example 4.12 for the service limit states. The bridge cross section is shown in Figure Ex4.14-1 and the plan view in Figure Ex4.14-2.
Load modifier $\eta = 1.0$

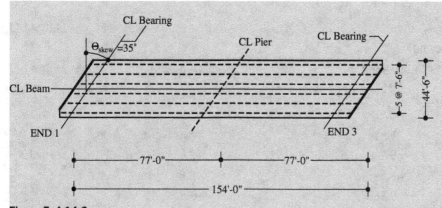

Figure Ex4.14-2
Plan view of bridge with skew.

Design Parameters

Same as Examples 4.12 and 4.13. The general information on the span is shown in Figure Ex4.14-3.

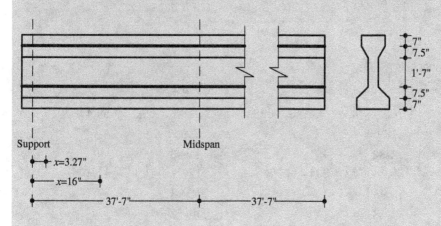

Figure Ex4.14-3
Design check sections and strand debonding.

Beam Top Stress Check for Concrete

The total stresses in the Tables are obtained using those stresses shaded in the Tables at the same cross section location.

Table Ex4.14-1
Beam top stress checking

Cross Section	Debond	Midspan
M_{NC} (kft)	$266.9 + 43.8 + 372.3 = 683$	$398.3 + 65.3 + 555.4 = 1019$
S_{NC} (in.^{3})	5070	5070
σ_{NC} (k/in.^{2})	−1.61	−2.41
$M_{composite}$ (kft)	$85.2 + 103 + 804.4 = 992.6$	$127.1 + 153.6 + 1{,}150.5 = 1431.2$
$S_{composite}$ (in.^{3})	45,890	45,890
$\sigma_{composite}$ (k/in.^{2})	−0.26	−0.37
σ_{ps} (k/in.^{2})	+0.52	+0.85
Total σ	−1.35(C)	−1.93(C)
Stress limit (k/in.^{2})	3.90	3.90
Check	1.35 < 3.9 OK	1.93 < 3.9 OK

❏ **Beam Botttom Stress Check for Concrete**

Table Ex4.14-2
Beam bottom stress checking

Cross Section	Debond	Midspan
M_{NC} (kft)	$266.9 + 43.8 + 372.3 = 683$	$398.3 + 65.3 + 555.4 = 1019$
S_{NC} (in.^{3})	6189	6189
σ_{NC} (k/in.^{2})	+1.32	+1.98
$M_{composite}$ (kft)	$85.2 + 103 + 804.4 = 992.6$	$127.1 + 153.6 + 1150.5 = 1431.2$
$S_{composite}$ (in.^{3})	10,983	10,983
$\sigma_{composite}$ (k/in.^{2})	+1.08	+1.56
σ_{ps} (k/in.^{2})	−2.62	−3.59
Total σ	−0.22(C)	−0.05(C)
Stress limit (k/in.^{2})	3.90	3.90
Check	0.46 < 3.9 OK	0.05 < 3.9 OK

☐ **Deck Top Stress Check for Concrete**

Table Ex4.14-3
Deck top stress checking

Cross Section	Debond	Midspan
M_{NC} (kft)		
S_{NC} (in.^{3})		
σ_{NC} (k/in.^{2})		
$M_{composite}$ (kft)	$85.2 + 103 + 804.4 = 992.6$	$127.1 + 153.6 + 1150.5 = 1431.2$
$S_{composite}$ (in.^{3})	21,923	21,923
$\sigma_{composite}$ (k/in.^{2})	−0.54	−0.78
σ_{ps} (k/in.^{2})	—	—
Total σ	$-0.54(0.832) = -0.45(C)$	$-0.78(0.832) = -0.65(C)$
Stress limit (k/in.^{2})	2.70	2.70
Check	0.45 < 2.70 OK	0.65 < 2.7 OK

Example 4.13 illustrates the design of prestressed concrete I-beam bridge spans of simple support to meet this limit state requirement.

4.8.5 Strength I Limit State Design for Moment and Shear

MOMENT

Strength Limit State
For the flange section shown in Figure 4.8-14, the strength I limit state design for flexure is given in the AASHTO specifications to satisfy

5.7.3.2.1
$$M_u \le \varphi M_n \tag{4.8-23}$$

where M_u is the combined factored load effects according to the strength limit state in Section 3.4.1 and

5.7.3.2.2, 5.5.4.2
$$\varphi = 1.0 \tag{4.8-24}$$

5.7.3.2.2
$$M_n = A_{ps}f_{ps}\left(d_p - \frac{a}{2}\right) + A_s f_s \left(d_s - \frac{a}{2}\right) - A'_s f'_s \left(d'_s - \frac{a}{2}\right)$$
$$+ 0.85 f'_c (b - b_w) h_f \left(\frac{a}{2} - \frac{h_f}{2}\right) \tag{4.8-25}$$

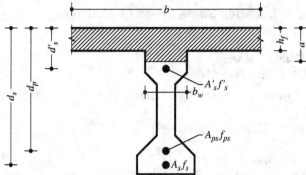

Figure 4.8-14
Parameters for strength of prestressed I beam with
flange section.

where (see Figure 4.8-14) A_{ps} = area of prestressing steel

f_{ps} = average stress in prestressing steel

d_p = distance from extreme compression fiber to centroid of prestressing tendons

a = depth of equivalent stress block $= c\,\beta_1$

c = distance measured from neutral axis

β_1 = stress block factor $0.85 - 0.05\,(f'_c - 4)$ (f'_c in k/in.^{2}) subject to $0.65 < \beta_1 < 0.85$ *5.7.2.2*

A_s = area of non-prestressed tension reinforcement

f_s = stress in mild steel tension reinforcement at nominal flexural resistance

d_s = distance from extreme compression fiber to centroid of nonprestressed tensile reinforcement

A'_s = area of compression reinforcement

f'_s = stress in non-prestressed compression reinforcement at nominal flexural resistance

d'_s = instance from extreme compression fiber to centroid of compression reinforcement

f'_c = 28-day concrete compressive strength

b = width of compression face of member; for flange section in compression, effective width of flange *4.6.2.6*

b_w = web width or diameter of circular section

h_f = compression flange depth

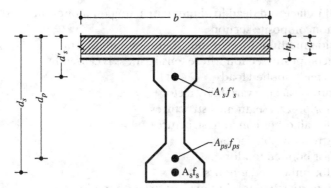

Figure 4.8-15
Parameters for strength of prestressed I beam with rectangular section.

The nominal capacity in Eq. 4.8-25 is derived based on the force equilibrium on the cross section at ultimate failure.

For a rectangular section, with the neutral axis in the concrete deck, the nominal strength is reduced to (see Figure 4.8-15)

5.7.3.2.2
$$M_n = A_{ps}f_{ps}\left(d_p - \frac{a}{2}\right) + A_s f_s \left(d_s - \frac{a}{2}\right) - A_s' f_s' \left(d_s' - \frac{a}{2}\right)$$
(4.8-26)

Reinforcement Limits
The current AASHTO specifications have the maximum reinforcement limit deleted for the flexural strength consideration.

The minimum reinforcement limit of the specifications requires that, at any section of a flexural component, the amount of prestressed and non-prestressed tensile reinforcement shall be adequate to exceed or equal the lesser of $1.33M_u$ and a factored flexural cracking resistance M_r:

5.7.3.3.2
$$\min\left(1.33M_u, M_{cr}\right) \le \varphi M_n$$
(4.8-27)

where

5.7.3.3.2, 5.5.4.2
$$\varphi = 1.0$$
(4.8-28)

5.7.3.3.2
$$M_{cr} = \gamma_3 \left[S_c \left(\gamma_1 f_r + \gamma_2 f_{cpe}\right) - M_{ndc}\left(\frac{S_c}{S_{ndc}} - 1\right)\right]$$
(4.8-29)

where S_c = section modulus for extreme fiber of composite section where tensile stress is caused by externally applied loads
f_r = concrete rupture modulus = $0.24\sqrt{f_c'}$ *5.4.2.6, 5.7.3.3.2*
f_{cpe} = compressive stress in concrete due to effective prestress forces only (after allowance for all prestress losses) at extreme fiber of section where tensile stress is caused by externally applied loads

M_{dnc} = total unfactored dead-load moment acting on monolithic or noncomposite section

S_{nc} = section modulus for extreme fiber of monolithic or noncomposite section where tensile stress is caused by externally applied loads

γ_1 = flexural cracking variability factor

= 1.2 for precast segmental structures

= 1.6 for all other concrete structures

γ_2 = prestress variable factor

= 1.1 for bonded tendors

= 1.0 for unbonded tendors

γ_3 = ratio of specified minimum yield strength to ultimate tensile strength of the reinforcement

= 0.67 for A615, Grade 60 reinforcement

= 0.75 for A706 Grade 60 reinforcement

= 1.00 for prestressed concrete structures

SHEAR

The nominal shear resistance V_n is required to satisfy

5.5.4.2 $$V_u \leq \varphi_v V_n \qquad \varphi_v = 0.9$$ (4.8-30)

where V_u is the factored and combined load effects according to the Strength I limit state for the superstructure component here:

3.4.1 $$V_u = 1.25 \, DC + 1.5 \, DW + 1.75 \, LL \, (1 + IM)$$

The nominal shear resistance is given in the AASHTO specifications as

5.7.3.3.2 $$V_n = \min \left(V_c + V_s + V_p, 0.25 f_c' b_v d_v + V_p \right)$$ (4.8-31)

The first of the two terms from which the lesser is to be taken as the nominal resistance is shown as a sum of three resistances: V_c relying on tensile stresses in the concrete, V_s relying on tensile stresses in the transverse reinforcement (stirrup), and V_p being the vertical component of the prestressing force. The second of the two terms in Eq. 4.8-31 represents the maximum shear resistance of the cross section and ensure that the concrete in the web will not crush prior to yield of the transverse reinforcement.

When shear is being designed for, the cross section and the flexure-controlled prestressing steel usually have been determined. Therefore, V_c, $0.25 f_c' \, b_v \, d_v$, and V_p in Eq. 4.8-31 can be determined accordingly. Thus, the design for shear can proceed as follows:

If $V_u \leq \varphi_v \min \left(V_c + V_p, 0.25 f_c' b_v d_v + V_p \right)$ is true without transverse steel V_s, then no transverse steel V_s is needed. Otherwise, if $V_c + V_p > 0.25 f_c' b_v d_v + V_p$ is true, then V_s is still not needed since $0.25 f_c' b_v d_v + V_p$

controls the resistance. However, if $V_c + V_p < 0.25 f_c' b_v d_v + V_p$, then V_s is needed, which can be designed as follows.

Find the required $V_{s,\,required}$ as

$$V_{s,\,required} = \frac{V_u}{\varphi_v} - V_c - V_p \qquad (4.8\text{-}32)$$

Then use $V_{s,\,required}$ to design the transverse steel:

5.8.3.3
$$V_s = \frac{A_v f_y d_v \left(\cot \theta + \cot \alpha \right) \sin \alpha}{s} \qquad (4.8\text{-}33)$$

where A_v = area of transverse non-prestressed steel
f_y = yield strength of non-prestressed steel
d_v = effective shear depth of cross section
θ = angle of inclination of diagonal compressive stresses
α = angle of inclination of transverse reinforcement to longitudinal axis
s = spacing of transverse steel

More details of this approach are provided below regarding determination of V_c as a contribution from the concrete.

Determination of Concrete Shear Resistance V_c: Option 1
The AASHTO specifications offer two optional approaches to determining the nominal resistance V_c. One of them is presented here:

5.8.3.3
$$V_c = 0.0316 \beta \sqrt{f_c'}\, b_v d_v \qquad (4.8\text{-}34)$$

where f_c' = 28-day concrete compressive strength (k/in.^{2})
b_v = effective web width taken as minimum web width within depth d_v
d_v = effective shear depth between resultants of tensile and compressive forces due to flexure
β = factor indicating ability of diagonally cracked concrete to transmit tension and shear, whose estimation procedure is explained below 5.8.3.4

For sections containing at least the minimum amount of transverse reinforcement, the values of β and θ shall be as specified in Table 4.8-3 (*B5.2*). In using this table, ε_x shall be taken as the calculated longitudinal strain at the middepth of the member when the section is subjected to M_u, N_u, and V_u:

B5.2
$$\varepsilon_x = \frac{|M_u|/d_v + 0.5 N_u + 0.5 \left(V_u - V_p \right) \cot \theta - A_{ps} f_{p0}}{2 \left(E_s A_s + E_p A_{ps} \right)} \qquad (4.8\text{-}35)$$

An iterative procedure is needed here to arrive at a set of values of ε_x, θ, and β satisfying this equation and Table 4.8-3 (B5.2). According to the AASHTO specifications, this process starts with an assumed $\varepsilon_x < 0.001$ as an input to Table 4.8-3, uses the resulting θ and β values in Table 4.8-3 to update ε_x, and continues until convergence is reached.

For sections containing less transverse reinforcement than specified in Article 5.8.2.5, the values of β and θ shall be as specified in Table 4.8-4 (B5.2). Using this table, ε_x shall be taken as the largest calculated longitudinal strain which occurs within the web of the member when the section is subjected to N_u, M_u, and V_u:

B5.2
$$\varepsilon_x = \frac{|M_u|/d_v + 0.5N_u + 0.5\left(V_u - V_p\right)\cot\theta - A_{ps}f_{p0}}{E_s A_s + E_p A_{ps}}$$

(4.8-36)

The same iteration procedure used for Eq. 4.8-35 above is applicable here, except that the AASHTO specifications give a requirement for $\varepsilon_x < 0.002$ as the initial value.

However, when Eq. 4.8-35 or 4.8-36 gives a negative strain ε_x, the following equation should be used to recompute the strain:

Table 4.8-3
Values of θ and β for sections with transverse reinforcement

$\frac{V_u}{f_c'}$	$\varepsilon_x \times 1000$								
	≤-0.20	≤-0.10	≤-0.05	≤ 0	≤ 0.125	≤ 0.25	≤ 0.50	≤ 0.75	≤ 1.00
≤ 0.075	22.3	20.4	21.0	21.8	24.3	26.6	30.5	33.7	36.4
	6.32	4.75	4.10	3.75	3.24	2.94	2.59	2.38	2.23
≤ 0.100	18.1	20.4	21.4	22.5	24.9	27.1	30.8	34.0	36.7
	3.79	3.38	3.24	3.14	2.91	2.75	2.50	2.32	2.18
≤ 0.125	19.9	21.9	22.8	23.7	25.9	27.9	31.4	34.4	37.0
	3.18	2.99	2.94	2.87	2.74	2.62	2.42	2.26	2.13
≤ 0.150	21.6	23.3	24.2	25.0	26.9	28.8	32.1	34.9	37.3
	2.88	2.79	2.78	2.72	2.60	2.52	2.36	2.21	2.08
≤ 0.175	23.2	24.7	25.5	26.2	28.0	29.7	32.7	35.2	36.8
	2.73	2.66	2.65	2.60	2.52	2.44	2.28	2.14	1.96
≤ 0.200	24.7	26.1	26.7	27.4	29.0	30.6	32.8	34.5	36.1
	2.63	2.59	2.52	2.51	2.43	2.37	2.14	1.94	1.79
≤ 0.225	26.1	27.3	27.9	28.5	30.0	30.8	32.3	34.0	35.7
	2.53	2.45	2.42	2.40	2.34	2.14	1.86	1.73	1.64
≤ 0.250	27.5	28.6	29.1	29.7	30.6	31.3	32.8	34.3	35.8
	2.39	2.39	2.33	2.33	2.12	1.93	1.70	1.58	1.50

Source: AASHTO LRFD Bridge Design Specifications, 2012. Used by permission.

Table 4.8-4
Values of θ and β for sections with less than minimum transverse reinforcement

S_{xe} (in.)	$\varepsilon_x \times 1000$										
	≤-0.20	≤-0.10	≤-0.05	≤0	≤0.125	≤0.25	≤0.50	≤0.75	≤1.00	≤1.50	≤2.00
≤5	25.4	25.5	25.9	36.4	27.7	28.9	30.9	32.4	33.7	35.6	37.2
	6.36	6.06	5.56	5.15	4.41	3.91	3.26	2.86	2.58	2.21	1.96
≤10	27.6	27.6	28.3	39.3	31.6	33.5	36.3	38.4	40.1	42.7	44.7
	5.78	5.78	5.38	4.89	4.05	3.52	2.88	2.50	2.23	1.88	1.65
≤15	29.5	29.5	29.7	31.1	34.1	36.5	39.9	42.4	44.4	47.4	49.7
	5.34	5.34	5.27	4.73	3.82	3.28	2.64	2.26	2.01	1.68	1.46
≤20	31.2	31.2	31.2	32.3	36.0	38.8	42.7	45.5	47.6	50.9	53.4
	4.99	4.99	4.99	4.61	3.65	3.09	2.46	2.09	1.85	1.52	1.31
≤30	34.1	34.1	34.1	34.2	38.9	42.3	46.9	50.1	52.6	65.3	59.0
	4.46	4.46	4.46	4.43	3.39	2.82	2.19	1.84	1.60	1.30	1.10
≤40	36.6	36.6	36.6	36.6	41.2	45.0	50.2	53.7	56.3	60.2	63.0
	4.06	4.06	4.06	4.06	3.20	2.62	2.00	1.66	1.43	1.14	0.95
≤60	40.8	40.8	40.8	40.8	44.5	49.2	55.1	58.9	61.8	65.8	68.6
	3.50	3.50	3.50	3.50	2.92	2.32	1.72	1.40	1.18	0.92	0.75
≤80	44.3	44.3	44.3	44.3	47.1	52.3	58.7	62.8	65.7	69.7	72.4
	3.10	3.10	3.10	3.10	2.71	2.11	1.52	1.21	1.01	0.76	0.62

Source: AASHTO LRFD Bridge Design Specifications, 2012. Used by permission.

B5.2
$$\varepsilon_x = \frac{|M_u|/d_v + 0.5N_u + 0.5\left(V_u - V_p\right)\cot\theta - A_{ps}f_{p0}}{2\left(E_c A_c + E_s A_s + E_p A_{ps}\right)} \qquad (4.8\text{-}37)$$

where A_c = area of concrete on flexural tension side of member B5.2

A_{ps} = area of prestressing steel on flexural tension side of member B5.2

A_s = area of non prestressed steel on flexural tension side of member at section under consideration B5.2 . In calculating A_s for use in this equation, bars terminated at a distance less than their development length from the section under consideration should be ignored.

f_{p0} = parameter taken as modulus of elasticity of prestressing tendons multiplied by locked-in difference in strain between prestressing tendons and surrounding concrete. For the usual levels of prestressing, a value of $0.7f_{pu}$ will be appropriate for both pretensioned and posttensioned members.

M_u = factored moment, not to be taken less than $V_u d_v$

N_u = factored axial force, taken as positive if tensile and negative if compressive

V_u = factored shear force

4.8.6 Service Limit State Check and Constructability Check

Deflection

As discussed in Section 4.8.4 for steel beam design, live-load deflection control is optional in the AASHTO specifications. If done, the criteria in Table 4.7-1 is recommended to be used. The dead-load deflection is also treated in the same way by camber.

Stress Control

As discussed earlier in Section 4.8.4, stress control is particularly critical for pretressed concrete members to prevent concrete cracking. Therefore, it is exercised early in design, particularly for stresses that can become critical in fabrication and construction of the concrete members. Tables 4.8-1 and 4.8-2 give the general requirements for stress control at prestress transfer in fabrication and service after construction. There can be other stages in construction when stresses in the prestressed concrete beam become close to the thresholds. Checking for these stages can also be viewed as part of the constructability check to complete the design.

Cracking Control

For crack control, Eq. 4.6-6 *(5.7.3.4)* needs to be checked to ensure the spacing of the steel.

References

American Association of State and Highway Transportation Officials (AASHTO) (2012), *LRFD Bridge Design Specifications*, 6th ed., AASHTO, Washington, DC.

American Association of State and Highway Transportation Officials (AASHTO) (2011), *Manual for Bridge Evaluation*, 2nd ed., AASHTO, Washington, DC.

American Institute of Steel Construction (AISC) (2011), *Steel Construction Manual* 14th ed.

Slutter, R. G., and Fisher, J. W. (1966a, Jan.), "Fatigue Strength of Shear Connectors," Report No. 316.2, Fritz Engineering Laboratory.

Slutter, R. G., and Fisher, J. W. (1966b, Mar.), "A Proposed Procedure for the Design of Shear Connectors in Composite Beams," Report No. 316.4, Fritz Engineering Laboratory.

Problems

4.1   Consider a reinforced concrete deck supported by five prestressed concrete I beams spaced at 8 ft 6 in. The deck is 9.5 in. thick with 1.5 in. of wearing surface for a northern state. The top-flange width of the beams is 18 in. Design the deck's interior bay reinforcement for positive and negative moments according to the AASHTO specifications. Make reasonable assumptions if needed.

4.2 Design the interior bays of the same reinforced concrete bridge deck in Problem 4.1 using the empirical design method in the AASHTO specifications. Use reasonable assumptions if needed. Compare your result with that of Problem 4.1. Comment on your comparison.

4.3 Consider the same reinforced concrete deck in Problem 4.1. Search the Internet or use other information sources to identify an acceptable/crash-tested reinforced concrete barrier. Design the deck overhang's transverse steel using the selected concrete barrier. Use reasonable assumptions when needed. Indicate the source of information for the barrier you have selected.

4.4 Design the interior steel beam of the superstructure in Problem 3.7 for the Strength I limit state. Prepare the cross-sectional drawing to scale.

4.5 Design the exterior steel beam of the superstructure in Problem 3.7 for the Strength I limit state, including preparation of the cross-sectional drawing to scale.

4.6 Change the span arrangement of Problem 3.3 to two identical simply supported spans. Design the interior steel beam of the superstructure for the Strength I limit state, including preparation of the cross-sectional drawing to scale.

4.7 Change the span arrangement of Problem 3.3 to two identical simply supported spans. Design the exterior steel beam of the superstructure for the Strength I limit state, including preparation of the cross-sectional drawing to scale.

4.8 Design the interior steel beam of the superstructure in Problem 3.1 for the Strength II limit state using a permit truck load that you are required to find via an Internet search. Also prepare the cross-sectional drawing to scale.

4.9 Design the exterior steel beam of the superstructure in Problem 3.1 for the strength II limit state using a permit truck load that you are required to find via an Internet search, including preparation of the cross-sectional drawing to scale.

5 Bearing Design

5.1 Introduction

5.1 Introduction

Bearings represent an important and distinct component in highway bridges. They are often treated as part of the substructure. However, it is preferred here that bearings be viewed differently from substructure components, and therefore they are dealt with separately in this chapter. The main difference between bearings and other bridge components is that they accommodate movement and thus reduce the forces transferred through. Therefore, when bearings have lost their capability of accommodating movement, they need to be replaced to avoid large and undesired stresses that may cause more serious issues. For example, when a bearing freezes and becomes unable to rotate as required, extra bending moment will develop, possibly causing concrete members to be subjected to the moment and therefore crack. This can subsequently introduce water leakage through cracked concrete and accelerate deterioration of other bridge components.

Of course, there are bridges that do not use bearings, such as those with so-called integral abutments. As the name indicates, integral abutment bridges use abutments integrated with the superstructure. As a result, forces induced by restraining movement, such as those due to thermal expansion and contraction, are transferred through solid materials not to be relieved by bearings. Apparently, the span length for those bridges without bearing needs to be relatively short so that the induced forces will not become too large to withstand. For relatively longer spans, such temperature-induced stresses can become excessive and bearings accordingly become necessary. Nevertheless, bridges without bearings, such as integral abutment bridges, have the advantage of not having joints where bearings would be. This offers the advantage of avoiding water leakage through joints, especially for those bridges in cold-climate areas where deicing chemicals are used to increase transportation safety, which are also corrosive and damaging.

When bearings are needed, they are required to meet some of the following requirements:

❑ Able to transfer vertical forces from the superstructure at its support points

❑ Able to accommodate horizontal translation along the bridge's longitudinal axis and/or transverse axis due to thermal effects and other load effects

❑ Able to accommodate rotation about the bridge's transverse axis

❑ Able to secure the superstructure in contact with the substructure to prevent uplifting. Such a function may require a tie-down system.

5.2 Types of Bridge Bearing

There are quite a few different types of bridge bearing, perhaps as analogically as different types of bridges. The mechanisms these bearings use can be quite distinct in accommodating movement and even their sizes significantly vary more or less depending on the size of the span begin supported. For example, Figure 5.2-1 shows an elastomeric bearing about 4 in. thick supporting a steel girder. Figure 5.2-2 displays a much larger bearing for contrast. Figure 5.2-3 exhibits a further larger bearing for a longer bridge span. On the other hand, like those commonly used bridge span types, a small group of bearings are used more often. They are briefly discussed in Section 5.3. Then in Section 5.4, the design of elastomeric pad bearings is presented in detail, since they have become overwhelmingly popular in the United States over the past three to four decades. They not only are used in new bridges, but also are often the first choice for bearing replacement in bridge rehabilitation projects.

5.2.1 Bearing Types According to Function

According to their functions, bearings can be categorized as expansion and fixed bearings. By these names, expansion bearings allow superstructure movement without creating unwanted stresses and/or failure. In general, expansion bearings refer to those bearings that are able to accommodate movement in the bridge longitudinal direction and in the vertical plane

Figure 5.2-1
Elastomeric bearing supporting steel girder.

Figure 5.2-2
Large guided pot bearing (Courtesy of the D. S. Brown Company, North Baltimore, OH).

of the beam or component being supported. The movement may include those due to thermal expansion and contraction of the superstructure being supported and those due to horizontal load along the longitudinal axis of the bridge such as vehicle braking force. Several examples of expansion bearing follow.

EXPANSION BEARINGS

Rocker bearings are commonly seen in existing highway bridges as expansion bearings. One example is seen in Figure 5.2-4 supporting a continuous beam with a pin-and-hanger connection. Figure 5.2-5 illustrates the

Figure 5.2-3
Elastomeric fixed bearing with
plan area of 98 in. × 158 in.
(Courtesy of the D. S. Brown
Company, North Baltimore, OH).

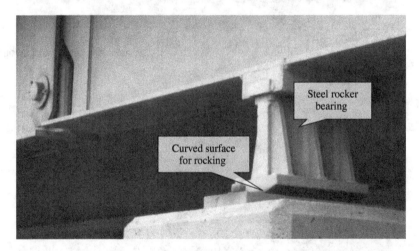

Steel rocker
bearing

Curved surface
for rocking

Figure 5.2-4
Steel rocker bearing.

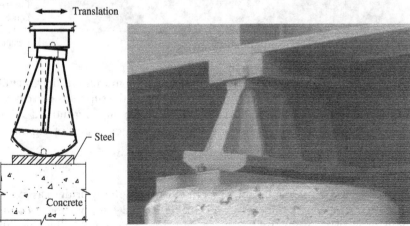

Translation

Steel

Concrete

Figure 5.2-5
Components and
mechanism of steel
rocker bearing.

mechanism and components of the rocker bearing. The connection between the sole plate and a pin on the top of the bearing and the pin and seat at the bottom of the bearing will allow an amount of rotation of the bearing when the beam moves in the longitudinal direction. When the real rotation and translation are within the range that the bearing is capable of accommodating, the bearing will satisfy the movement requirement for expansion (and contraction).

Elastomeric bearing (including both reinforced and plain pads) are also widely used as expansion bearings. Figure 5.2-6 shows such a bearing supporting a prestressed concrete I beam, and Figure 5.2-7 shows the mechanism of the bearing to accommodate vertical deformation, rotation, and translation. The most attractive feature of elastomeric bearings is that they do not have movable parts that require maintenance such as cleaning and/or lubrication. As a result, their installation and replacement become simple, easy, and of low cost. For contrast, rocker bearings, shown

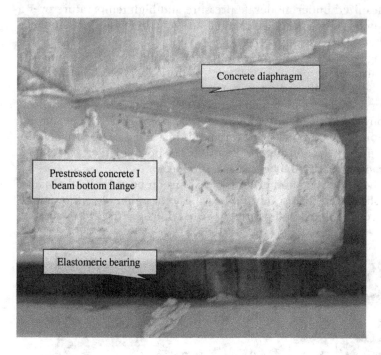

Figure 5.2-6
Steel-reinforced elastomeric bearing supporting concrete girder.

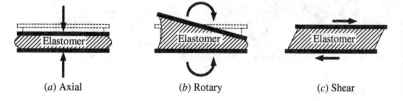

(*a*) Axial   (*b*) Rotary   (*c*) Shear

Figure 5.2-7
Mechanism of elastomeric bearing to accommodate (left) axial deformation, (middle) rotation, and (right) translation.

in Figure 5.2-5, have to move around the top and bottom pins. Water and/or moisture can remain in and/or around these moving parts. Rust can accumulate and cause the bearing to not be able to move any longer, which is referred to as "frozen." The constrained movement has caused other parts of the bridge to be stressed not as designed. For instance, the pier supporting the bearing has been observed to crack due to this extra stress that was not designed for.

It is noteworthy to mention that the bearing manufacturing process is unique for each bearing with respect to its size and the corresponding dimensions of its components. Therefore, production line manufacturing is not considered suitable for bridge bearing fabrication because this wide variation in size and part type requires only a few identical bearings for a single bridge. There is no need for large-quantity production or it would be too expensive to develop a production line to manufacture a few bearings of a single kind.

For example, elastomeric bearings are made using synthetic or natural rubber vulcanized under moderate pressure and high temperature over a period of time. They are molded manually, as seen in Figure 5.2-8, for relatively smaller bearing sizes.

For sizing bearings, Figure 5.2-8 (left) shows molds using aluminum bars or plates partitioning to the right spaces for the right sizes of bearing. The spaces will be filled with raw elastomers to the right weight calculated from the bearing sizes. The work bench is seen in Figure 5.2-8 (right) on wheels so as to conveniently slide into the oven to be pressurized and heated for vulcanization. For larger bearings, the molds may need to be specially fabricated to meet the requirements. Figure 5.2-9 shows a larger elastomeric bearing ready to receive the beam to be supported.

Sliding bearings are another kind of expansion bearing. By the name, there is a sliding mechanism in the bearing to accommodate movement, translation, or rotation. For the former, the bearing has a flat sliding surface.

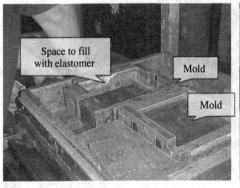 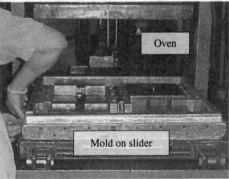

Figure 5.2-8
(Left) Mold preparation; (Right) Oven for vulcanizing elastomeric bearings.

Figure 5.2-9
Elastomeric expansion bearing installed prior to erecting superstructure (Courtesy of the D. S. Brown Company, North Baltimore, OH).

For the latter, a spherical or cylindrical surface is often provided. The two surfaces sliding against one another have used similar or different materials. For example, bronze sliding against steel with lubrication was an option. Later, polytetrafluoroethylene (PTFE) was found to provide a very low friction coefficient and a high strength for compression load, making it perfect for application as a sliding surface for bridge bearing. It has largely replaced other sliding surfaces for sliding bearing. PTFE is known by the DuPont brand name Teflon.

FIXED BEARINGS

Fixed bridge bearings often refer to those that restrain translation movement but still are able to accommodate uniaxial or multiaxial rotation. Several examples of these bearings are given below.

Steel-reinforced elastomeric bearings can be used for both expansion and fixed functions. The flexibility these bearings offers bridge designers an attractive option when selecting bearing type. Elastomeric bearings with a dowling system restricting translation have been used as fixed bearings, as shown in Figure 5.2-10 as an example.

Spherical bearings refer to those that have a convex surface for sliding. They are often used as fixed bearings while allowing multiaxial rotation. Figure 5.2-11 shows the spherical surface of a sliding bearing used as a fixed bearing. Notice the four steel guide plates that will restrict translations of the superstructure.

Steel pinned bearings are another type of fixed bearing. An example is shown in Figure 5.2-12. As seen, the pin needs to be adequately lubricated

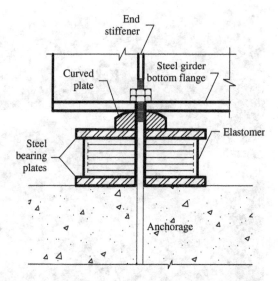

Figure 5.2-10
Steel-reinforced elastomeric bearing as fixed bearing.

Figure 5.2-11
Convex spherical surface of fixed bearing with 62 in diameter (Courtesy of the D. S. Brown Company, North Baltimore, OH).

for the bearing to allow rotation or extra and undesired stress will develop in the beam and other components, which has not been considered in design.

5.2.2 Bearing Types According to Material

According to the material used, bearings are referred to as steel bearings, elastomeric bearings, and so on. As discussed above, steel bearings can be both fixed and expansion bearings, and so can elastomeric bearings. Figure 5.2-13 compares the two. As seen, elastomeric bearings are shorter and thus more stable. Elastomeric bearings also do not have the moving parts that steel bearings do, which require (parts) maintenance. Elastomeric bearings (including nonreinforced elastomeric pads) are becoming much more popular in U.S. bridge construction, particularly for short- and

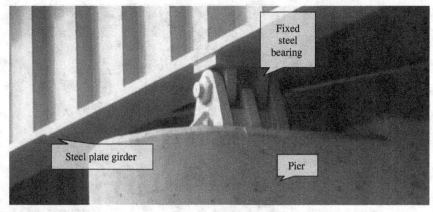

Figure 5.2-12
Steel pinned bearing as fixed bearing on pier.

Figure 5.2-13
Contrast of (left) steel and (right)
elastomeric bridge bearings.

medium-span highway bridges. The main reason is that elastomeric bearings do not have moving components requiring maintenance. This advantage is obviously translated to lower cost for the life cycle.

Sometimes bearings are also referred to according to their internal characteristic materials. For example, fiberglass-reinforced elastomeric pads (abbreviated as FGP in the AASHTO specifications) use fiberglass plates instead of steel plates as reinforcement. Cotton-duck fabric reinforced elastomeric pads represent another example (abbreviated in the AASHTO specifications as CDP). These bearings are typically for lower vertical reaction forces and thus more suitable for shorter bridge spans.

Figure 5.2-14
Elastomeric bridge bearing pad.

5.2.3 Bearing Types According to Characteristic Shape

Bridge bearings also acquire their distinct names referring to their geometric shape, obviously for convenience in practice. Examples are bearing pad, pot bearing, spherical bearing, and cylindrical bearing.

"Bearing pads" has been loosely used to refer to elastomeric bearings shaped as rectangles or circular solids, as shown in Figure 5.2-14. In the AASHTO specifications, "pads" is used to exclusively refer to bearings made of elastomer without steel reinforcement. In other words, elastomeric bearings without reinforcement (made of plain elastomer only) and reinforced using fiberglass plates or cotton-duck fabrics are all referred to as "pads" in the specifications. Apparently these bearing pads possess a lower load-carrying capacity without steel reinforcement. As a result, they are typically used in shorter bridge spans.

Pot bearings resemble a pot, holding an elastomeric disk and a metal piston in the pot, as seen in Figure 5.2-15. The piston needs to be able to move to accommodate rotation in a fixed bearing and both rotation and translation in an expansion bearing. These moving parts do need maintenance and sometimes need to be replaced when not functioning as designed.

Spherical and cylindrical bearings refer to the geometric shape of their respective sliding surface. A spherical bearing as a fixed bearing is shown in Figure 5.2-11.

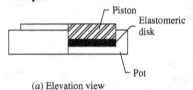

(a) Elevation view

(b) Plan view

Figure 5.2-15
Pot bearing. Adapted from *AASHTO LRFD Bridge Design Specifications*, 2012. Used by permission.

5.3 Appropriate Selection of Bearings

The bearings chosen for a particular bridge and its surrounding conditions need to have appropriate load and movement capabilities. Often these bearings need to provide more than one function. For instance, a steel-reinforced elastomeric expansion bearing needs to adequately permit translation and rotation as well as transfer the reaction force in the vertical direction. A spherical fixed bearing may be required to accommodate multiaxial rotation and constrain translation. To facilitate appropriate selection, the AASHTO specifications provide some general information as to the capability of commonly used bearing types. Table 5.3-1 (*14.6.2*) includes that information taken from the specifications, which also note that the information "is based on general judgment and observation, and there will obviously be some exceptions."

This table addresses bearing resistance to load in three orthogonal directions: longitudinal, transverse, and vertical. It also covers consideration to rotations about axes in these three directions and translations in the longitudinal and transverse directions. This table helps form a global view about the functions required of bearings. Four levels of appropriateness are used in this table for bearings to be evaluated, as defined in the footnote to the table. Note that an example of additional elements is seen in Figure 5.2-11, where four plates as additional elements are welded around the spherical

Table 5.3-1
Bearing suitability   *14.6.2*

Type of Bearing	Movement		Rotation about Bridge Axis Indicated			Resistance to Loads		
	Lg	Tr	Lg	Tr	Vt	Lg	Tr	Vt
Plain elastomeric pad	S	S	S	S	L	L	L	L
Fiberglass-reinforced pad	S	S	S	S	L	L	L	L
Cotton-duck-reinforced pad	U	U	U	U	U	L	L	S
Steel-reinforced elastomeric bearing	S	S	S	S	L	L	L	S
Plane sliding bearing	S	S	U	U	S	R	R	S
Curved sliding spherical bearing	R	R	S	S	S	R	R	S
Curved sliding cylindrical bearing	R	R	U	S	U	R	R	S
Disc bearing	R	R	S	S	L	S	S	S
Double cylindrical bearing	R	R	S	S	U	R	R	S
Pot bearing	R	R	S	S	L	S	S	S
Rocker bearing	S	U	U	S	U	R	R	S
Knuckle pinned bearing	U	U	U	S	U	S	R	S
Single-roller bearing	S	U	U	S	U	U	R	S
Multiple-roller bearing	S	U	U	U	U	U	U	S

Note: S = suitable; U = unsuitable; L = suitable for limited applications; R = may be suitable but requires special considerations or additonal elements such as sliders or guideways; Lg = about longitudinal axis; Tr = about transverse axis; Vt = about vertical axis.
Source: AASHTO LRFD Bridge Design Specifications, 2012. Used by permission.

bearing to restrict unwanted translation. Bearings listed as suitable for a specific application are likely to be so with little or no effort of the design engineer other than good design and detailing practice. Bearings listed as unsuitable are likely to be marginal, even if the design engineer makes extraordinary efforts to make the bearing work properly. Bearings listed as suitable for limited application may work if the load and rotation requirements are not excessive.

5.4 Design of Elastomeric Bearings

This section has a focus on the design of elastomeric bridge bearings. They may be reinforced or not reinforced. Figure 5.4-1 offers a typical procedure for designing elastomeric bearings. Steel-reinforced elastomeric bearings can be designed using either of the two methods specified in the AASHTO

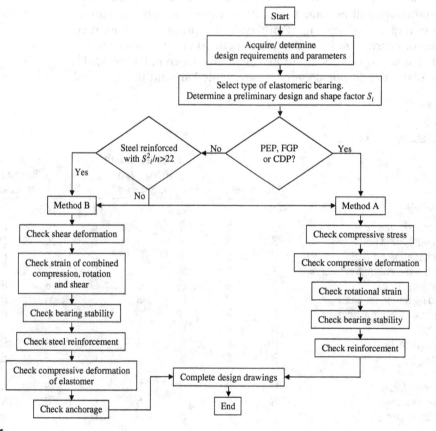

Figure 5.4-1
Flowchart for elastomeric bearing design.

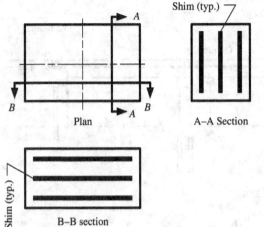

Figure 5.4-2
Cross section of steel-reinforced elastomeric bearing.

design code. They are referred to as Method A and Method B, respectively. Method B is intended to be used for steel-reinforced elastomeric bearings, which is presented next. Method A can be used for steel-reinforced elastomaric bearing meeting certain requirements as well as elastomeric pads without reinforcement and with reinforcement of fiberglass and cotton-duck fabric.

Design Method B in the AASHTO specifications is intended to be used for designing steel-reinforced elastomeric bearings only. Example 5.1 presents application of Method B to a prestressed concrete beam bridge superstructure, whose design is given in Examples 4.12 to 4.14. Steel-reinforced elastomeric bearings consist of alternate layers of steel reinforcement and bonded elastomer of constant thickness. In addition to internal reinforcements, those bearings may have external steel load plates bonded to either or both the upper and lower elastomer layers. Figure Ex5.1-2 shows two cross sections of a typical rectangular bearing of this kind in two perpendicular directions.

5.4.1 Method B

Example 5.1 Elastomeric Bearing Design
(for Examples 4.12 to 4.14)

❑ **Design Requirement**
For the prestressed concrete I-beam bridge in Examples 4.12, 4.13, and 4.14, design the elastomeric bearings for the beams. The bridge elevation view is shown in Figure Ex5.1-1.
Load modifier $\eta = 1.0$

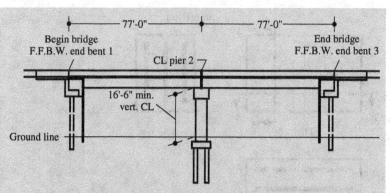

Figure Ex5.1-1
Elevation view of bridge.

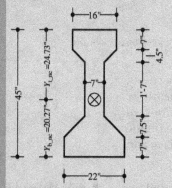

Figure Ex5.1-2
AASHTO Type III prestressed I-beam section.

☐ **Design Parameters**

Span length $L_{span} = 77$ ft
Number of beams, $N_{beams} = 6$
$\theta_{skew} = 35°$
Number of design traffic lanes per roadway = 3
Distance from centerline pier to centerline bearing,
    $K = 11$ in.
Distance from end of beam to centerline of bearing, $J = 6$ in.
Beam length $L_{beam} = L_{span} - 2(K - J) = 77$ ft $- 2(11$ in. $- 6$ in.$)$
    $= 76$ ft 2 in.
Beam design length $L_{design} = L_{span} - 2k = 77$ ft $- 2(11$ in.$)$
    $= 75$ ft 2 in.
Noncomposite beam section (Type III) as seen in Figure Ex5.1-3.
Compressive strength of concrete for prestressed beams,
    $f'_{c,\,beam} = 6.5$ k/in.^{2}

Modulus of elasticity for prestressed concrete:

$$E_{beam} = 1820\sqrt{f'_{c,\,beam}} = 1820 \times \sqrt{6.5 \text{ k/in.}^2}$$

C 5.4.2.4

$$= 4640 \text{ k/in.}^2$$

Shear modulus of elastomer for hardness 60 at 73°F: $G = 0.13$ k/in.^{2} *14.7.6.2.1*

☐ **Trial Design as Shown in Figure Ex5.1-3.**

Bearing length and width: $L = 16$ in., $W = 12$ in.
Internal elastomer thickness $h_{ri} = 0.5$ in. (two layers)
Shape factor:

14.7.5.1 $S_i = \dfrac{LW}{2\,h_{ri}\,(L + W)} = \dfrac{16\,(12)}{2\,(0.5)\,(16 + 12)} = 6.86$

Use $\frac{1}{8}$-in.-thick shims.

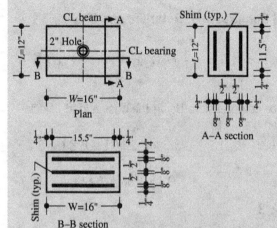

Figure Ex5.1-3
Preliminary design of elastomeric bearing.

❑ Shear Deformation Check

Find maximum shear deformation of elastomer

$\Delta_{s,\text{thermal}}$ = (thermal expansion coefficient)(expected temperature change)(span length)

= 0.000006/°F(80°F − 25°F)75.17 ft(12 in./ft) = 0.3 in.

Conservatively assuming lowest construction temperature of 25°F.
5.4.2.2

$\Delta_{s,\text{shrinkage}}$ = (shrinkage strain $\varepsilon_{\text{shrinkage}}$)(span length)

$$\varepsilon_{\text{shrinkage}} = k_s k_{hs} k_f k_{td}(0.48 \times 10^{-3}) = \left[\frac{t/\left(26e^{0.36(V/S)} + t\right)}{t/(45 + t)}\right]\left(\frac{1064 - 94(V/S)}{923}\right)$$

$$\times (2 - 0.014H)\frac{5}{1 + f'_c}\left(\frac{t}{61 - 4f'_c + t}\right)(0.48 \times 10^{-3})$$

For a relatively large t, $V/S = 6$ and $H = 75$ for the assumed Detroit, Michigan, area,

$$\varepsilon_{\text{shrinkage}} = 1\left(\frac{1064 - 94(6)}{923}\right)[2 - 0.014(75)]\frac{5}{1 + 6.5}(1)\frac{0.48}{1000}$$

$$= 0.00016$$

5.4.2.3.3 $\Delta_{s,\text{shrinkage}} = 0.00016(75, 17\ \text{ft})\,12\ \text{in/ft.} = 0.14\ \text{in.}$

14.7.5.3.2 h_{rt} = 2 layers each 0.5 in. = 2(0.5 in.) = 1 in. > $2\Delta_s$

$$= 2(0.3\ \text{in.} + 0.14\ \text{in.}) = 0.88\ \text{in.}$$

OK for shear deformation.

☐ **Check for Combined Compression, Rotation, and Shear Requirements**

$$\gamma_{a,st} + \gamma_{r,st} + \gamma_{s,st} + 1.75\,(\gamma_{a,cy} + \gamma_{r,cy} + \gamma_{s,cy}) \leq 5$$

14.7.5.3.3
$$\gamma_{a,st} \leq 3$$

Calculation for γ_a for compression: Using the design load calculation results in Example 4.12:

$$\text{Static } \gamma_{a,st} = D_a \frac{\sigma_s}{GS_i} = 1.4 \frac{\dfrac{21.2\ k + 3.5\ k + 29.6\ k + 6.8\ k + 8.2\ k}{16\ \text{in. (12 in.)}}}{0.13\ k/\text{in.}^2\ (6.86)}$$

$$= 0.57 < 3$$

Service I limit state load combination.
   OK for static compressive strain. *14.7.5.3.3*

$$\text{Cyclic } \gamma_{a,cy} = D_a \frac{\sigma_s}{GS_i} = 1.4 \frac{95\ k/[16\ \text{in.(12 in.)}]}{0.13\ k/\text{in.}^2\ (6.86)} = 0.78$$

Calculation for γ_r for rotation: Using design load calculation results in Examples 4.12 and 4.13, rotation θ_s for static load is given as

$$\theta_s \text{ for static load} = \frac{w\,(\text{span length})^3}{24EI}$$

$$= \frac{(0.56 + 0.09 + 0.79 + 0.18 + 0.22)}{24\,(4640\ k/\text{in.}^2)\,398{,}785\ \text{in.}^4}$$

$$= 0.0025\ \text{rad}$$

$$\theta_s \text{ due to prestressing caused camber} = \frac{M\,(\text{span length})}{2EI}$$

$$M = \text{moment due to prestressing}$$

$$= (\text{area of strands})(\text{prestress after total loss})(\text{eccetricity})$$

$$= -(0.192\ \text{in.}^2)19(202.5\ k/\text{in.}^2 - 42.67\ k/\text{in.}^2)$$

$$\times (20.27\ \text{in.} - 7.42\ \text{in.}) = -7492\ \text{kin.}$$

$$\theta_s = -\frac{7492\ \text{kin.}(75.17\ \text{ft})12\ \text{in./ft}}{2\,(4640\ k/\text{in.}^2)\,125{,}390\ \text{in.}^4} = -0.0058\ \text{rad}$$

Service I limit state load combination:

$$\text{Static } \gamma_{r,st} = D_r \left(\frac{L}{h_{ri}}\right)^2 \frac{\theta_s}{n} = 0.5 \left(\frac{12 \text{ in.}}{0.5 \text{ in.}}\right)^2$$

$$\times \frac{0.0025 + 0.0025 - 0.0058}{2}$$

$$= -0.12$$

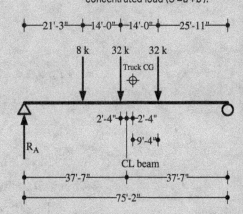

Figure Ex5.1-4
Support rotation due to concentrated load ($S = a + b$).

Set $\gamma_{r,st} = 0$ conservatively. An allowance of 0.0025 rad is included for uncertainty. *14.4.2.1*

Using design load calculation results in Examples 4.12 to 4.14, estimate θ_s for cyclic load: For a concentrated load P on a simply supported beam as shown in Figure Ex5.1-4, the rotation at the left support is

$$\theta_{\text{left}} = \frac{Pb}{6 \, S \, EI} \left(S^2 - b^2\right)$$

Assume that the HL93 truck's position inducing the maximum rotation at the left support is the same as the one causing the maximum moment as shown in Figure Ex5.1-5. The cyclic rotation due to HL93 truck is estimated as follows:

Figure Ex5.1-5
HL93 truck position for maximum rotation.

$$\theta_s \text{ for HL93 truck} = \frac{\sum P_i b_i \left(S^2 - b_i^2\right)}{6 \, S \, EI}$$

$$= \frac{\left[\begin{array}{c} 8(53.92)(75.17^2 - 53.92^2) + 32 \text{ k}(39.92)(75.17^2 - 39.92^2) \\ +32 \text{ k}(25.92)(75.17^2 - 25.92^2) \end{array}\right]}{\dfrac{6 \, (75.17 \text{ ft}) \, 4640 \text{ k/in.}^2 \left(398{,}785 \text{ in.}^4\right)}{(12 \text{ in./ft})^2}}$$

$$= 0.0018 \text{ rad}$$

$$\theta_s \text{ for HL93 lane} = \frac{wS^3}{24EI}$$

$$= \frac{0.64 \text{ k/ft}(75.17 \text{ ft})^3 (12 \text{ in./ft})^2}{24 \, (4640 \text{ k/in.}^2) \, 398{,}785 \text{ in.}^4} = 0.00088 \text{ rad}$$

$$\text{Cyclic } \gamma_{r,cy} = D_r \left(\frac{L}{h_{ri}}\right)^2 \frac{\theta_s}{n} = 0.5 \left(\frac{12 \text{ in.}}{0.5 \text{ in.}}\right)^2$$

$$\times \frac{0.0018(1.33) + 0.00088 + 0.0025}{2} = 0.83$$

An allowance of 0.0025 rad is included for uncertainty. 14.4.2.1

Calculation for γ_s for shear

Using the design load calculation results in above and Examples 4.12 to 4.14

$$\text{Static } \gamma_{s,st} = \frac{\Delta_s}{h_{rt}} = \frac{0.44 \text{ in.}}{2(0.5 \text{ in.})} = 0.44$$

$$\text{Cyclic } \gamma_{s,cy} = 0$$

Check for combined compression, rotation, and shear

$$\gamma_{a,st} + \gamma_{r,st} + \gamma_{s,st} + 1.75(\gamma_{a,cy} + \gamma_{r,cy} + \gamma_{s,cy})$$

$$= 0.57 + 0 + 0.44 + 1.75(0.78 + 0.83 + 0) = 3.83 < 5$$

OK for combined strain. *14.7.5.3.3*

☐ **Check Stability**

$$A = \frac{1.92(h_{rt}/L)}{\sqrt{1 + 2L/W}} = \frac{1.92(2(0.5 \text{ in.})/12 \text{ in.})}{\sqrt{1 + 2(12 \text{ in.})/16 \text{ in.}}} = 0.10$$

$$B = \frac{2.67}{(S_i + 2)[1 + L/(4W)]} = \frac{2.67}{(6.86 + 2)\{1 + 12/[4(16)]\}} = .25$$

$$2A = 0.2 < B = 0.25$$

OK for stability *14.7.5.3.4*

☐ **Check Reinforcement Thickness**

Shim thickness $h_s = 0.125$ in. > 0.0625 in.

OK for reinforcement thickness. *14.7.5.3.5*

☐ **Check Reinforcement for Service Limit State**

Thickenss $h_s = 0.125$ in.

$$> \frac{3h_{ri}\sigma_s}{F_y} = \frac{3(0.5 \text{ in.})\dfrac{21.2 + 3.5 + 29.6 + 6.8 + 8.2 + 95}{16 \text{ in.}(12 \text{ in.})} \text{ k}}{36 \text{ k/in.}^2} = 0.036 \text{ in.}$$

OK for reinforcement thickness for service limit. *14.7.5.3.5*

Check Reinforcement for Fatigue Limit State

Thickenss $h_s = 0.125$ in.

$$> \frac{2h_{ri}\sigma_L}{\Delta F_{TH}} = \frac{2(0.5 \text{ in.})\dfrac{95 \text{ k}}{16 \text{ in.}(12 \text{ in.})}}{24 \text{ k/in.}^2} = 0.021 \text{ in.}$$

OK for reinforcement thickness for fatigue limit.
14.7.5.3.5

Check Anchorage

$$\frac{\theta_s}{n} = \frac{0.0025 + 0.0025 - 0.0058 + 1.75 \begin{bmatrix} 0.0018(1.33) + 0.00088 + 0.0025 \end{bmatrix}}{2}$$

$$= 0.0047 < \frac{3\varepsilon_a}{S_i} = \frac{3(0.05)}{6.86} = 0.021$$

OK for anchorage check. No restraint system is needed. *14.7.5.4*

Check Live-Load Compressive Deformation

$$\delta_L = \sum \varepsilon_{Li} h_{ri} = 2 \text{ strain due to stress } \frac{95 \text{ k}}{12 \text{ in.}(16 \text{ in.})}(0.5 \text{ in.})$$

$$= 2(0.021)0.5 \text{ in.} = 0.021 \text{ in.} < 0.125 \text{ in.}$$

Strain is found using Figure Ex5.1-6.
OK for compressive deflection. *C14.7.5.3.6*

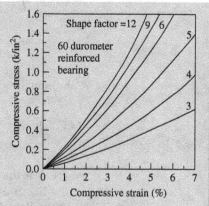

Figure Ex5.1-6
Elastomer stress–strain relation graph in Figure Ex5.4-6. *C14.7.6.3.3.* From *AASHTO LRFD Bridge Design Specifications*, 2012. Used by permission.

Tapered elastomer layers are not allowed in the AASHTO specifications for this type of elastomeric bearings. The top and bottom cover layers are restricted to being no thicker than 70% of the internal layers. Tapered layers can cause larger shear strains and bearings made with them fail prematurely due to delamination or rupture of the reinforcement. In addition, internal layers should have the same thickness because the strength and stiffness of the bearing in resisting compressive load are controlled by the thickest layer that deforms most and induces largest strain.

The so-called shape factor S_i of a layer i of an elastomeric bearing is an important parameter that significantly influences its behavior and capacity. It is defined in the AASHTO specifications for rectangular bearing pads as

14.7.5.1
$$S_i = \frac{WL}{2h_{ri}(W+L)} \tag{5.4-1}$$

where $W =$ width of bearing parallel to axis of bearing rotation, usually parallel to bridge transverse axis

L = length of bearing perpendicular to axis of bearing rotation, usually along bridge longitudinal axis

h_{ri} = thickness of ith internal elastomer layer, namely thickness of each internal layer

For circular bearing pads, the shape factor is

14.7.5.1
$$S_i = \frac{D}{4h_{ri}}$$
(5.4-2)

where D is the diameter of the projection of the bearing's loaded surface in the horizontal plane. As seen in Eqs. 5.4-1 and 5.4-2, the shape factors are actually a dimensionless ratio of the bearing horizontal cross sectional area to the total side area of a layer, which is the area that bulges under a vertical load on the bearing horizontal area. Elastomers in a bearing bulge as shown later in Figure 5.4-4.

ELASTOMERIC MATERIAL PROPERTIES

The mechanical properties of elastomeric material are sensitive to temperature, much more so than such conventional structural material as steel or concrete. This variability is also much more noticeable. The AASHTO specifies the shear modulus of the elastomer at 73°F as the basis for design. In addition, the elastomer is required to have a specified shear modulus G between 0.080 and 0.175 k/in.[2] The shear modulus is one of the most important material properties used in the design of elastomeric bearings and it is the primary means of specifying the elastomer. Hardness has been used widely for that purpose in the past and is still used in Method A, to be discussed below, because the test for it is simple and quick. However, the test results for elastomeric hardness show that they correlate only loosely with shear modulus G.

For the purposes of elastomeric bearing design, a bridge site is classified in the AASHTO specifications as being in one of the temperature Zones A, B, C, D, or E. The definitions for these temperature zones are given in Table 5.4-1, taken from the AASHTO specifications. The table also gives the minimum grade of elastomer required for each low-temperature zone. However, the required information on low temperature for the site may not always be available to the bridge design engineer. In the absence of the precise temperature information needed to use this table, the AASHTO specifications also include a U.S. map to approximately determine to which zone a bridge site may belong.

It should be noted that steel-reinforced elastomeric bearings designed using Method B in the AASHTO specifications need to be physically tested in accordance with relevant specifications. These bearings are designed to resist relatively high stresses. Their integrity depends on good quality control during manufacture, which needs to be ensured by rigorous testing.

Table 5.4-1
Low-temperature zones and minimum grades of elastomer   *14.7.5.2*

	Zone A	Zone B	Zone C	Zone D	Zone E
50-yr low temperature (°F)	0	−20	−30	−45	<−45
Maximum number of consecutive days when temperature does not rise above 32°F	3	7	14	N/A	N/A
Minimum low-temperature elastomer grade	0	2	3	4	5
Minimum low-temperature elastomer grade when special force provisions are incorporated	0	0	2	3	5

Source: AASHTO LRFD Bridge Design Specifications, 2012. Used by permission.

REQUIREMENT ON SHEAR DEFORMATIONS

According to the AASHTO specifications, the maximum horizontal displacement of the bridge superstructure is required to be controlled as half of the total thickness of the deformable part of the elastomeric bearing:

14.7.5.3.2 $$h_{rt} \geq 2\Delta_s$$ (5.4-3)

where h_{rt} = total elastomer thickness
Δ_s = maximum total shear deformation of elastomer from service load limit state

The shear deformation is limited to $\pm 0.5 h_{rt}$, as stated in Eq. 5.4-3, in order to avoid rollover at the edges and delamination due to fatigue. This deformation is illustrated in Figure 5.4-3.

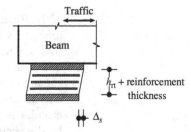

Figure 5.4-3
Shear deformation of elastomeric bearing.

REQUIREMENT ON STRAIN DUE TO COMBINED COMPRESSION, ROTATION, AND SHEAR

Elastomers are almost incompressible. For example, when a steel-laminated bearing is loaded in compression, the elastomer in an internal layer significantly expands laterally due to significant Poisson effect typically associated with elastomers. That expansion is partially restrained by the steel plates to which the elastomer layer is chemically bonded. The restraint results in bulging of the layer between the plates, as illustrated in Figure 5.4-4. This bulging induces shear stresses at the bonded interface between the elastomer and the steel. If these stresses become large enough, they can cause shear failure of the bond or failure of the elastomer adjacent to it. This is the most common form of damage in steel-laminated elastomeric bearings. Elastomeric bearing design is required therefore to control shear

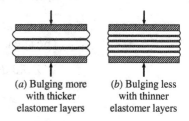

(a) Bulging more with thicker elastomer layers   (b) Bulging less with thinner elastomer layers

Figure 5.4-4
Elastomer bulging constrained by bonded reinforcements.

strain in the elastomer. This is also the reason nonuniform thicknesses of elastomers in a bearing are not allowed, as the thickest layer, bulging more, will control bearing design. Thus uniform layer thickness is the optimal design.

The AASHTO specifications require the following strain control for combined effects of axial load, rotation, and shear deformation at the service limit state:

14.7.5.3.3
$$\left(\gamma_{a,\text{st}} + \gamma_{r,\text{st}} + \gamma_{s,\text{st}}\right)$$

$$+ 1.75\left(\gamma_{a,\text{cy}} + \gamma_{r,\text{cy}} + \gamma_{s,\text{cy}}\right) \leq 5 \tag{5.4-4}$$

14.7.5.3.3
$$\gamma_{a,\text{st}} \leq 3 \tag{5.4-5}$$

where　　γ = strain
　　　　　a = axial load
　　　　　r = rotation
　　　　　s = shear deformation
　　　　　st = static load effect
　　　　　cy = cyclic load effect

Traffic-induced live load should be included as cyclic load, and all other loads may be considered static. In rectangular bearings, the shear strains shall be evaluated for rotation about the axis parallel to the transverse axis of the bridge in the plane of bearing length L defined earlier. Evaluation of shear strains for rotation in the perpendicular direction (about the axis parallel to the longitudinal axis of the bridge) should also be considered. For circular bearings, the total rotation needs to be considered, which should be the vector sum of the rotations about two primary orthogonal axes.

The shear strains γ_a, γ_r, and γ_s need to be estimated in the design process to satisfy Eqs. 5.4-4 and 5.4-5. A precise determination of these strains in the bearing may require advanced analysis techniques. The following approximations are acceptable in the AASHTO specifications. Figure 5.4-5 provides an intuitive understanding for these three terms respectively.

Shear Strain γ_a due to Axial Load
The shear strain due to axial load may be taken as

14.7.5.3.3
$$\gamma_a = D_a \frac{\sigma_a}{GS_i} \tag{5.4-6}$$

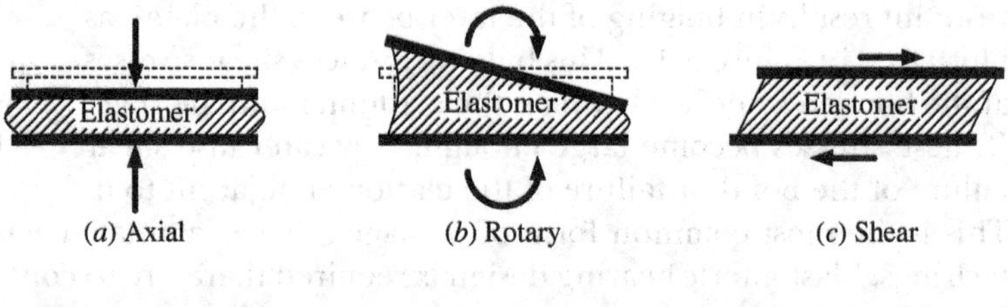

Figure 5.4-5
Axial, rotary, and shear deformations.

(a) Axial　　　　(b) Rotary　　　　(c) Shear

where D_a = dimensionless coefficient for calculation of strain due to axial load = 1.4 for rectangular bearings and 1.0 for circular bearings

σ_a = average compressive satress due to static or cyclic load

G = shear modulus of elastomer

S_i = shape factor of bearing

Shear Strain γ_r due to Rotation

The shear strain due to rotation for elastomeric bearings may be taken as

14.7.5.3.3
$$\gamma_r = D_r \left(\frac{D_b}{h_{ri}} \right)^2 \frac{\theta_s}{n}$$
(5.4-7)

where D_r = dimensionless coefficient for rorational shear strain calculation = 0.5 for rectangular bearings and 0.375 for circular bearings

D_b = characteristic bearing dimension = length L for rectangular bearings and diameter D for circular bearings

θ_s = total rotation due to static load or cyclic load under the service limit state

n = number of internal layers of elastomer, defined as those layers bonded on both faces. When the thickness of the exterior layer of the elastomer is equal to or greater than one-half the thickness of an interior layer, n may be increased by one-half for each such exterior layer.

h_{ri} = thickness of ith internal elastomeric layer

Shear Strain γ_s due to Shear Deformation

According to the AASHTO specifications, the shear strain due to shear deformation in a bearing may be taken as

14.7.5.3.3
$$\gamma_s = \frac{\Delta_s}{h_{rt}}$$
(5.4-8)

where Δ_s = maximum total shear deformation due to static or cyclic load under service limit state

h_{rt} = total elastomer thickness

This calculation is similar to that in Eq. 5.4-3 except that now static and cyclic loads are separately calculated to be used in Eq. 5.4-4 with different weights.

REQUIREMENT ON STABILITY OF ELASTOMERIC BEARING

Elastomeric bearings also need to be investigated in design for possible instability to prevent such failure under the service limit state. Bearings satisfying the following requirement in Eq. 5.4-9 are considered to be stable, and no further investigation of stability is required. Otherwise, a stress check is required as discussed later.

14.7.5.3.4 $$2A \leq B \tag{5.4-9}$$

where

14.7.5.3.4 $$A = \frac{1.92 \left(h_{rt}/L\right)}{\sqrt{1 + 2L/W}} \tag{5.4-10}$$

14.7.5.3.4 $$B = \frac{2.67 \left(h_{rt}/L\right)}{\left(S_i + 2\right) \left[1 + L/\left(4W\right)\right]} \tag{5.4-11}$$

The symbols used here have been defined earlier in this chapter. These two equations are for rectangular bearings, explicitly using their dimension parameters L and W. When $L > W$, the bearing stability is required to be investigated by interchanging L and W in these equations. For circular bearings, these equations may still be used with $L = W = 0.8D$, where D is the bearing diameter as defined above in this chapter.

For rectangular bearings not satisfying Eq. 5.4-9, the stress due to the total load is required to satisfy the following equation:

❑ If the bridge deck is free to translate horizontally,

14.7.5.3.4 $$\sigma_i = \frac{GS_i}{2A - B} \tag{5.4-12}$$

❑ If the bridge deck is fixed against horizontal translation,

14.7.5.3.4 $$\sigma_i = \frac{GS_i}{A - B} \tag{5.4-13}$$

REQUIREMENTS ON STEEL REINFORCEMENT

The AASHTO specifications require that the minimum steel reinforcement thickness h_s be 0.0625 in. If there are holes in the steel reinforcement, the minimum thickness needs to be increased by a factor equal to twice the gross width divided by the net width with the holes excluded. The thickness is also required to satisfy:

❑ At the service limit state

14.7.5.3.5 $$h_s \geq \frac{3h_{ri}\sigma_s}{F_y} \tag{5.4-14}$$

❑ At the fatigue limit state

14.7.5.3.5 $$h_s \geq \frac{2h_{ri}\sigma_L}{\Delta F_{TH}} \tag{5.4-15}$$

where σ_s = average compressive stress due to total load at applicable service load

σ_L = average compressive stress at service limit state due to
  live load

ΔF_{TH} = constant-amplitude fatigue threshold for Category A
  in Table 4.7-3 *(6.6.1.2)* = 24 k/in.^{2}

F_y = yield strength of steel reinforcement

These requirements target at the steel reinforcement yield strength and
fatigue strength, respectively.

REQUIREMENT ON COMPRESSIVE DEFLECTION OF ELASTOMER

According to the AASHTO specifications, deflections of elastomeric bearings induced by dead load and instantaneous live load alone are required to
be considered separately. The loads considered here shall be at the service
limit state with all load factors equal to 1.

Limiting live-load deflection in bearings is important to ensure that deck
joints and seals are not damaged by trucks crossing them. Excessively flexible bearings in compression could cause a small elevation difference in the
road surface at a deck joint when a truck crosses it, giving rise to additional
impact loading. A maximum relative live-load deflection across a joint is suggested at 0.125 in. in the AASHTO specifications. Joints and seals that are
sensitive to relative deflections may require a smaller gap than that.

It should be emphasized that elastomers have a nonlinear load–
deflection relation in the compression service load range. It is quite
different from conventional structural materials used in construction such
as steel and Portland cement concrete. In the absence of information
specific to the particular elastomer to be used, the following equation or
Figure 5.4-6 may be used for calculating the dead- and live-load compressive
strains needed below:

C14.7.5.3.6
$$\varepsilon \geq \frac{\sigma}{4.8GS^2}$$
(5.4-16)

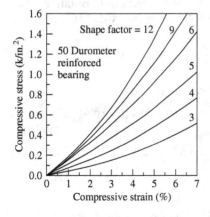

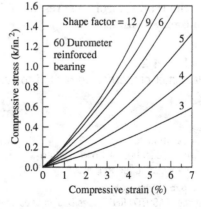

Figure 5.4-6
Stress–strain relation for
typical elastomers. From
*AASHTO LRFD Bridge
Design Specifications*,
2012. Used by permission.

where ε = strain in individual layer of elastomer
       σ = instantaneous live-load compressive stress or dead-load
          compressive stress in individual elastomer layer
       G = shear modulus of elastomer
       S = shape factor of individual elastomer layer

In addition, long-term dead-load deflections should be addressed in design where the joints and seals between sections of the bridge rest on bearings of different design and when estimating redistribution of forces in continuous bridges caused by settlement.

Loadings considered in this section should be at the service limit state with all load factors equal to 1. Instantaneous live-load deformation δ_L is taken in the AASHTO specifications as

14.7.5.3.6 $$\delta_L = \sum_i \varepsilon_{Li} h_{ri} \qquad (5.4\text{-}17)$$

where ε_{Li} is the instantaneous live-load compressive strain in the ith elastomer layer, h_{ri} is the thickness of the ith elastomeric layer. The initial dead-load deflection δ_d is given in the AASHTO specifications as

14.7.5.3.6 $$\delta_d = \sum_i \varepsilon_{di} h_{ri} \qquad (5.4\text{-}18)$$

where ε_{di} is the initial dead-load compressive strain in the ith elastomer layer.

Long-term dead-load deflection δ_{lt}, including the effects of creep, is given in the AASHTO specifications as a sum of both the initial and creep-included deflections:

14.7.5.3.6 $$\delta_{lt} = \delta_d + a_{cr}\delta_d \qquad (5.4\text{-}19)$$

where a_{cr} is the creep deflection divided by the initial dead-load deflection. Values for ε_{Li} and ε_{di} are required to be determined from test results or by analysis. Creep deformation should be determined from information relevant to the elastomeric compound used. If it is elected not to obtain a value for the ratio a_{cr} from test results, Table 5.4-2 from the AASHTO specifications may be used.

Table 5.4-2
Elastomer properties correlated to hardness   *14.7.6.2*

	Hardness (Shore A)		
	50	60	70[a]
Shear modulus at 73°F (k/in.²)	0.095–0.130	0.130–0.200	0.200–0.300
Creep deflection at 25 yr divided by initial deflection, a_{cr}	0.25	0.35	0.45

[a]For PEP and FGP only.
Source: AASHTO LRFD Bridge Design Specifications, 2012. Used by permission.

REQUIREMENT ON ANCHORAGE FOR BEARINGS WITHOUT BONDED EXTERNAL PLATES

In bearings without externally bonded steel plates, a restraint system shall be used to secure the bearing against horizontal movement if

14.7.5.4
$$\frac{\theta_s}{n} \geq \frac{3\varepsilon_a}{S_i} \qquad\qquad (5.4\text{-}20)$$

where n = number of interior layers of elastomer, where interior layers are defined as those layers that are bonded on both faces. When the thickness of the exterior layer of elastomer is equal to or greater than one-half the thickness of an interior layer, n may be increased by one-half for each such exterior layer.

S_i = shape factor of ith internal layer as defined in Eq. 5.4-1

ε_a = total of static and cyclic average axial strain taken as positive for compression in which the cyclic component is multiplied by 1.75 from applicable service load combinations

θ_s = total of static and cyclic service limit state design rotation angles of the elastomer in radians, in which the cyclic component is multiplied by 1.75 as indicated in Eq.5.4-4 *14.7.5.3.3.*

Method A in the AASHTO specifications is to be applied to the design of the following cases:

5.4.2 Method A

❑ Plain elastomeric pads (PEP)

❑ Pads reinforced with discrete layers of fiberglass (FGP)

❑ Steel-reinforced elastomeric bearings with $S_i^2/n < 22$ and primary rotation about the axis parallel to the transverse axis of the bridge, where n is the number of interior layers of elastomer, where interior layers are defined as those layers that are bonded on each face. When the thickness of the exterior layer of elastomer is equal to or greater than one-half the thickness of an interior layer, the parameter n may be increased by one-half for each such exterior layer; S_i is the shape factor of the ith internal layer defined in Eq. 5.4-1.

❑ Cotton-duck pads (CDP) with closely spaced layers of cotton duck and manufactured and tested under compression

While FGP may have different layer thicknesses in the same bearing, steel-reinforced elastomeric bearings designed using Method A are required to have the same thickness for each internal layer and the cover layers have no more than 70% of the thickness of the internal layers. The shape factor for PEP, FGP, and steel-reinforced elastomeric bearings is the same as that for

Method B in Eq. 5.4-1 *(14.7.5.1)*. The shape factor for CDP is based upon the total pad thickness.

REQUIREMENTS ON COMPRESSIVE STRESS

At the service limit state, according to the AASHTO specifications, the average compressive stresses σ_s and σ_L in any layer are required to satisfy:

❑ For PEP

$$\sigma_s \leq 1.0GS_i$$

14.7.6.3.2 $\sigma_s \leq 0.80 \text{ k/in.}^2$ (5.4-21)

❑ For FGP

$$\sigma_s \leq 1.25GS_i$$

14.7.6.3.2 $\sigma_s \leq 1.0 \text{ k/in.}^2$ (5.4-22)

❑ For CDP

14.7.6.3.2 $\sigma_s \leq 3.0 \text{ k/in.}^2$ (5.4-23)

14.7.6.3.2 $\sigma_L \leq 2.0 \text{ k/in.}^2$ (5.4-24)

❑ For steel-reinforced elastomeric bearings

14.7.6.3.2 $\sigma_s \leq 1.25GS_i$ (5.4-25)

14.7.6.3.2 $\sigma_L \leq 1.25 \text{ k/in.}^2$ (5.4-26)

where σ_s = average compressive stress due to total load from applicable service load

σ_L = average compressive stress at service limit state (load factor = 1.0) due to live load

For FGP, the value of S_i used should be based upon an h_{ri} layer thickness that equals the greatest distance between midpoints of two fiberglass reinforcement layers.

The above stress limits may need to be increased by 10% where shear deformation is prevented.

REQUIREMENT ON COMPRESSIVE DEFLECTION

In addition to the requirements on compressive deflection in Method B for steel-reinforced elastomeric bearing design, more specific requirements of the specifications are presented here.

For Steel-Reinforced Elastomeric Bearings   The initial compressive deflection of an internal layer of a steel-reinforced elastomeric bearing at the service limit state without impact is required to not exceed $0.09 \, h_{ri}$, where h_{ri} is the thickness of an internal layer. *(14.7.6.3.3)*

For FGP In lieu of using specific product data, the compressive deflection of a FGP should be taken as 1.5 times the deflection estimated for steel-reinforced bearings of the same shape factor as in Method B.

For PEP The initial compressive deflection of a PEP at the service limit state without impact is required not to exceed $0.09h_{ri}$, where h_{ri} is the thickness of a PEP. *(14.7.6.3.3)*

For CDP The computed compressive strain ε_s may be taken as

14.7.6.3.3 $$\varepsilon_s = \frac{\sigma_s}{E_c}$$ (5.4-27)

where E_c = uniaxial compressive stiffness of CDP bearing pad. It may be taken as 30 k/in.^{2} in lieu of pad-specific test data.

σ_s = average compressive stress due to total load from applicable service load limit states

REQUIREMENT ON SHEAR DEFORMATION

The maximum horizontal superstructure displacement Δ_s to be accommodated by the bearing is required to be computed under the service limit state. It may be reduced to account for the pier flexibility and modified for construction procedures, as appropriate. The computation required in Eq. 5.4-3 is applicable here, while the requirements for individual cases of bearing type are given as follows:

For PEP, FGP, and steel-reinforced elastomeric bearings

14.7.6.3.4 $$h_{rt} \geq 2\Delta_s$$ (5.4-38)

For CDP

14.7.6.3.4 $$h_{rt} \geq 10\Delta_s$$ (5.4-29)

where h_{rt} = smaller of total elastomer or bearing thickness

Δ_S = maximum total shear deformation from applicable service load limit state

REQUIREMENT ON ROTATION

According to the AASHTO specifications, the rotation here is taken as the maximum sum of the effects of initial lack of parallelism or rotation (e.g., due to prestress-induced beam camber) and subsequent girder end rotation due to imposed loads and movements (such as live load). Stress shall be the maximum stress associated with the load conditions inducing the maximum rotation.

The AASHTO specifications require that the maximum compressive strain due to combined compression and rotation of CDP at the service limit state ε_t not exceed

14.7.6.3.5b
$$\varepsilon_t = \varepsilon_c + \frac{\theta_s L}{2t_p} \leq 0.2 \qquad (5.4\text{-}30)$$

where

14.7.6.3.5b
$$\varepsilon_c = \frac{\sigma_s}{E_c} \qquad (5.4\text{-}31)$$

Also the maximum rotation is required to be limited as follows:

14.7.6.3.5b
$$\theta_s \leq 0.8 \frac{2t_p \varepsilon_c}{L} \qquad (5.4\text{-}32)$$

14.7.6.3.5b
$$\theta_L \leq 0.2 \frac{2t_p \varepsilon_c}{L} \qquad (5.4\text{-}33)$$

where E_c = uniaxial compressive stiffness of CDP bearing pad; may be taken as 30 k/in.^{2} in lieu of pad-specific test data
L = length of CDP bearing pad in plane of rotation (in.)
t_p = total thickness of CDP pad (in.)
ε_c = maximum uniaxial strain due to compression under total load from applicable service load combinations
ε_t = maximum uniaxial strain due to combined compression and rotation from applicable service load combinations
σ_s = average compressive stress due to total load associated with maximum rotation from applicable service load combinations
θ_L = maximum rotation of CDP pad at service limit state (load factor 1.0) due to live load
θ_s = maximum rotation of CDP pad from applicable service load combinations

REQUIREMENT ON STABILITY
To ensure stability, the total thickness of the pad shall not exceed the least of $L/3$, $W/3$, or $D/4$.

REQUIREMENT ON REINFORCEMENT
The fiberglass reinforcement in FGP is required to have a strength in each plan direction of at least $2.2h_{ri}$ in k/in. If the layers of elastomer are of different thicknesses, h_{ri} is taken as the mean thickness of the two layers of the elastomer bonded to the same reinforcement. If the fiberglass

reinforcement contains holes, its strength is increased over the minimum value specified herein by twice the gross width divided by net width.

References

American Association of State and Highway Transportation Officials (AASHTO) (2012), *LRFD Bridge Design Specifications*, 6th ed., AASHTO, Washington, DC.

American Association of State and Highway Transportation Officials (AASHTO) (2011), *Manual for Bridge Evaluation*, 2nd ed., AASHTO, Washington, DC.

Roeder, C.W. (2000, Mar.), "LRFD Design Criteria for Cotton Duck Pad Bridge Bearing," Prepared for National Cooperative Highway Research Program, Transportation Research Board, NCHRP Web Document 24 (Project 20-07[99]): Contractor's Final Report, Department of Civil Engineering, University of Washington, Seattle, WA.

Roeder, C. W. (2002, Apr.), "Thermal Movement Design Procedure for Steel and Concrete Bridges," Report Presented to the National Cooperative Highway Research Program, NCHRP 20-07/106, Department of Civil Engineering, University of Washington, Seattle, WA.

Problems

5.1 Design an elastomeric bearing for the interior beam in Problems 4.6 and 3.3. Use reasonable assumptions if needed. They should be consistent with what have been used in Problems 4.6 and 3.3.

5.2 Design an elastomeric bearing for the exterior beam in Problems 4.7 and 3.3. Use reasonable assumptions if needed. They should be consistent with what have been used in Problems 4.7 and 3.3. Comment on the differences and similarities of your results of this problem and Problem 5.1, if any. Comment on the differences in the requirements here and Problem 5.1 if any.

6 Substructure Design

6.1 Introduction

Bridge substructure here refers to those structural components supporting the bearings or the superstructure if bearings are not used. Typical examples of substructure components are piers, abutments, and foundations such as spread footings and piles. Figures 6.1-1 and 6.1-2 show examples of piers for the typical short and medium bridge spans commonly seen in the United States. Figures 6.1-3 and 6.1-4 provide examples of highway bridge abutment.

As discussed earlier, construction of bridge structures starts with the supporting components. Typical pier construction starts with the completed foundation, such as the piles shown in Figure 6.1-5 followed by column forming as shown in Figure 6.1-6 and pier cap forming as shown in Figure 6.1-7. The completed pier structure, seen in Figure 6.1-8, can receive beams, as seen in Figure 6.1-9.

Bridge structures, like all other things on the earth, age and deteriorate. So the design engineers should envision what may be expected years

Figure 6.1-1
Highway bridge piers with multiple concrete columns.

Figure 6.1-2
Bridge piers for crossing highways.

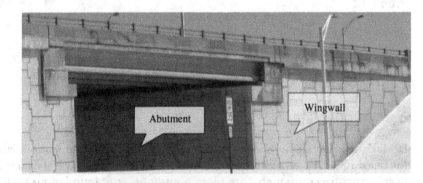

Figure 6.1-3
Highway bridge abutment with
orthogonal wingwall.

Figure 6.1-4
Highway bridge abutment with parallel wingwall.

Figure 6.1-5
Piles for pier in median of two-way separated highway.

Figure 6.1-6
Forming for pier columns.

Figure 6.1-7
Forming for pier cap after completing columns.

Figure 6.1-8
Completed bridge pier.

Figure 6.1-9
Pier supporting spread prestressed
concrete box beams.

or decades down the road and consider them in the design. Figure 6.1-10 shows a photograph of a two-column pier experiencing rehabilitation. Spalled and loose concrete has been removed down to material of good condition. Steel reinforcement has been examined for its adequate cross-sectional area. New concrete will need to be placed to protect the steel

Figure 6.1-10
Concrete pier being
rehabilitated.

Figure 6.1-11
Abutment footing top with anchored epoxy-coated
steel reinforcement.

Figure 6.1-12
Forming for abutment stem wall.

and renew this pier structure. It is critical in the original design of the structure to arrange and design features that can mitigate or eliminate such deterioration. While the particular causes of the observed deterioration are not necessarily applicable, a general attention in design to future possible deterioration is advised and measures taken to prevent premature failure and/or reduce deterioration rate.

Abutments represent another group of important and necessary components for highway bridges. Their main functions are to retain the earth material at the connection between the road and the bridge and transfer the superstructure loads to the foundation system. Figure 6.1-11 shows an abutment footing ready to be formed for a reinforced concrete abutment. The steel reinforcements are epoxy coated as seen and provide an anchor for the abutment wall to be cast. Figure 6.1-12 exhibits ribbed forms prepared to cast the abutment backwall, and Figure 6.1-13 shows the completed backwall and wingwall.

6.2 Piers

Piers are structures that provide intermediate supports to spans, continuous or simply supported, as opposed to abutments, which provide only end supports of the entire bridge. There are several commonly used piers for short- and medium-span highway bridges. They are presented here to provide more in-depth understanding before detailed design procedures and quantitative requirements are presented in Section 6.4 for reinforced concrete multicolumn piers.

The most commonly used highway bridge piers are made of reinforced concrete, taking advantage of its self-weight for stability, ease in sizing, and

Figure 6.1-13
Completed abutment and wingwall.

cost effectiveness. Reinforced concrete piers, can be single-column, multi-column, and wall piers. These will be discussed next.

Single-column piers sometimes are referred to as T columns and hammerhead columns depending on the shape or appearance. Examples are shown in Figures 6.2-1 and 6.2-2. They either may be integral with the superstructure without bearings or may provide independent support through bearings for the superstructure. These piers are sometimes supported by a spread footing. They may also be supported by drilled-shaft- or pile-supported foundation. Their main cross section can be of various shapes (rectangle, circle, polygon, etc.) largely depending on considerations of construction cost and aesthetic appearance. They can be prismatic or flared to form the pier cap to support the superstructure. They also can be blended with the geometric configuration of the superstructure system and/or components.

    Another main advantage of single-column piers is that they require less ground space or land. In areas where land is limited and costly, such as urban regions, single-column piers have been widely used to save space. This can also avoid skewed spans, which impose additional requirements on the bridge span. Single-column piers, however, have no other members to form a redundant system for support, and they may need more material to ensure their stability and strength compared with multicolumn piers.

6.2.1 Single-Column Piers

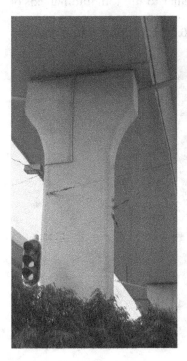 

Figure 6.2-1
Single-column piers of rectangular cross section (left) with independent support through bearings and (right) with integral support without bearings.

Figure 6.2-2
Single-column piers of (left) circular section and (right) polygonal section with (left) independent support through bearings or (right) integral support without bearings.

6.2.2 Multicolumn Piers

When space is available, multicolumn piers or bents are often selected. Two- or three-column systems may be most popular, as seen in Figures 6.2-3 to 6.2-6. The columns also have been seen to have various cross sections. As the appearance may suggest, multicolumn piers or bents are designed as frames in the plane of the pier (perpendicular to the longitudinal axis of the bridge if there is no skew), with moment-resisting connections between the cap and each column. Figures 6.2-3 to 6.2-6 show piers with a pier cap

Figure 6.2-3
Three-column pier supporting prestressed concrete spread box beams.

Figure 6.2-4
Two-column pier supporting concrete unibox beam.

Figure 6.2-5
Five-column pier.

Figure 6.2-6
Eight-column pier.

as independent supports to the superstructure. However, piers can also be monolithically cast and structurally connected with the superstructure, especially for bridges in seismically active areas. Multicolumn piers or bents are usually fixed at the base of the pier or supported on a spread footing or pile-supported foundation or a solid-wall shaft. Multicolumn piers are apparently more stable compared with single-column piers. When there are three or more columns, there is a significant amount of redundancy built in the system. For example, if one of the columns is severely damaged, the system may still be stable and not collapse, which will allow for repair or replacement.

6.2.3 Wall Piers

Wall piers use a wall to support the superstructure between abutments at the ends of a bridge. An example is given in Figure 6.2-7. They are not as popularly used as single- or multicolumn piers, but they are an option when significant superstructure load needs to be transmitted to the ground and/or when the load needs to be distributed more evenly to the earth.

Wall piers are analyzed and designed as columns for forces and moments acting out of the plane of the wall (i.e., about its weak axis). They are also analyzed and designed as piers for load effects in the plane of the wall (i.e., about the strong axis). Wall piers may be pinned or fixed or free at the top and are conventionally fixed at the base. Short and stubby wall piers are often pinned at the base to eliminate the high moments experienced if otherwise fixed.

Besides reinforced concrete, steel also has been used to construct bridge piers but much less commonly. Figures 6.2-8 through 6.2-10 show three

Figure 6.2-7
Solid-wall pier.

Figure 6.2-8
Steel truss bridge pier.

Figure 6.2-9
Steel frame bridge pier.

Figure 6.2-10
Steel box bridge piers.

examples. The first one displays a spatial steel truss tower used as a bridge pier. The second one is an old bridge supported by a steel frame pier. The last one shows steel box frames as bridge piers. They may be analyzed and designed using truss and frame models.

6.3 Abutments

A bridge abutment is the structure supporting the end of end spans connecting to the road and retaining the soil behind it. Therefore there are usually two abutments for two ends of the entire bridge. When the bridge has only one span the two abutments support the two ends of the single span. Another important function of abutments is to transmit the load from the superstructure to the earth, as well as the loads directly applied to the abutment itself, such as earth pressure behind the abutment and live-load surcharge through the soil. Chapter 3 includes an introduction to those loads to abutments. Figures 6.3-1 and 6.3-2 show example abutment structures for highway bridges.

Abutments may be categorized in different ways. The AASHTO design specifications use a way differentiating them according to their relation to the approach fill or embankment as follows:

❏ **Stub Abutments** These are typically located at or near the top of approach fills, usually supported on a pile foundation. Compared with full-depth abutments, stub abutments make the bridge span longer. An example of stub abutment is shown in Figure 6.3-1.

Figure 6.3-1
Stub abutment parallel
wingwall in service.

Figure 6.3-2
Full-height abutment with return wingwall with superstructure to be erected: (left) front view; (right) back view.

❑ **Full-Depth (or Full-Height) Abutments** These are typically located at the approximate front toe of the approach embankment, restricting the opening under the bridge and making the bridge span shorter. Figure 6.3-2 shows an example of full-depth abutment.

❑ **Partial-Depth (or Partial-Height or Semistub) Abutments** These are located approximately at middepth of the front slope of the approach embankment. In other words, they are between stub and full-depth abutments with respect to their relation to embankment. The backwall and wingwalls may retain fill material or the embankment slope may continue behind the backwall. An example of partial-depth abutment is shown in Figure 6.3-3.

Another group of abutments has deviated from the traditional concept of abutment, which is referred to as integral abutment. The term *integral abutment* is not as explicit about its location with respect to the embankment as those discussed above. Instead it refers to its unique and different feature

Figure 6.3-3
Partial-depth abutment supporting
steel superstructure.

compared to other traditional abutments: integral with superstructure. In other words, integral abutments are rigidly connected to their superstructure, as opposed to through bearings or joints, as are traditional abutments. With no joint on the roadway surface, there will be no water leakage through the joint and no or very little associated deterioration. Bridges with integral abutments are thus also referred to as jointless bridges.

On the other hand, rigidly connected abutments and superstructures impose a new challenge: extra stresses at and around the rigid connection induced by thermal expansion/contraction movement. Therefore, integral abutments need to be designed with adequate coverage of these extra stresses. When span length becomes long, these additional stresses may become excessive. Accordingly, integral abutments are suitable only for relatively shorter bridge spans.

Abutments may or may not have wingwalls. Abutments without wingwalls are not popularly used in the United States. The main type is the buried abutment which has earth fill on both the front and back sides. The following discussion focuses on abutments with wingwalls.

The word "abutment" in bridge design strictly refers to the part of the bridge end support directly subjected to the superstructure loads. The word "wingwall" refers to the structure extending from the abutment, as seen in Figures 6.3-1 and 6.3-2. As shown, wingwalls are not directly subjected to the superstructure loads.

Wingwalls may be subdivided into three groups according to their relation to the abutment: straight, splayed, and return wingwalls. The first group refers to the wingwalls continuing from the main abutment structure, as illustrated in Figure 6.3-1. The third group has the wingwalls extending at a right angle from the main abutment structure, as shown in Figure 6.3-2. Abutments with return wingwalls are often used when the embankment is steep, and they are also called U abutments, referring to the abutment–wingwall shape in the plan. The second group of wingwalls is between the first and third groups with respect to the turning angle extending from the main abutment.

Wingwalls are used mainly to retain the soil and thereby transmit live load. In contrast, the abutment structure needs to function not only as a retaining wall but also as a structure to transmit superstructure loads to the ground as well as to provide access of vehicle traffic and sometimes also pedestrian traffic. Note that the wingwall may or may not be structurally continuous with the main abutment. Usually it does not need to be as thick as the stem wall in the abutment because it carries much less or no load from the superstructure.

6.4 Foundations

Foundations are important components of bridge systems. They transmit all loads to the ground. It should be emphasized that foundation design requires expertise in both structural engineering and geotechnical engineering. While bridge engineers are typically well educated and trained in the area of structural engineering, geotechnical engineers need to be the decision makers in foundation design with assistance from structural engineers.

In general, three kinds of foundation systems are commonly used as bridge foundations for piers or abutments: spread footings, driven piles, and drilled shafts. Spread footings are also referred to as shallow foundations and driven piles and drilled shafts as deep foundations. Some examples are seen in Figures 6.4-1 and 6.4-2. Foundations are usually not readily visable after construction is completed.

A subsurface investigation, including borings and soil tests, is required according to AASHTO design specifications to provide pertinent and sufficient information for the design of substructure units. The type and cost of foundations should also be considered in the economic and aesthetic studies for location and bridge alternate selection.

It should be emphasized that earth masses move and so do water channels, although often slowly. Over decades, if not centuries, of the expected bridge life, this continues to take place. It is important to understand the history of such movements and use that information to determine what to expect over the life of the new bridge being designed. In the AASHTO design specifications, the current topography of the bridge site is required to be established via contour maps and photographs. Such studies need to include the history of the site in terms of movement of earth masses, soil

Figure 6.4-1
Foundation support to twin-arch bridges.

Figure 6.4-2
Foundation support to steel
box arch bridge.

and rock erosion, and meandering of waterways. Corresponding measures can then be developed to accommodate future movements if expected and significant.

6.4.1 Spread Footings

Spread footings refer to shallow foundations supported on soil or rock (see Figure 6.4-3). Some bridge owners require them to be supported on rock. This type of foundation seems to be not as widely used as deeper foundations of driven piles or drilled shafts for bridge structures. It is perhaps

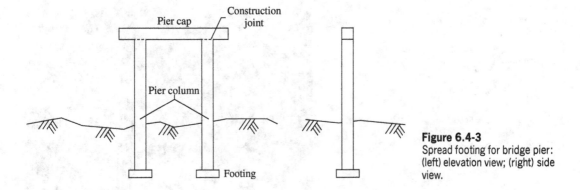

Figure 6.4-3
Spread footing for bridge pier:
(left) elevation view; (right) side
view.

because bridge design has been conservatively practiced and also the loads to be transferred to the earth through the foundations are significant.

Spread footings consist of a footing to distribute the force and moment from the substructure of the bridge to the foundation material of soil or rock. The footing's footprint area needs to be large enough to eliminate tensile contact stress on the bottom surface of the footing. Namely, on the soil or rock there should be no tensile stress because the contact they will have would not be able to develop such a mechanical relation. In other words, the foundation material is considered to have a zero tensile capacity for spread footings.

Driven piles are commonly used as bridge foundations. They are typically slender and deep units usually partly embedded in the ground. They may be installed using a number of techniques. Figure 6.4-4 shows an example of pile-driving equipment for bridge pier foundation. Figures 6.4-5 and 6.4-6 show piles as pier and abutment foundation, respectively. Piles derive their load-carrying capacity from the surrounding soil and/or from the soil or rock strata below their tip. As a result they may possess a capability for sustaining tensile (pulling) force. Consequently a pile group can have a resistance to moment in addition to axial compressive force because some of the piles in the group can carry tensile force and some compression force, forming a resisting moment.

6.4.2 Driven Piles

Drilled shafts are similar to driven piles as deep foundations and are also commonly used as bridge foundations. The major difference between drilled shafts and driven piles is in the construction methods. Piles are manufactured before being placed in the ground. Drilled shafts are formed in place. They are constructed using fresh concrete placed in a drilled hole with or without steel reinforcement. Drilled shafts derive their capacity from the surrounding soil and/or from the soil or rock strata below their tip, similar to driven piles. Drilled shafts are also commonly referred to as

6.4.3 Drilled Shafts

Figure 6.4-4
Pile driving for bridge pier foundation.

Figure 6.4-5
Driven piles for bridge
pier foundation.

Figure 6.4-6
Driven piles as bridge
abutment foundation.

caissons, drilled caissons, and bored piles. They also may have a capability
to resist moment in addition to axial compressive force.

6.5 Design of Piers

As briefly introduced earlier in this chapter, piers provide intermediate support to highway bridges between abutments. Besides vertical loads, they are
also usually subjected to lateral loads from the superstructure if constraint is
provided by the pier to the superstructure. For example, if the pier is rigidly
connected to the superstructure, these loads will apply lateral forces to the
pier: braking force BR, wind load on vehicle WL, wind load on structure WS,
and so on. When fixed bearings are used on the pier, the pier will also be
subjected to lateral force. In fact, bearings with some stiffness in the lateral
direction on the pier will also induce such forces. For example, elastomeric
bearings are also capable of transferring lateral forces to the pier. Design
of piers is required to satisfy the criteria for the service limit state and the
strength limit state, which are discussed in this section.

Piers are required to be designed with respect to at least these loads according to the AASHTO design specifications (Figure 6.5-1):

**6.5.1 Loads on
Piers**

❑ Self-weight of pier

❑ Loads applied from bridge superstructure and transmitted to pier

❑ Earth load on pier

❑ Earthquake loads as result of ground motion

For stability check, the earth loads shall be multiplied by the maximum
and/or minimum load factors given in Chapter 3, whichever govern the

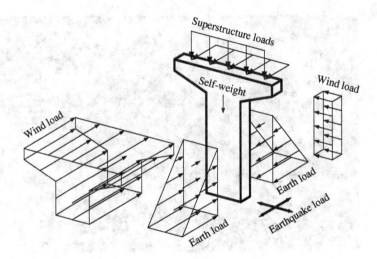

Figure 6.5-1
Loads on a pier.

design. An earth load may act as a resistance in the stability check, and thus a minimum factor for that load should be applied in order to check the most unfavorable condition. In general, the design needs to be investigated for the load combination producing the most severe condition to the structure.

6.5.2 Limit States for Piers

STRENGTH LIMIT STATE

Piers are required to be investigated at the strength limit states for the following failure modes:

❑ Bearing resistance failure if on spread footing

❑ Lateral sliding for stability

❑ Excessive loss of base contact (tensile stress)

❑ Pier structural failure

For piers supported in deep foundations, some of these failure modes are also relevant to the foundation.

SERVICE LIMIT STATE

In the AASHTO design specifications, piers are required to be investigated for excessive vertical and lateral displacement and overall stability at the service limit state. If supported on a deep foundation with driven piles or drilled shafts, the displacement and stability investigations will need to cover the foundation as well. Overall stability for piers on spread footings shall be evaluated using limit state equilibrium methods of analysis.

For footings on soil, the location of the resultant of the reaction forces shall be within the middle two-thirds of the base width. For footings on rock, the location of the resultant of the reaction forces shall be within the middle nine-tenths of the base width. This concept is based on the use of plastic bearing pressure distribution for the limit state. *11.6.3.3*

Table 6.5-1
Resistance factors for spread footing foundations   *10.5.5.2.2*

Method Soil Condition			Resistance Factor φ
Bearing resistance:	φ_b	Soils	0.45–0.50
		Footings on rock	0.45
		Plate load test	0.55
Sliding resistance	φ_t	Precast concrete placed on sand	0.90
		Cast-in-place concrete on sand	0.80
		Cast-in-place or precast concrete on clay	0.85
	φ_{ep}	Passive earth pressure component of sliding resistance	0.50

Table 6.5-2
Resistance factors for reinforced concrete piers and abutments for structural failure modes   *5.5.4.2.1*

Component	φ
Tension-controlled reinforced concrete	0.90
Tension-controlled prestressed concrete	1.00
Shear and torsion	
Normal-weight concrete	0.90
Lightweight concrete	0.70
Compression-controlled sections with or ties	0.75
Bearing on concrete	0.70
Compression in strut-and-tie models	0.70

RESISTANCE FACTORS

For piers supported on spread footing, the AASHTO specifications specify the resistance factors shown in Table 6.5-1 for the two overall strength limit states: bearing resistance and sliding.

For structural failure modes, the resistance factors for the specific materials apply. For example, for reinforced concrete as the most popularly used material for modern highway bridge piers, Table 6.5-2 displays the AASHTO resistance factors.

Compared with other substructure components, piers are also subject to hazards unique to them: collision with trucks (load CT) and with vessels (load CV). Both have been touched upon in Chapter 3.

When the possibility of collision exists from highway or waterway traffic, an appropriate risk analysis should be made to determine the capacity of impact resistance and/or the appropriate protection system. How to determine collision load CV for design has been discussed in Chapter 3.

6.5.3 Pier Protection

For collision with trucks, the AASHTO design specifications include the following general requirements. Consideration is required to be given to safe passage of vehicles both on and under the bridge. The hazard of errant vehicles inducing collision and structural failure should be minimized. This may be accomplished using three approaches. (1) Locate obstacles at a safe distance from the travel lanes. (2) Keep a safe distance between the travel lanes and the structure component of concern. (3) Design the structural component to withstand the collision load. The first distance is specified as shorter than the second, as discussed below. These approaches require different levels of cost depending on a number of site-specific factors. They should be compared and evaluated before a decision is made. Note that input from highway or safety engineers and structural engineers should be included in the assessment and decision process.

Obstacles such as barriers should be structurally independent from the pier or abutment being protected. Namely, the barriers or guardrails will not transmit loads to the bridge structure if a collision occurs. When barriers are used for this purpose, the AASHTO specifications identify the difference between 42- and 54-in.-high barriers. Full-scale crash tests have shown that some vehicles have a greater tendency to lean over or partially cross over the lower barrier than the higher one. However, if the component is more than about 10 ft away from the barrier, the difference is no longer important. *2.3.2.2.1*

Piers and abutments located within a distance of 30 ft from the edge of the roadway or 50 ft to the centerline of a railway track should be investigated for collision. The bridge owner may also decide the specific site condition.

Truck collision load may also be addressed by providing adequate structural strength, as guided in the AASHTO specifications. The pier or abutment is then required to be designed for an equivalent static force of 600 kips, acting in direction of 0 to 15 degrees with the pavement edge in a horizontal plane at a distance of 5 ft above ground (*3.6.5.1*).

In summary, where the design choice is to redirect or absorb the collision load, protection should consist of one of the following:

❑ An embankment

❑ A structurally independent, crashworthy ground-mounted 54-in.-high barrier located within 10 ft from the component being protected

❑ A 42.0-in.-high barrier located more than 10 ft from the component being protected

The barriers are also required to be structurally and geometrically capable of surviving the crash test for test level 5 (TL-5), as given in Table 3.3-3. *3.6.5*

Scour is also a serious hazard to piers and abutments if they are in or near a water stream. The scour potential should be determined and the design developed to minimize failure from this condition. Expertise in hydrology and hydraulics is required to perform the analysis and design. Therefore bridges engineers typically trained in structural engineering need to work with hydraulic engineers on this issue. It should be noted that more bridge failure cases have been attributed to scour than any other causes. Adequate attention is required on this hazard.

Ice load is another load unique to piers in water. The AASHTO specifications require that the pier nose be designed to effectively break up or deflect floating ice or drift. Accordingly, facing the nose with steel (plates or angles) or granite has been an option to protect the pier and extend its life.

6.6 Design of Abutments

Abutment design is in many ways similar to pier design. A subsurface investigation, including borings and soil tests, is required in the AASHTO specifications to provide pertinent and sufficient information for the design. The type and cost of foundations should be considered in the economic and aesthetic studies for location and bridge alternate selection. Typically, when a bridge design project involves both abutments and piers, this study is included in one task, addressing both structural components.

Topography of the bridge site needs to be established via contour maps and photographs. Topographic studies are also required to include the history of the site in terms of movement of earth masses, soil and rock erosion, and so on. These studies will provide an understanding or expectation for possible future developments. Particularly when a waterway is part of the relevant topography, such as for bridges over a stream, creek, or river, it is critical to understand the the evolution of the waterway. Design of the abutment needs to be put in the perspective of viewing the bridge life as part of a long history since the topography evolves with time.

It is noted that the AASHTO specifications also state that backfill materials for abutments should be granular and free draining. Where walls retain in situ cohesive soils, drainage should be provided to reduce hydrostatic water pressure behind the wall.

Abutments are required to be designed with respect to at least these loads according to the AASHTO design specifications (Figure 6.6-1):

6.6.1 Loads on Abutments

- ❑ Lateral earth and water pressures, including any live- and dead-load surcharge
- ❑ Self-weight of abutment
- ❑ Loads applied from bridge superstructure and transmitted to abutment
- ❑ Temperature and shrinkage deformation effects
- ❑ Earthquake loads as a result of ground motion such as an earthquake

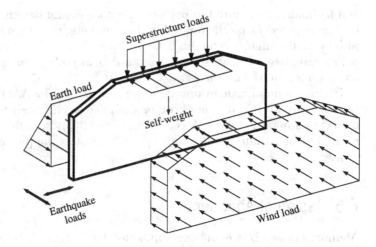

Figure 6.6-1
Loads on an abutment.

Example 6.1 illustrates the loads on a typical bridge abutment located at a seismically not very active site.

Example 6.1 Abutment Design (Loads)

❏ **Design Requirement**
   Design an abutment for the steel superstructure designed in Examples 4.9 to 4.11. Assume that the abutment is supported on a pile foundation.
   Load modifier $\eta = 1.0$

❏ **Material Properties**   *3.5.1*
   Concrete density $W_c = 0.145$ k/ft^3
   Concrete 28-day compressive strength $f'_c = 4$ k/ft^2
   Reinforcement strength $f_y = 60$ k/ft^2

❏ **Reinforcing Steel Cover Requirements**   *5.12.3*
   Backwall back cover $C_{backwall} = 2.5$ in.
   Stem back cover $C_{stem} = 2.5$ in.
   Footing top cover $C_{footing\ top} = 2.0$ in.
   Footing bottom cover $C_{footing\ bottom} = 3.0$ in.

❏ **Relevant Superstructure Data**
   Girder spacing $S = 106$ in. $= 8$ ft 10 in.
   Number of girders, $N_g = 5$
   Span length $L_{span} = 115$ ft
   Parapet height $H_{parapet} = 42$ in. $= 3.5$ ft
   Out-to-out deck width $w_{deck} = 43$ ft 2 in. $= 43.17$ ft

☐ Abutment Dimensions

The preliminary dimensions of the abutment are shown in Figures Ex6.1-1 through Ex6.1-3.

A full-depth reinforced concrete cantilever abutment is chosen for the site conditions.

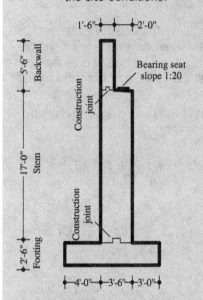

Figure Ex6.1-1
Side view of reinforced concrete cantilever abutment.

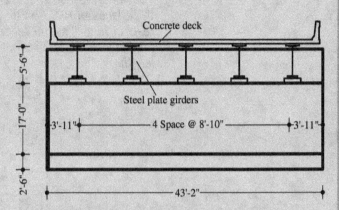

Figure Ex6.1-2
Front view of reinforced concrete cantilever abutment.

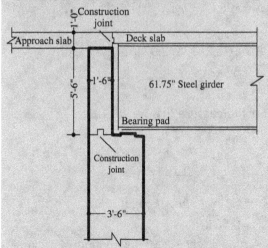

Figure Ex6.1-3
Details of abutment and superstructure.

The dimensions are based on previous designs and past experience.

☐ **Loads on Backwall**
Dead Load

$$DC_{backwall} = 5.5 \text{ ft } (1.5 \text{ ft}) \, 0.145 \text{ k/ft}^3 = 1.20 \text{ k/ft}$$

Live Load

The backwall live-load effects are computed by placing one, two, or three lanes of HL93 on the backwall, including impact and the respective multiple presence factors. This load acts at the front (bridge side) of the backwall and is distributed to the entire wall length.

$$LL_{3lanes\ backwall} = m_3 \frac{6 (16 \text{ k}) (1 + IM) + 3 (0.64 \text{ k/ft}) \, 1.5 \text{ ft}}{L_{abutment}}$$

$$= 0.85 \frac{6 (16 \text{ k}) \, 1.33 + 3 (0.64 \text{ k/ft}) \, 1.5 \text{ ft}}{43.17 \text{ ft}}$$

$$= 2.57 \text{ k/ft}$$

3.6.1.1.2

$$LL_{2lanes\ backwall} = 1.0 \frac{4 (16 \text{ k}) \, 1.33 + 2 (0.64 \text{ k/ft}) \, 1.5 \text{ ft}}{43.17 \text{ ft}}$$

$$= 2.02 \text{ k/ft}$$

$$LL_{1lane\ backwall} = 1.2 \frac{2 (16 \text{ k}) \, 1.33 + (0.64 \text{ k/ft}) \, 1.5 \text{ ft}}{43.17 \text{ ft}}$$

$$= 1.21 \text{ k/ft}$$

$$LL_{backwall} = \max (LL_{1\ lane\ backwall}, \ LL_{2lanes\ backwall},$$
$$LL_{3\ lanes\ backwall})$$
$$= \max (1.21 \text{ k/ft}, \ 2.02 \text{ k/ft}, \ 2.57 \text{ k/ft})$$
$$= 2.57 \text{ k/ft}$$

Lateral Earth Pressure EH   *3.11.5*

At bottom of backwall,

3.11.5.1 $p = k\gamma_s z = 0.3 \, (0.11 \text{ k/ft}^3) \, 5.5 \text{ ft}$
$$= 0.18 \text{ k/ft}^2$$

$$EH_{backwall} = \frac{1}{2} \left(0.18 \text{ k/ft}^3 \right) 5.5 \text{ ft}$$
$$= 0.50 \text{ k/ft}$$

at height 1.83 ft from bottom of backwall

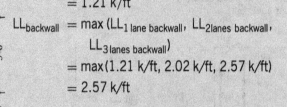

Figure Ex6.1-4
Loads on abutment backwall.

Live-Load Surcharge LS   *3.11.6.4*

$$\text{Surcharge pressure } \Delta_p = k\gamma_s h_{eq}$$

Using $k = k_a = 0.3$, $\gamma_s = 0.11$ k/ft^3, and $h_{eq} = 2$ ft, γ_s as average of loose and compacted gravel.   *3.5.1*

Abutment Height (ft)	h_{eg}(ft)
5	4
10	3
$\geq$20	2

3.11.6.4 $\Delta_p = k_a \gamma_s h_{eq} = 0.3\,(0.11\text{ k/ft}^3)\,2\text{ ft} = 0.066\text{ k/ft}^2$

The lateral load due to the live-load surcharge is

$$LS_{backwall} = \Delta_p h_{backwall} = 0.066\text{ k/ft}^2\,(5.5\text{ ft}) = 0.36\text{ k/ft}$$

Loads on Stem

If calculations are not shown, refer to "Loads on backwall" above.

Dead Loads from Superstructure

From fascia girders

$$R_{DC\,fascia} = 70.15 + 12.12 = 82.3\text{ k} \qquad R_{DW\,fascia} = 13.5\text{ k}$$

From interior girders

$$R_{DC\,interior} = 73.72 + 12.12 = 85.8\text{ k} \qquad R_{DW\,interior} = 13.5\text{ k}$$

The superstructure dead-load reactions are distributed to the abutment.

$$DC_{girders} = \frac{2R_{DC\,fascia} + 3R_{DC\,interior}}{L_{abutment}} = \frac{2\,(82.3\text{ k}) + 3\,(85.8\text{ k})}{43.17\text{ ft}} = 9.78\text{ k/ft}$$

$$DW_{girders} = \frac{2R_{DW\,fascia} + 3R_{DW\,interior}}{L_{abutment}} = \frac{2\,(13.5\text{ k}) + 3\,(13.5\text{ k})}{43.17\text{ ft}} = 1.56\text{ k/ft}$$

Dead Load from Backwall

$$DC_{backwall} = 1.20\text{ k/ft}$$

Stem Dead Load (Figure Ex6.1-5)

$$DC_{stem} = 17\text{ ft }(3.5\text{ ft})\,0.145\text{ k/ft}^3 = 8.63\text{ k/ft}$$

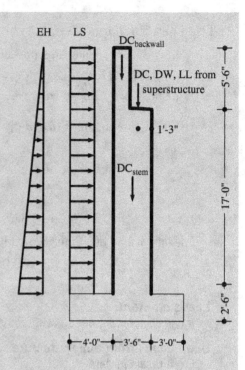

Figure Ex6.1-5
Loads on abutment stem.

Live Loads

$$LL_{girders} = V_{HL93\ truck}(1 + IM) + V_{HL93\ lane}$$

$$= 66.16\ k\,(1 + 0.33) + 36.8\ k = 124.8\ k/lane$$

$$LL_{stem} = \frac{3\ m_3\ LL_{beams}}{L_{abutment}} = \frac{3\ lanes\ (0.85)\ 124.8\ k/lane}{43.17\ ft}$$

$$= 7.37\ k/ft$$

Lateral Earth Pressure EH   *3.11.5*
At bottom of stem,

$$p = k\gamma_s z = 0.3\,(0.11\ k/ft^3)\,22.5\ ft = 0.74\ k/ft^2 \qquad 3.11.5.1$$

$$EH_{backwall} = \frac{1}{2}\left(0.74\ k/ft^2\right)22.5\ ft$$

$$= 8.3\ k/ft \qquad \text{at height 7.5 ft from bottom of stem}$$

Live-Load Surcharge LS   *3.11.6.4*

$$LS_{stem} = LS_{backwall} = 0.36\ k/ft$$

Braking Force BR

25% design truck $= 0.25(72\,k) = 18\,k$

25% design tandem $= 0.25(50\,k) = 12.5\,k$

5% [design truck $+$ design lane $(115\,ft)] = 0.05[72\,k + 0.64\,k/ft\,(115\,ft)] = 7.3\,k$

5% [design tandem $+$ design lane $(115\,ft)] = 0.05\,[50\,k + 0.64\,k/ft\,(115\,ft)] = 6.2\,k$

$$BR = \frac{\max\,(18,\ 12.5,\ 7.3,\ 6.2)}{L_{abutment}} = \frac{18\,k}{43.17\,ft} = 0.42\,k/ft$$

Acting 6 ft above roadway.   *3.6.4*

Wind Load on Superstructure WS
For usual girder and slab bridges with span length < 125 ft and height < 30 ft above ground.   *3.8.1.2.2.*
Transverse wind load of $0.05\,k/ft^2$ is ignored because it will not contribute to control stem design.
Longitudinal wind load $= 0$

Wind Load on Substructure (Abutment) WS   *3.8.1.2.3*
In the longitudinal direction: $0.04\,k/ft(17\,ft) = 0.68\,k/ft$
In the transverse direction: WS is not calculated because it will not contribute to govern stem design.

Wind Load on Vehicles WL
For usual girder and slab bridges with span length < 125 ft and height < 30 ft above ground. Acting 6 ft above roadway.   *3.8.1.3*
In the longitudinal direction, $0.04\,k/ft\,(115\,ft) = 4.6\,k/lane$

$$WL = \frac{4.6\,k/lane\,(3\,lanes)\,0.85}{L_{abutment}} = \frac{11.7\,k}{43.17\,ft} = 0.27\,k/ft$$

In the transverse direction, WL is not calculated because it will not contribute to govern stem design.

Vertical Wind Load WS
Strength III limit state only.   *3.8.2*

Wind pressure (plane area of bridge) $= 0.02\,k/ft^2\,(115\,ft/2)$

$$= 1.15\,k/ft$$

Earthquake Load EQ

Assume the bridge to be in seismic zone I with an acceleration coefficient of 0.03 and soil type I. Thus no seismic analysis is required except providing the minimum connection.

$$N = (8 + 0.02L + 0.08H)(1 + 0.000125 S^2)$$

$$= [8 + 0.02(115 \text{ ft}) + 0](1 + 0) = 10.3 \text{ in.} < 15 \text{ in. provided.}$$

OK for earthquake load.

Temperature Load TU

Assume steel girder setting temperature $T_{setting} = 40\,°F$. The temperature range is −30 to 120°F for cold climate.   *3.12.2.1*

Expansion will cause a force toward the abutment, canceling the effects of EH and LS and creating a noncritical loading condition. Thus it is not computed.

Contraction:

$$\Delta T_{fall} = T_{setting} - (-30) = 40°F - (-30°F) = 70°F$$

6.4.1
$$\begin{aligned}\Delta_{expansion} &= \varepsilon(\Delta T_{fall})L_{span} \\ &= 6.5 \times 10^{-6}\,\text{in./in./°F} \,(70°F)\, 115 \text{ ft} \,(12 \text{ in./ft}) \\ &= 0.63 \text{ in.}\end{aligned}$$

14.6.3.1
$$TU_{bearing} = GA\frac{\Delta_{expansion}}{h_{rt}} = 0.095 \text{ k/in}^2 \left(182 \text{ in.}^2\right) \frac{0.63 \text{ in.}}{3.5 \text{ in.}}$$

$$= 3.1 \text{ k/bearing}$$

Assumed elastomer shear modulus G, area A, and total thickness h_{rt}
14.7.5.2

$$\begin{aligned}TU &= \frac{5\ TU_{bearing}}{L_{abutment}} \\ &= \frac{5 \text{ bearings } (3.1 \text{ k/bearing})}{43.17 \text{ ft}} \\ &= 0.36 \text{ k/ft}\end{aligned}$$

☐ Loads on Footing

See "Loads on backwall" and/or "Loads on stem" above if more details are desired.

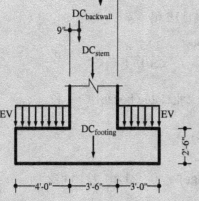

Figure Ex6.1-6
Loads on abutment footing.

Dead Loads

$$DC_{footing} = 10.5 \text{ ft } (2.5 \text{ ft}) \, 0.145 \text{ k/ft}^3 = 3.84 \text{ k/ft}$$

DC from stem and backwall:

$$DC_{stem} = 8.63 \text{ k/ft} \qquad DC_{backwall} = 1.20 \text{ k/ft}$$

DC from superstructure:

$$DC_{girders} = 9.78 \text{ k/ft} \qquad DW_{girders} = 1.56 \text{ k/ft}$$

See "loads on stem."

Earth Load EV

$$\text{Using } \gamma_S = 0.11 \text{ k/ft}^3 :$$

Average of loose and compacted gravel.   *3.5.1*

$$DC_{earth \, back} = 22.5 \text{ ft } (4 \text{ ft}) \, 0.11 \text{ k/ft}^3 = 9.90 \text{ k/ft}$$

$$DC_{earth \, front} = 1 \text{ ft } (3 \text{ ft}) \, 0.11 \text{ k/ft}^3 = 0.33 \text{ k/ft}$$

Live Loads from Superstructure

Maximum unfactored live load used for abutment footing design:

$$IM = 0.33 \, (1 - 0.125 D_E) = 0.33 \, [1 - 0.125 \, (1 \text{ ft})] = 0.29$$

Assume 1 ft of minimum depth of earth cover for footing.
    3.6.2.2

$$LL_{girder} = V_{HL93 \, truck}(1 + IM) + V_{HL93 \, lane}$$

$$= 66.16 \text{ k } (1 + 0.29) + 36.8 \text{ k} = 122.1 \text{ k/lane}$$

$$LL_{footing} = \frac{3 \, m_3 LL_{girder}}{L_{abutment}} = \frac{3 \text{ lanes } (0.85) \, 122.1 \text{ k/lane}}{43.17 \text{ ft}}$$

$$= 7.21 \text{ k/ft}$$

Lateral Earth Pressure EH   *3.11.5*

At bottom of footing,

$$p = k \gamma_s z = 0.3 \, (0.11 \text{ k/ft}^3) \, 25 \text{ ft} = 0.83 \text{ k/ft}^2 \qquad \textit{3.11.5.1}$$

$$EH_{backwall} = \frac{1}{2} \left(0.83 \text{ k/ft}^2 \right) 25 \text{ ft} = 10.3 \text{ k/ft}$$

at height 8.3 ft from footing bottom

Live-Load Surcharge LS   *3.11.6.4*

$$LS_{stem} = LS_{backwall} = LS_{footing} = 0.36 \text{ k/ft}$$

Braking Force BR

$$BR = 0.42 \, k/ft$$

Acting 6 ft above roadway. See "loads on stem."   3.6.4

Wind Load on Substructure (Abutment) WS   3.8.1.2.3

In the longitudinal direction, WS = 0.9 k/ft.

In the transverse direction, WS is not calculated because it will not control footing design.

Wind Load on Vehicles WL

For usual girder and slab bridges with span length < 125 ft and height < 30 ft above ground. Acting 6 ft above roadway.
3.8.1.3

In the longitudinal direction, WL = 0.27 k/ft.

See "Loads on stem."

In the transverse direction, WL is not calculated because it will not contribute to govern footing design.

Vertical Wind Load WS

Strength III limit state only.   3.8.2

Wind pressure = 1.15 k/ft

(See "Loads on stem.")

Temperature Load TU

$$TU = 0.36 \, k/ft$$

See "Loads on Stem."

To check stability, the earth loads shall be multiplied by the maximum and/or minimum load factors given in Chapter 3, whichever govern the design. An earth load may act as a resistance in the stability check, and thus a minimum load factor for that load should be applied in order to check the most unfavorable condition. In general, the design needs to be investigated for the combination producing the most unfavorable condition to the structure.

To compute load effects in abutments, the weight of the filling material directly over an inclined or stepped rear face or over the base of a reinforced concrete spread footing may be considered as part of the effective weight of the abutment. Example 6.2 shows a typical process of analyzing load effects for the same abutment in Example 6.1 as well as their combinations for design. Examples 6.3 and 6.4 continue the design of the same abutment including wingwalls.

Example 6.2 Abutment Design (Load Effects and Combinations)

❑ Design Requirement

Analyze the load effects and their combinations for the abutment in Example 6.1 for the steel superstructure design example 4.9. The load factors are shown in Table Ex6.2-1.

Load modifier $\eta = 1.0$

Table Ex6.2-1
Load Factors

Load	Strength I	Strength III	Strength V	Service I
DC	1.25	1.25	1.25	1.00
DW	1.50	1.50	1.50	1.00
EV	1.35	1.35	1.35	1.00
LL	1.75	—	1.35	1.00
EH	1.50	1.50	1.50	1.00
LS	1.75	—	1.35	1.00
WS	—	1.40	0.40	0.30
WL	—	—	1.00	1.00
TU	0.50	0.50	0.50	1.00
BR	1.75	—	1.35	1.00

❑ Abutment Dimensions

The preliminary dimensions of the abutment are shown in Figure Ex6.2-1.

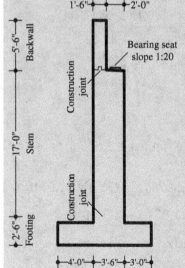

Figure Ex6.2-1
Reinforced concrete cantilever abutment.

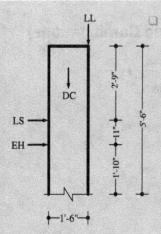

Figure Ex6.2-2
Loads on abutment backwall
(see Example 6.1 for load
magnitudes).

☐ Backwall Bottom Cross Section

Vertical Forces

Load combination is summarized in Table Ex6.2-2 using Figure Ex6.2-2.

$$F_{DC\,backwall} = 1.20 \text{ k/ft} \qquad F_{LL\,backwall} = 2.57 \text{ k/ft}$$

From Example 6.1.

Table Ex6.2-2
Combined vertical forces for backwall bottom

Limit State	Load Combination	Total Load Effect
Strength I	1.25(1.2) + 1.75(2.57)	6.0 k/ft
Strength III	1.25(1.2)	1.5 k/ft
Strength V	1.25(1.2) + 1.35(2.57)	5.0 k/ft
Service I	1(1.2) + 1(2.57)	3.8 k/ft

Shear

As summarized in Table Ex6.2-3 using Figure Ex6.2-2

$$V_{EH} = 0.5 \text{ k/ft} \qquad V_{LS} = 0.36 \text{ k/ft}$$

From Example 6.1.

Table Ex6.2-3
Combined shear forces for backwall bottom

Limit State	Load Combination	Total Load Effect
Strength I	1.5(0.5) + 1.75(0.36)	1.4 k/ft
Strength III	1.5(0.5)	0.8 k/ft
Strength V	1.5(0.5) + 1.35(0.36)	1.2 k/ft
Service I	1(0.5) + 1(0.36)	0.9 k/ft

Moment

Load combination is summarized in Table Ex6.2-4 based on Figure Ex6.2-2

$$M_{DC} = 0 \qquad M_{LL} = 2.57 \text{ k/ft } (0.75 \text{ ft}) = 1.9 \text{ kft/ft}$$

$$M_{EH} = 0.5 \text{ k/ft } (1.83 \text{ ft}) = 0.9 \text{ kft/ft}$$

$$M_{LS} = 0.36 \text{ k/ft } (2.75 \text{ ft}) = 1.0 \text{ kft/ft}$$

See Figure Ex6.2-2.

Table Ex6.2-4
Combined moments for backwall bottom

Limit State	Load Combination	Total Load Effect
Strength I	$1.5(0.9) + 1.75(1.9) + 1.75(1)$	6.4 kft/ft
Strength III	$1.5(0.9)$	1.4 kft/ft
Strength V	$1.5(0.9) + 1.35(1.9) + 1.35(1)$	5.3 kft/ft
Service I	$1(0.9) + 1(1.9) + 1(1)$	3.8 kft/ft

Bottom Cross Section of Stem
Vertical Forces

$$F_{DC\ backwall} = 1.20\ k/ft \qquad F_{DC\ stem} = 8.63\ k/ft \qquad F_{DC\ girder} = 9.78\ k/ft$$
$$F_{DW\ girder} = 1.56\ k/ft \qquad F_{LL\ girder} = 7.37\ k/ft \qquad F_{WS\ vertical} = 1.15\ k/ft$$

From Example 6.1. Load combination is summarized in Table Ex6.2-5. Also see Figure Ex6.2-3.

Table Ex6.2-5
Combined vertical forces for stem bottom

Limit State	Load Combination	Total Load Effect
Strength I	$1.25(1.2 + 8.63 + 9.78) + 1.5(1.56) + 1.75(7.37)$	39.8 k/ft
Strength III	$1.25(1.2 + 8.63 + 9.78) + 1.4(1.15)$	26.1 k/ft
Strength V	$1.25(1.2 + 8.63 + 9.78) + 1.5(1.56) + 1.35(7.37)$	36.8 k/ft
Service I	$1(1.2 + 8.63 + 9.78) + 1(1.56) + 1(7.37)$	28.5 k/ft

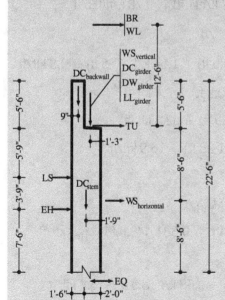

Figure Ex6.2-3
Loads on abutment stem (see Example 6.1 for load magnitudes).

Vertical WS is for strength III and service IV limit states only. 3.8.2.

Shears

From Example 6.1. Load Combination is summarized in Table Ex6.2-6. Also see Figure Ex6.2-3.

$$V_{EH} = 8.3 \text{ k/ft} \qquad V_{LS} = 0.066 \text{ k/ft}^2 \ (22.5 \text{ ft}) = 1.5 \text{ k/ft}$$

$$V_{BR} = 0.42 \text{ k/ft} \qquad V_{WL} = 0.27 \text{ k/ft}$$

$$V_{WS} = 0.68 \text{ k/ft} \qquad V_{TU} = 0.36 \text{ k/ft}$$

Table Ex6.2-6
Combined shear forces for stem bottom

Limit State	Load Combination	Total Load Effect
Strength I	$1.5(8.3) + 1.75(1.5) + 1.75(0.42) + 0.5(0.36)$	16.0 k/ft
Strength III	$1.5(8.3) + 1.4(0.68) + 0.5(0.36)$	13.6 k/ft
Strength V	$1.5(8.3) + 1.35(1.5) + 0.4(0.68) + 1(0.27) + 1.35(0.42) + 0.5(0.36)$	15.8 k/ft
Service I	$1(8.3) + 1(1.5) + 0.3(0.68) + 1(0.27) + 1(0.42) + 1(0.36)$	11.1 k/ft

Moments

$$M_{DC} = -F_{DC \text{ backwall}}(1 \text{ ft}) + F_{DC \text{ girder}}(0.5 \text{ ft})$$

$$= -1.2 \text{ k/ft}(1 \text{ ft}) + 9.78 \text{ k/ft}(0.5 \text{ ft})$$

$$= 3.7 \text{ kft/ft}$$

$$M_{DW} = F_{DW \text{ girder}} (0.5 \text{ ft}) = 1.56 \text{ k/ft} (0.5 \text{ ft}) = 0.8 \text{ kft/ft}$$

$$M_{LL} = 7.37 \text{ k/ft} (0.5 \text{ ft}) = 3.7 \text{ kft/ft}$$

$$M_{EH} = 8.3 \text{ k/ft} (7.5 \text{ ft}) = 62.3 \text{ kft/ft}$$

$$M_{LS} = 1.5 \text{ k/ft} (11.25 \text{ ft}) = 16.9 \text{ kft/ft}$$

For Strength III limit state:

$$M_{WS} = M_{WS \text{ vertical}} + M_{WS \text{ horizontal}}$$

$$= 1.15 \text{ k/ft}(0.5 \text{ ft}) + 0.68 \text{ k/ft}(8.5 \text{ ft}) = 6.4 \text{ kft/ft}$$

For other limit states:

$$M_{WS} = M_{WS \text{ horizontal}} = 0.68 \text{ k/ft} (8.5 \text{ ft}) = 5.8 \text{ kft/ft}$$

$$M_{WL} = 0.27 \text{ k/ft } (29.5 \text{ ft}) = 8.0 \text{ kft/ft}$$

$$M_{TU} = 0.36 \text{ k/ft } (17 \text{ ft}) = 6.1 \text{ kft/ft}$$

$$M_{BR} = 0.42 \text{ k/ft } (29.5 \text{ ft}) = 12.4 \text{ kft/ft}$$

See Example 6.1. Load combination is summarized in Table Ex6.2-7. Also see Figure Ex6.2-3.

Table Ex6.2-7

Combined moments for stem bottom

Limit State	Load Combination	Total Load Effect
Strength I	$1.25(3.7) + 1.5(0.8) + 1.75(3.7) + 1.5(62.3) + 1.75(16.9)$ $+ 0.5(6.1) + 1.75(12.4)$	160.1 kft/ft
Strength III	$1.25(3.7) + 1.5(0.8) + 1.5(62.3) + 1.4(6.4) + 0.5(6.1)$	111.3 kft/ft
Strength V	$1.25(3.7) + 1.5(0.8) + 1.35(3.7) + 1.5(62.3) + 1.35(16.9)$ $+ 0.4(5.8) + 1(8.0) + 0.5(6.1) + 1.35(12.4)$	157.2 kft/ft
Service I	$1(3.7) + 1(0.8) + 1(3.7) + 1(62.3) + 1(16.9) + 0.3(5.8)$ $+ 1(8.0) + 1(6.1) + 1(12.4)$	115.6 kft/ft

Vertical WS is for Strength III and Service IV limit states only. 3.8.2.

Bottom Cross Section of Footing
Vertical Forces

$$F_{DC \text{ backwall}} = 1.20 \text{ k/ft} \qquad F_{DC \text{ stem}} = 8.63 \text{ k/ft} \qquad F_{DC \text{ girder}} = 9.78 \text{ k/ft}$$

$$F_{DW \text{ girder}} = 1.56 \text{ k/ft} \qquad F_{LL \text{ girder}} = 7.37 \text{ k/ft} \qquad F_{WS \text{ vertical}} = 1.15 \text{ k/ft}$$

$$F_{EV} = 9.90 \text{ k/ft} + 0.33 \text{ k/ft} = 10.2 \text{ k/ft} \qquad F_{DC \text{ footing}} = 3.84 \text{ k/ft}$$

From Example 6.1. Load combination is summarized in Table Ex6.2-8. Also see Figure Ex6.2-4.

Table Ex6.2-8

Combined vertical forces for footing bottom

Limit State	Load Combination	Total Load Effect
Strength I	$1.25(1.2 + 8.63 + 9.78 + 3.84) + 1.35(10.2) + 1.5(1.56) + 1.75(7.37)$	58.3 k/ft
Strength III	$1.25(1.2 + 8.63 + 9.78 + 3.84) + 1.35(10.2) + 1.4(1.15)$	44.7 k/ft
Strength V	$1.25(1.2 + 8.63 + 9.78 + 3.84) + 1.35(10.2) + 1.5(1.56) + 1.35(7.37)$	55.4 k/ft
Service I	$1(1.2 + 8.63 + 9.78 + 3.84) + 1(10.2) + 1(1.56) + 1(7.37)$	42.6 k/ft

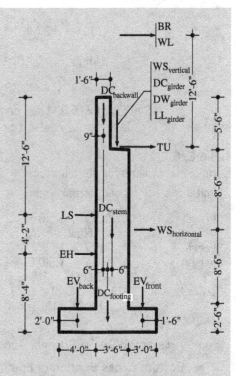

Figure Ex6.2-4
Loads on abutment footing (see Example 6.1 for load magnitudes).

Vertical WS is for Strength III and Service IV limit states only. 3.8.2.

Shears:

From Example 6.1. Load combination is summarized in Table Ex6.2-9. Also see Figure Ex6.2-4.

$$V_{EH} = 10.3 \text{ k/ft} \quad V_{LS} = 0.066 \text{ k/ft}^2 \text{ (25 ft)} = 1.65 \text{ k/ft}$$
$$V_{BR} = 0.42 \text{ k/ft} \quad V_{WL} = 0.27 \text{ k/ft}$$
$$V_{WS} = 0.68 \text{ k/ft} \quad V_{TU} = 0.36 \text{ k/ft}$$

Table Ex6.2-9
Combined shear forces for footing bottom

Limit State	Load Combination	Total Load Effect
Strength I	$1.5(10.3) + 1.75(1.65) + 1.75(0.42) + 0.5(0.36)$	19.3 k/ft
Strength III	$1.5(10.3) + 1.4(0.68) + 0.5(0.36)$	16.6 k/ft
Strength V	$1.5(10.3) + 1.35(1.65) + 0.4(0.68) + 1(0.27) + 1.35(0.42) + 0.5(0.36)$	19.0 k/ft
Service I	$1(10.3) + 1(1.65) + 0.3(0.68) + 1(0.27) + 1(0.42) + 1(0.36)$	13.2 k/ft

Moments:

$$M_{DC} = -F_{DC\ backwall}(0.5\ ft) + F_{DC\ stem}(0.5\ ft) + F_{DC\ girder}(1\ ft)$$

$$= -1.2\ k/ft(0.5\ ft) + 8.63\ k/ft(0.5\ ft) + 9.78\ k/ft(1\ ft) = 13.5\ kft/ft$$

$$M_{DW} = F_{DW\ girder}\ (1\ ft) = 1.56\ k/ft\ (1\ ft) = 1.56\ kft/ft$$

$$M_{LL} = 7.37\ k/ft\ (1\ ft) = 7.4\ kft/ft$$

$$M_{EH} = 10.3\ k/ft\ (8.3\ ft) = 85.5\ kft/ft$$

$$M_{LS} = 1.65\ k/ft\ (12.5\ ft) = 20.6\ kft/ft$$

For strength III limit state:

$$M_{WS} = M_{WS\ vertical} + M_{WS\ horizontal}$$

$$= 1.15\ k/ft(1\ ft) + 0.68\ k/ft(11\ ft) = 8.6\ kft/ft$$

For other limit states:

$$M_{WS} = M_{WS\ horizontal} = 0.68\ k/ft\ (11\ ft) = 7.5\ kft/ft$$

$$M_{WL} = 0.27\ k/ft\ (32\ ft) = 8.6\ kft/ft$$

$$M_{TU} = 0.36\ k/ft\ (19.5\ ft) = 7.0\ kft/ft$$

$$M_{BR} = 0.42\ k/ft\ (32\ ft) = 13.4\ kft/ft$$

See Example 6.1. Load combination is summarized in Table Ex6.2-10. Also see Figure Ex6.2-4.

Table Ex6.2-10
Combined moments for footing bottom

Limit State	Load Combination	Total Load Effect
Strength I	1.25(13.5) + 1.5(1.56) + 1.75(7.4) + 1.5(85.5) + 1.75(20.6) + 0.5(7.0) + 1.75(13.4)	223.4 kft/ft
Strength III	1.25(13.5) + 1.5(1.56) + 1.5(85.5) + 1.4(8.6) + 0.5(7.0)	163.0 kft/ft
Strength V	1.25(13.5) + 1.5(1.56) + 1.35(7.4) + 1.5(85.5) + 1.35(20.6) + 0.4(7.5) + 1(8.6) + 0.5(7.0) + 1.35(13.4)	218.5 kft/ft
Service I	1(13.5) + 1(1.56) + 1(7.4) + 1(85.8) + 1(20.6) + 0.3(7.5) + 1(8.6) + 1(7.0) + 1(13.4)	159.8 kft/ft

Vertical WS is for Strength III and Service IV limit states only. 3.8.2.

6.6.2 Limit States and Resistance Factors for Design

Design of abutments is required to satisfy the criteria for the service limit state and strength limit state. As discussed above, abutments should be designed to withstand loads transmitted through the superstructures, lateral earth and water pressures with any live and dead load surcharge if applicable, the self-weight of the wall, temperature and shrinkage effects, and earthquake loads. These loads need to be combined according to the limit states defined in the specifications as discussed in Chapter 3. Several significant and unique aspects of this combination are emphasized below. After load effect analysis, Example 6.2 shows how these load effects should be combined for corresponding limit states and specific components.

STRENGTH LIMIT STATE

Abutments are required to be investigated at the strength limit states at least for the following failure modes:

- ❏ Bearing resistance failure
- ❏ Lateral sliding
- ❏ Excessive loss of base contact
- ❏ Pullout failure of anchors or soil reinforcements
- ❏ Abutment component structural failure

SERVICE LIMIT STATES

Abutments also are required to be investigated for excessive vertical and lateral displacement and overall stability at the service limit state, according to the AASHTO specifications.

Similar to the requirements for piers on spread footing, for abutments on soil, the location of the resultant of the reaction forces should be within the middle two-thirds of the base width. For those on rock, the location of the resultant of the reaction forces should be within the middle nine-tenths of the base width. These requirements ensure that no tensile stress will develop at the bottom surface of the foundation against overturning. *11.6.3.3*

RESISTANCE FACTORS

For abutments not supported on deep foundation (driven piles or drilled shafts), the AASHTO specifications require the resistance factors in Table 6.5-1 for the two overall strength limit states: bearing resistance and sliding. For structural failure modes, the resistance factors for the specific materials apply. For example, for reinforced concrete as the most popularly used material for abutments, Table 6.5-2 displays the AASHTO resistance factors.

6.6.3 Section Propotioning and Reinforcement Design for Abutments

The AASHTO specifications also emphasize the requirement for controlling temperature and shrinkage cracks as follows:

5.10.8

$$A_s \geq \frac{1.3bh}{2(b+h)f_y}$$

(6.6-1)

This value of A_s is limited to

5.10.8
$$0.11 \leq A_s \leq 0.60 \qquad\qquad (6.6\text{-}2)$$

where A_s = area of reinforcement in each direction and each face
(in.^{2}/ft)
b = least width of component section (in.)
h = least thickness of component section (in.)
f_y = yield strength of reinforcing bars in k/in.^{2} $\leq$75 k/in.^{2}

When using the above equation, the calculated area of reinforcing steel must be equally distributed on both concrete faces. In addition, the maximum spacing of the temperature and shrinkage reinforcement must be the smaller of 3 times the deck thickness or 18 in.

Example 6.3 illustrates how these requirements are met as well as other aspects of a design case of reinforced concrete abutment, where estimations for loads and load effects have been performed in Examples 6.1 and 6.2.

Example 6.3 Abutment Design (Member Design)

❏ **Design Requirement**
   The abutment and wingwall properties as well as information about
      the superstructure that the abutment supports are required.
   Load modifier $\eta = 1.0$

❏ **Material Properties**   *3.5.1*
   Concrete density $W_c = 0.145$ k/ft^3
   Concrete 28-day compressive strength $f_c' = 4.0$ k/ft^2
   Reinforcement strength $f_y = 60$ k/ft^2

   Reinforcing Steel Cover Requirements   *5.12.3*
      Backwall back cover $C_b = 2.5$ in.
      Stem back cover $C_s = 2.5$ in.
      Footing top cover $C_{ft} = 2.0$ in.
      Footing bottom cover $C_{fb} = 3.0$ in.

   Other Cover Requirements   *5.12.3*
      Backwall back cover = 2.5 in.
      Stem cover = 2.5 in.
      Footing top cover = 2.0 in.
      Footing bottom cover = 3.0 in.

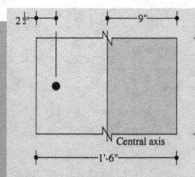

Figure Ex6.3-1
Abutment backwall cross section for calculation of M_{cr}.

☐ **Design Abutment Backwall**

Design for flexure for minimum reinforcement. Assume No. 4 bars, bar diameter = 0.5 in. bar area = 0.20 in.^{2}

The minimum reinforcement requirement will be met first. 5.7.3.3.2

For $M_{cr} = S_c f_r$,   5.7.3.3.2

$$S_c = \frac{bh^2}{6} = \frac{12\,(18)^2}{6} = 648 \text{ in.}^3$$

according to Figure Ex6.3-1.

5.4.2.6 $f_r' = 0.24\sqrt{f_c'} = 0.24\sqrt{4 \text{ k/in.}^2} = 0.48 \text{ k/in.}^2$

$M_{cr} = 648 \text{ in.}^3 \left(0.48 \text{ k/in.}^2\right) = 25.92 \text{ kft/ft}$

Use $M_{design} = \min\left((0.75)(1.6)M_{cr},\ 1.33M_u\right)$ to design steel spacing s:

$$1.2\,M_{cr} = 1.2\,(25.92 \text{ kft/ft}) = 31.10 \text{ kft/ft}$$

$$1.33\,M_{strength\ I\ backwall} = 1.33\,(6.4 \text{ kft/ft})$$

$$= 8.5 \text{ kft/ft} < 31.10 \text{ kft/ft} \qquad \text{controls}$$

See Example 6.6 for $M_{strength\ I\ stem}$.

So $M_{design} = 8.5 \text{ kft/ft}$.

Find the required amount of reinforcing steel A_s:

$$\text{Effective depth } d_e = h - C_{bot} - \frac{\text{bar diameter}}{2}$$

$$= 18 \text{ in.} - 2.5 \text{ in.} - \frac{0.5 \text{ in.}}{2} = 15.25 \text{ in.}$$

5.5.4.2.1 $R_n = \dfrac{M_{design}}{\varphi_f b d_e^2} = \dfrac{8.5 \text{ kft/ft}\,(12 \text{ in./ft})}{(0.9)\,12 \text{ in.}\,(15.25 \text{ in.})^2} = 0.041 \text{ k/in.}^2$

$$\rho = 0.85\frac{f_c'}{f_y}\left[1 - \sqrt{1 - \frac{2R_n}{0.85f_c}}\right]$$

$$= 0.85\frac{4 \text{ k/in.}^2}{60 \text{ k/in.}^2}\left[1 - \sqrt{1 - \frac{2\,(0.041 \text{ k/in.}^2)}{0.85\,(4 \text{ k/in.}^2)}}\right] = 0.00069$$

The above equations are for strength design of reinforced concrete.

$$A_s = \rho b d_e = 0.00069(12 \text{ in.})15.25 \text{ n.} = 0.13 \text{ in.}^2/\text{ft}$$

$$\text{Required bar spacing } s = \frac{\text{bar area}}{A_s} = \frac{0.20 \text{ in.}^2}{0.13 \text{ in.}^2/\text{ft}} = 1.54 \text{ ft} = 18.5 \text{ in.}$$

Use No. 4 bars at 18 in.

$$A_{s \text{ provided}} = \text{bar area}\frac{12 \text{ in.}}{\text{bar spacing}} = 0.2 \text{ in.}^2\frac{12 \text{ in./ft}}{18 \text{ in.}} = 0.13 \text{ in.}^2/\text{ft}$$

OK for minimum reinforcement and Strength I limit state for backwall.

Design Distribution Steel for Crack Control under Service I Limit State

$$\text{For} \quad s \leq \frac{700\gamma_e}{\beta_s f_{ss}} - 2d_c$$

$$D_c = 2.5 \text{ in.} \quad \gamma_e = 0.75$$

For substructure. 5.7.3.4

$$\beta_s = 1 + \frac{d_c}{0.7(h - d_c)} = 1 + \frac{2.5 \text{ in.}}{0.7(18 \text{ in.} - 2.5 \text{ in.})} = 1.23$$

$$f_{ss} = \frac{M_{\text{service I}}}{A_s j d_s}$$

where

$$A_s = 0.13 \text{ in.}^2/\text{ft} \qquad j = 1 - \frac{k}{3}$$

$$k = \sqrt{(\rho n)^2 + 2\rho n} - \rho n$$

$$= \sqrt{[0.00069(8)]^2 + 2(0.00069)8} - 0.00069(8)$$

$$= 0.10$$

$$j = 0.97$$

Thus,

$$f_{ss} = \frac{3.8 \text{ kft/ft}(12 \text{ in./ft})}{0.13 \text{ in.}^2/\text{ft}(0.97)15.25 \text{ in.}} = 23.7 \text{ k/in.}^2$$

$$s \leq \frac{700\gamma_e}{\beta_s f_s} - 2d_c = \frac{700(0.75)}{1.23(23.7)}\text{in.} - 2(2.5 \text{ in.})$$

$$= 13 \text{ in.}$$

Use No. 4 bars at 12. in superceding the 18 in. spacing based on Stregnth I limit state.

OK for moment Service I limit state.

OK also for minimum reinforcement and strength I limit state for backwall.

Design for Shear

The factored longitudinal shear force at the base of the backwall:

$$V_{u \text{ backwall strength I}} = 1.4 \text{ k/ft}$$

See Example 6.6 for $V_{u \text{ backwall strength I}}$.

The nominal shear resistance is the lesser of

5.8.3.3 $\qquad\qquad V_c + V_s \quad \text{or} \quad 0.25 f'_c b_v d_v$

where

5.8.2.9 $\qquad d_v = \max[0.9 \, d_e, 0.72 \, h]$
$\qquad\qquad = \max[0.9\,(15.25 \text{ in.}), 0.72\,(18 \text{ in.})]$
$\qquad\qquad = \max(13.73 \text{ in. } 12.96 \text{ in.}) = 13.73 \text{ in.}$

$$V_c = 0.0316 \, \beta \sqrt{f'_c} b_v d_v$$

$$= 0.0316\,(2)\sqrt{4 \text{ k/in.}^2}\,12 \text{ in./ft}\,(13.73 \text{ in.})$$

$$= 20.8 \text{ k/ft}$$

$$V_s = \frac{A_v f_y d_v \,(\cot\theta + \cot\alpha)\sin\alpha}{S}$$

is neglected for this abutment.

$$0.25\,f'_c\,b_v d_v = 0.25\,\left(4 \text{ k/in.}^2\right)\,12 \text{ in./ft}\,(13.7 \text{ in.})$$

$$= 164.9 \text{ k/ft}$$

$$V_n = \min\,(20.8, 164.9) = 20.8 \text{ k/ft}$$

Factored shear resistance:

$$V_r = \phi_v V_n = 0.90\,(20.8 \text{ k/ft})$$

$$= 18.7 \text{ k/ft} > V_{u \text{ backwall strength I}}$$

$$= 1.4 \text{ k/ft}$$

OK for shear in backwall.

Shrinkage and Temperature Reinforcement for Front Sur-face Exposed to Daily Temperature Variation

$$A_s \geq \frac{1.3\,bh}{2\,(b+h)\,f_y} = \frac{1.3\,(66\ \text{in.})\,18\ \text{in.}}{2\,(66\ \text{in.} + 18\ \text{in.})\,60}$$

$$= 0.15\ \text{in.}^2/\text{ft}$$

and A_s needs to be $0.11 \leq A_s \leq 0.6$.

Use No. 4 bars at 12 in. for the front face and each direction and $0.11 < 0.15 < A_{s\ provided} = 0.2\ \text{in.}^2/\text{ft} < 0.6$.

OK for shrinkage and temperature steel.

❑ **Design Abutment Stem**

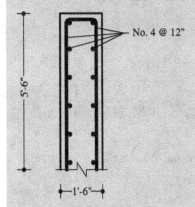

Figure Ex6.3-2
Reinforcement for abutment backwall.

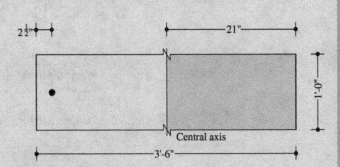

Figure Ex6.3-3
Abutment stem cross section for calculation of M_{cr}.

❑ **Design for Minimum Flexure Reinforcement**

Assume No. 8 bars, bar diameter = 1 in., bar area = 0.79 in.^{2}
The minimum reinforcement requirement will be met first.
5.7.3.3.2

For $M_{cr} = \gamma_3\gamma_1 S_c F_r$ 5.7.3.3.2

$$S_c = \frac{bh^2}{6} = \frac{12\,(42)^2}{6} = 3528\ \text{in.}^3$$

5.4.2.6 $f_r = 0.24\sqrt{f_c'} = 0.24\sqrt{4\ \text{k/in.}^2} = 0.48\ \text{k/in.}^2$

$$M_{cr} = 0.75(1.6)3528\ in.^3\left(0.48\ \text{k/in.}^2\right) = 169.3\ \text{kft/ft}$$

Use $M_{design} = \min(M_{cr}, 1.33M_u)$ to design steel spacing s:

$$1.33\,M_{strength\,I\,stem} = 1.33\,(160.1\text{ kft/ft}) = 212.9\text{ kft/ft}$$

$$> 169.3\text{ kft/ft} \quad \text{so } M_{cr} \text{ controls}$$

See Example 6.2 for $M_{strength\,I\,stem}$.
So $M_{design} = 169.3$ kft/ft.
Find the required amount of reinforcing steel A_s:

$$\text{Effective depth } d_e = h - C_{bot} - \frac{\text{bar diameter}}{2}$$

$$= 42\text{ in.} - 2.5\text{ in.} - \frac{1\text{ in.}}{2} = 39\text{ in.}$$

5.5.4.2.1 $$R_n = \frac{M_{design}}{\varphi_f b d_e^2} = \frac{169.3\text{ kft/ft }(12\text{ in./ft})}{(0.9)\,12\text{ in.}\,(39\text{ in.})^2} = 0.12\text{ k/in.}^2$$

$$\rho = 0.85\frac{f'_c}{f_y}\left[1 - \sqrt{1 - \frac{2R_n}{0.85f_c}}\right]$$

$$= 0.85\frac{4\text{ k/in.}^2}{60\text{ k/in.}^2}\left[1 - \sqrt{1 - \frac{2\,(0.12\text{ k/in.}^2)}{0.85\,(4\text{ k/in.}^2)}}\right]$$

$$= 0.002$$

The above equations are for strength design of reinforced concrete.

$$A_s = \rho b d_e = 0.002(12\text{ in.})39\text{ in.} = 0.953\text{ in.}^2/\text{ft}$$

Required bar spacing s using No. 8 bars $= \dfrac{\text{bar area}}{A_s}$

$$= \frac{0.79\text{ in.}^2}{0.953\text{ in.}^2/\text{ft}}$$

$$= 0.83\text{ ft} = 9.9\text{ in.}$$

Use No. 8 bars at 9 in.

$$A_{s,\,provided} = \text{bar area}\frac{12\text{ in.}}{\text{bar spacing}}$$

$$= 0.79\text{ in.}^2\frac{12\text{ in.ft}}{9\text{ in.}}$$

$$= 1.05\text{ in.}^2/\text{ft} > \text{required } 0.953\text{ in.}^2/\text{ft}$$

$M_{design} = 169.3$ kft/ft $> M_u = 160.1$ kft/ft for Strength I limit State.

OK for minimum reinforcement and strength I limit state for back-wall.

Design Distribution Steel for Crack Control under Service I Limit State

$$\text{For} \quad s \leq \frac{700\gamma_e}{\beta_s f_{ss}} - 2d_c$$

$$D_c = 2.5 \text{ in.} \qquad \gamma_e = 0.75$$

For substructure.   *5.7.3.4.*

$$\beta_s = 1 + \frac{d_c}{0.7(h - d_c)} = 1 + \frac{2.5 \text{ in.}}{0.7(42 \text{ in.} - 2.5 \text{ in.})} = 1.09$$

$$f_{ss} = \frac{M_{service}}{A_s j d_s}$$

where

$$A_s = 1.05 \text{ in.}^2/\text{ft} \quad j = 1 - \frac{k}{3}$$

$$k = \sqrt{(\rho n)^2 + 2\rho n} - \rho n$$

$$= \sqrt{[0.0026(8)]^2 + 2(0.0026)8} - 0.0026(8)$$

$$= 0.18$$

$$j = 0.94$$

Thus,

$$f_{ss} = \frac{115.6 \text{ kft/ft}(12 \text{ in./ft})}{1.05 \text{ in.}^2/\text{ft}(0.94)\,39 \text{ in.}} = 36.0 \text{ k/in.}^2$$

See Example 6.2 for $M_{service\,I}$.

$$s \leq \frac{700\gamma_e}{\beta_s f_s} - 2d_c = \frac{700(0.75)}{1.09(36)}\text{in.} - 2(2.5 \text{ in.}) = 8.4 \text{ in.} \; < \; 9 \text{ in.}$$

Use No. 8 bars at 8 in. based on the requirements for minimum steel and Stregnth I limit state.

OK for minimum reinforcement and Strength I limit state for stem I.

Design for Shear

The factored longitudinal shear force at the base of the stem:

$$V_{strength\ I\ stem} = 16\ k/ft$$

See Example 6.2 for $V_{strength\ I\ stem}$.

The nominal shear resistance is the lesser of

5.8.3.3 $V_c + V_s$ or $0.25f'_c b_v d_v$

where

5.8.2.9 $d_v = \max[0.9d_e,\ 0.72\ h]$

$$= \max[0.9(39\ in.),\ 0.72(42\ in.)]$$

$$= \max(35.10\ in.,\ 30.24\ in.) = 35.1\ in.$$

$$V_c = 0.0316\ \beta\sqrt{f'_c}\,b_v d_v$$

$$= 0.0316(2)\sqrt{4\ k/in.^2}\,12\ in./ft\ (35.1\ in.)$$

$$= 53.2\ k/ft$$

$$V_s = \frac{A_v f_y d_v\ (\cot\theta + \cot\alpha)\ \sin\alpha}{S}$$

is neglected for this abutment

$$0.25f'_c b_v d_v = 0.25\left(4\ k/in.^2\right)12\ in./ft\ (35.1\ in.)$$

$$= 421.2\ k/ft$$

$$V_n = \min(53.2,\ 421.2) = 53.2\ k/ft$$

Factored shear resistance:

$$V_r = \phi_v V_n = 0.90\ (53.2\ k/ft)$$

$$= 47.9\ k/ft > V_{u,\ stem\ strength\ I} = 16\ k/ft$$

OK for shear in stem.

Shrinkage and Temperature Reinforcement for Exposed Surface

$$A_s \geq \frac{1.3\ bh}{2(b+h)f_y} = \frac{1.3(204\ in.)42\ in.}{2(204\ in. + 42\ in.)60}$$

$$= 0.38\ in.^2/ft$$

and A_s needs to be $0.11 \leq A_s \leq 0.6$.

Use No. 6 bars at 12 in. in the transverse direction in each face and vertical in the front (bridge) face and $0.11 < 0.38 < A_{s, provided} = 0.44$ in.^{2}/ft < 0.6.

OK for shrinkage and temperature steel.

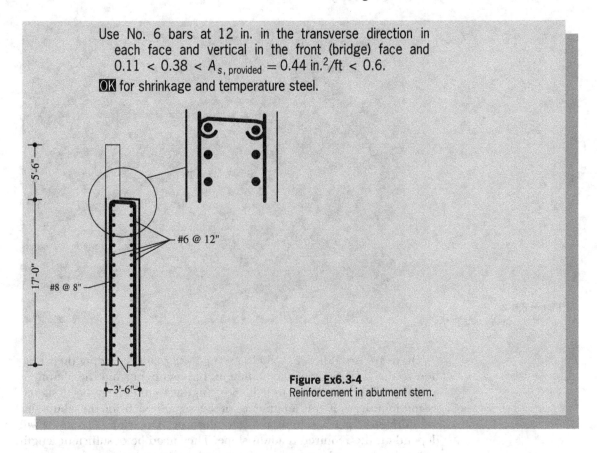

Figure Ex6.3-4
Reinforcement in abutment stem.

Wingwalls appear to be part of the abutment system, which may mislead that they belong to the abutment monolithically. In the AASHTO specifications, wingwalls are indeed not listed as independent components but under the heading of abutment. Nevertheless, when wingwalls are focused, the word "wingwall" is used with differentiation from the word "abutment." Figure 6.6-2 shows a wingwall not extending from the main body of abutment but making a right angle to it. When the entire abutment system is referred to, wingwalls are often meant to be part of the abutment system, although the word "system" is usually omitted. When specific analysis or design steps are dealt with, wingwalls are separated from the main body of the abutment system (usually consisting of the backwall, the stem wall, and the footing (or pile cap), which is also referred to as "abutment" without the words "main body" or "system." Understanding the omitted words will make it easier to understand the relevant provisions in the specifications as well as discussions in this chapter and other relevant chapters. In this section, wingwall is referred to as a local component of the entire abutment system, and so is 'abutment' in the entire system.

6.6.4 Design of Wingwalls

Figure 6.6-2
Wingwall in bridge abutment.

There are two different ways to arrange wingwalls regarding their relation to the main body of the abutment: independent from the abutment and the other associated with it. So wingwalls either may be designed as monolithic with the abutments or be separated from the abutment wall with an expansion joint and standing free. The wingwall lengths will depend on the required roadway slope. They need be of sufficient length to retain the roadway embankment and to furnish protection against erosion.

To be monolithic with the abutment's main body, steel-reinforcing bars or suitable reinforcement is required to be spaced across the junction between the wingwall and the main body of the abutment to tie them together. Such bars shall extend into the masonry on each side of the joint far enough to develop the strength of the bar as specified for bar reinforcement and shall vary in length so as to avoid planes of weakness in the concrete at their ends. To be independent from the main body of abutment, an expansion joint needs to be provided and the wingwall shall be keyed into the main body of the abutment.

Loads on wingwalls are similar to those on the main body of the abutment, except that the superstructure loads are assumed to not be carried by the wingwalls whether or not they are monolithic to the main body of the abutment.

Example 6.4 designs a wingwall for the same abutment in Examples 6.1 through 6.3. It includes load identification, load effect analysis, and member design.

Example 6.4 Wingwall Design

❑ **Design Requirement**
Design a wingwall structure for the abutment designed in Examples 6.1, 6.2, and 6.3. Assume that the wingall is also supported on piles.
Load modifier $\eta = 1.0$

❑ **Material Properties**   *3.5.1*
Concrete density $W_c = 0.145$ k/ft^3
Concrete 28-day compressive strength $f'_c = 4$ k/ft^2
Reinforcement strength $f_y = 60$ k/ft^2

❑ **Reinforcing Steel Cover Requirements**   *5.12.3*
Backwall back cover $C_{backwall} = 2.5$ in.
Stem back cover $C_{stem} = 2.5$ in.
Footing top cover $C_{footing\ top} = 2.0$ in.
Footing bottom cover $C_{footing\ bottom} = 3.0$ in.

❑ **Wingwall Dimensions**
The preliminary dimensions of the wingwall are shown in Figures Ex6.4-1 to Ex6.4-3. Assume that the abutment designed in Examples 6.1, 6.2, and 6.3 has an identical wingwall on each side being designed here.
A full-depth reinforced concrete wingwall is chosen for the site conditions.

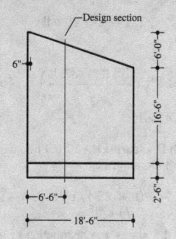

Figure Ex6.4-1
Elevation view of reinforced concrete wingwall.

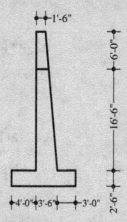

Figure Ex6.4-2
Side view of reinforced concrete wingwall.

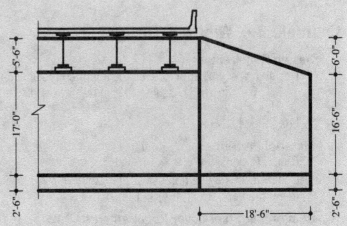

Figure Ex6.4-3
Front view of reinforced concrete abutment
with wingwall.

The dimensions are based on previous designs and past experience.

□ **Loads on Wingwall**

For sloped wingwalls, the design section can be taken at a distance of one-third down from the high end, as indicated in Figure Ex6.4-1. The one-third distance from the start of the slope is

$$\frac{18.5 - 0.5}{3} = 6 \text{ ft}$$

Thus, the section is located at 6.5 ft from the left end as shown in Figure Ex6.4-1. The design height of the wingwall is

$$h_{\text{design}} = 22.5 - 2 = 20.5 \text{ ft}$$

□ **Dead Load**

$$DC_{\text{wingwall}} = 20.5 \text{ ft} \left(\frac{1.5 \text{ ft} + 3.5 \text{ ft}}{2}\right) 0.145 \text{ k/ft}^3$$

$$= 7.43 \text{ k/ft}$$

□ **Lateral Earth Pressure EH**   *3.11.5*

At bottom of the wingwall,

3.11.5.1 $p = k_a \gamma_s z$
$$= 0.3(0.11 \text{ k/ft}^3)(20.5 \text{ ft} + 1 \text{ ft})$$
$$= 0.71 \text{ k/ft}^2$$

Figure Ex6.4-4 shows the distribution of earth load on the wingwall.

Figure Ex6.4-4
Earth pressure on wingwall.

$$EH = \frac{1}{2}\left(0.71 \text{ k/ft}^3\right) 21.5 \text{ ft}$$

$$= 7.63 \text{ k/ft} \quad \text{at height 7.17 ft from bottom of wingwall}$$

The horizontal component to the wingwall is

$$EH_{horizontal} = EH \cos\left(15°\right) = 7.63 \cos\left(15°\right) = 7.37 \text{ k/ft}$$

The vertical component is

$$EH_{vertical} = EH \sin\left(15°\right) = 7.63 \sin\left(15°\right) = 1.97 \text{ k/ft}$$

❑ Live-Load Surcharge LS   *3.11.6.4*

$$\text{Surcharge pressure} \quad \Delta_p = k \gamma_s h_{eq}$$

Using $k = k_a = 0.3$, $\gamma_s = 0.11 \text{ k/ft}^3$, and $h_{eq} = 2 \text{ ft}$, γ_s as average of loose and compacted gravel.   *3.5.1*

Abutment Height (ft)	h_{eq} (ft)
5.0	4.0
10.0	3.0
≥20.0	2.0

$$\Delta_p = k_a \gamma_s h_{eq} = 0.3(0.11 \text{ k/ft}^3)2 \text{ ft} = 0.066 \text{ k/ft}^2 \quad \text{3.11.6.4}$$

The live-load surcharge is

$$LS = \Delta_p h_{wingwall} = 0.066 \text{ k/ft}^2 \left(21.5 \text{ ft}\right) = 1.42 \text{ k/ft}$$

The horizontal component to the wingwall is

$$LS_{horizontal} = LS \cos\left(15°\right) = 1.42 \cos\left(15°\right) = 1.37 \text{ k/ft}$$

The vertical component is

$$LS_{vertical} = LS \sin\left(15°\right) = 1.42 \sin\left(15°\right) = 0.37 \text{ k/ft}$$

❑ Load Effects in Wingwall Load Combinations

Superstructure-transferred load effects are irrelevant for this wingwall, such as LL, BR, WL, DW, and so on. For the final state of the wingwall, WS on the substructure is also ignored.

Based on comparison of these four limit states and their associated load factors in Table Ex6.4-1, the Strength I limit state will control in the group of strength limit states. Accordingly, only Strength I and Service I limit states need to be considered hereafter.

Table Ex6.4-1
Load factors for wingwall

Load	Strength I	Strength III	Strength V	Service I
DC	1.25	1.25	1.25	1.00
EH	1.50	1.50	1.50	1.00
LS	1.75	—	1.35	1.00

Vertical forces:

$$DC_{wingwall} = 7.43 \text{ k/ft} \quad EH_{vertical} = 1.97 \text{ k/ft}$$

$$LS_{vertical} = 0.37 \text{ k/ft}$$

Combined vertical forces:

Limit State	Load Combination	Total Load Effect
Strength I	1.25(7.43) + 1.50(1.97) + 1.75(0.37)	12.9 k/ft
Service I	1(7.43) + 1(1.97) + 1(0.37)	9.8 k/ft

Shears:

$$EH_{horizontal} = 7.37 \text{ k/ft} \qquad LS_{horizontal} = 1.37 \text{ k/ft}$$

Combined shears:

Limit State	Load Combination	Total Load Effect
Strength I	1.50(7.37) + 1.75(1.37)	13.5 k/ft
Service I	1(7.37) + 1(1.37)	8.7 k/ft

Moments (at center of bottom cross section)

$$EH_{moment} = 7.37 \text{ k/ft} (7.17 \text{ ft}) = 52.8 \text{ kft/ft}$$

$$LS_{moment} = 1.37 \text{ k/ft} (10.75 \text{ ft}) = 14.7 \text{ kft/ft}$$

Combined moments:

Limit State	Load Combination	Total Load Effect
Strength I	1.50(52.8) + 1.75(14.7)	104.9 kft/ft
Service I	1(52.8) + 1(14.7)	67.5 kft/ft

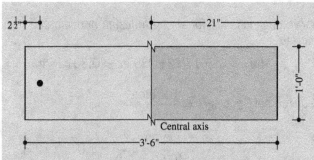

Central axis

Figure Ex6.4-5
Wingwall cross section for calculation of M_{cr}.

☐ Design Wingwall for Minimum Flexure Steel

Assume No. 8 bars, bar diameter = 1 in., bar area = 0.79 in.²
The minimum reinforcement requirement will be met first.
      5.7.3.3.2
For $M_{cr} = \gamma_3 \gamma_1 S_c f_r$,   *5.7.3.3.2*

$$S_c = \frac{bh^2}{6} = \frac{12\,(42)^2}{6} = 3528 \text{ in.}^3$$

5.4.2.6 $f_r = 0.24\sqrt{f_c'} = 0.24\sqrt{4 \text{ k/in.}^2} = 0.48 \text{ k/in.}^2$

$$M_{cr} = 0.75\,(1.6)\,3528 \text{ in.}^3 \left(0.48 \text{ k/in.}^2\right) = 169.3 \text{ kft./ft}$$

Use

$$M_{design} = \min(M_{cr},\ 1.33 M_u)$$

to design steel spacing s:

$$1.33 M_{strength\,I} = 1.33\,(104.9 \text{ kft/ft}) = 139.5 \text{ kft/ft} < 169.3 \text{ kft/ft}$$

Thus $1.33 M_{strength\,I}$ controls, $M_{design} = 139.5$ kft/ft, and the following design will cover Strength I limit state as well. Find the required reinforcing steel amount A_s:

Effective depth $d_e = h - C_{bot} - \dfrac{\text{bar diameter}}{2} = 42 \text{ in.} - 2.5 \text{ in.} - \dfrac{1 \text{ in.}}{2} = 39 \text{ in.}$

5.5.4.2.1 $R_n = \dfrac{M_{design}}{\varphi_f b d_e^2} = \dfrac{139.5 \text{ kft/ft}\,(12 \text{ in./ft})}{(0.9)\,12 \text{ in.}\,(39 \text{ in.})^2} = 0.1 \text{ k/in.}^2$

$$\rho = 0.85 \frac{f_c'}{f_y}\left[1 - \sqrt{1 - \frac{2 R_n}{0.85 f_c}}\right] = 0.85 \frac{4 \text{ k/in.}^2}{60 \text{ k/in.}^2}\left[1 - \sqrt{1 - \frac{2\,(0.1 \text{ k/in.}^2)}{0.85\,(4 \text{ k/in.}^2)}}\right]$$

$$= 0.0017$$

The above equations are for strength design of reinforced concrete.

$$A_s = \rho b d_e = 0.0017(12 \text{ in.})39 \text{ in.} = 0.80 \text{ in.}^2/\text{ft}$$

$$\text{Required bar spacing } s = \frac{\text{bar area}}{A_s}$$

$$= \frac{0.79 \text{ in.}^2}{0.80 \text{ in.}^2/\text{ft}} = 1.0 \text{ ft} = 12 \text{ in.}$$

Use No. 8 bars at 12 in.

$$A_{s, \text{ provided}} = \text{bar area} \frac{12 \text{ in.}}{\text{bar spacing}}$$

$$= 0.79 \text{ in.}^2 \frac{12 \text{ in./ft}}{12 \text{ in.}} = 0.79 \text{ in.}^2/\text{ft}$$

OK for Strength I limit state and minimum flexural steel requirements.

❏ **Check Steel distribution s for Crack Control under Service I Limit State**

$$\text{For required} \quad s \le \frac{700\,\gamma_e}{\beta_s f_{ss}} - 2d_c$$

$$D_c = 2.5 \text{ in} \qquad \gamma_e = 0.75$$

For substructure.   5.7.3.4

$$\beta_s = 1 + \frac{d_c}{0.7\,(h - d_e)} = 1 + \frac{2.5 \text{ in.}}{0.7(42 \text{ in.} - 2.5 \text{ in.})} = 1.09$$

$$f_{ss} = \frac{M_{\text{service I}}}{A_s j d_e}$$

where

$$A_s = 0.79 \text{ in.}^2/\text{ft}; \qquad j = 1 - \frac{k}{3}$$

$$k = \sqrt{(\rho n)^2 + 2\rho n} - \rho n$$

$$= \sqrt{[0.0017\,(8)]^2 + 2\,(0.0017)\,8} - 0.0017(8)$$

$$= 0.15$$

$$j = 1 - \frac{0.15}{3} = 0.95$$

Thus,

$$f_{ss} = \frac{67.5 \text{ kft/ft } (12 \text{ in./ft})}{0.79 \text{ in.}^2/\text{ft } (0.95) \, 39 \text{ in.}} = 27.7 \text{ k/in.}^2$$

$$s \le \frac{700\gamma_e}{\beta_s f_s} - 2d_c = \frac{700 \, (0.75)}{1.09 \, (27.7)} \text{in.} - 2(2.5 \text{ in.}) = 12.4 \text{ in.}$$

Keep No. 8 bars at 12 in. based on the stregnth I limit state requirement.

OK for flexural cracking under service I limit state.

☐ Design for Shear under Strength I Limit State

The factored shear force at the base of the wingwall is 13.5 k/ft. The nominal shear resistance is the lesser of

5.8.3.3 $V_c + V_s$ or $0.25 f'_c b_v d_v$

where

5.8.2.9 $d_v = \max[0.9 d_e, \, 0.72h]$

$$= \max[0.9 \, (39 \text{ in.}), \, 0.72 \, (42 \text{ in.})]$$

$$= \max \, (35.1 \text{ in.}, \, 30.2 \text{ in.}) = 35.1 \text{ in.}$$

$$V_c = 0.0316 \beta \sqrt{f'_c} b_v d_v$$

$$= 0.0316 (2) \sqrt{4 \text{ k/in.}^2} \, 12 \text{ in./ft } (35.1 \text{ in.})$$

$$= 53.2 \text{ k/ft}$$

$$V_s = \frac{A_v f_y d_v \, (\cot \theta + \cot \alpha) \sin \alpha}{S}$$

is neglected for this abutment.

$$0.25 f'_c \, b_v d_v = 0.25 \left(4 \text{ k/in.}^2\right) \, 12 \text{ in./ft } (35.1 \text{ in.})$$

$$= 421.2 \text{ k/ft}$$

$$V_n = \min \, (53.2, \, 421.1) = 53.2 \text{ k/ft}$$

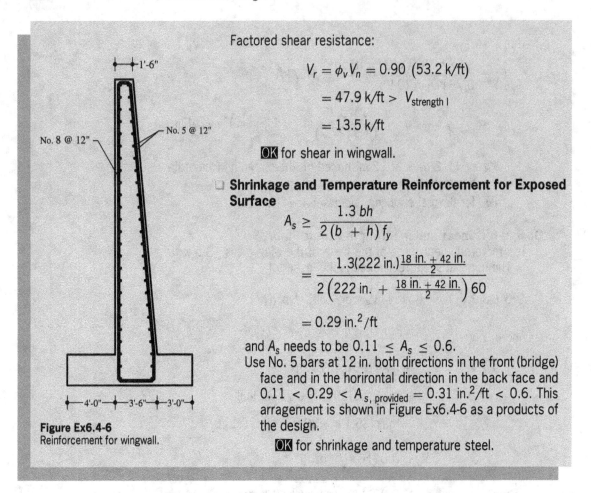

Factored shear resistance:

$$V_r = \phi_v V_n = 0.90 \ (53.2 \ \text{k/ft})$$

$$= 47.9 \ \text{k/ft} > V_{\text{strength I}}$$

$$= 13.5 \ \text{k/ft}$$

OK for shear in wingwall.

❑ **Shrinkage and Temperature Reinforcement for Exposed Surface**

$$A_s \geq \frac{1.3 \ bh}{2 \ (b + h) \ f_y}$$

$$= \frac{1.3(222 \ \text{in.})\frac{18 \ \text{in.} + 42 \ \text{in.}}{2}}{2 \left(222 \ \text{in.} + \frac{18 \ \text{in.} + 42 \ \text{in.}}{2} \right) 60}$$

$$= 0.29 \ \text{in.}^2/\text{ft}$$

and A_s needs to be $0.11 \leq A_s \leq 0.6$.
Use No. 5 bars at 12 in. both directions in the front (bridge) face and in the horirontal direction in the back face and $0.11 < 0.29 < A_{s, \ \text{provided}} = 0.31 \ \text{in.}^2/\text{ft} < 0.6$. This arragement is shown in Figure Ex6.4-6 as a products of the design.

OK for shrinkage and temperature steel.

Figure Ex6.4-6
Reinforcement for wingwall.

Labels in figure: 1'-6" · No. 5 @ 12" · No. 8 @ 12" · 4'-0" · 3'-6" · 3'-0"

6.6.5 Integral Abutments

As briefly commented on earlier, integral abutments are very different from routine abutments. No bearings are used in integral abutment bridges. Namely, the abutments are rigidly connected with the superstructure. Therefore the superstructure's deformations or movements are constrained by the substructure abutments. In other words, integral abutment bridges accommodate deformations or movements due to temperature, creep, shrinkage, and so on, using their material strength, not through bearings. As a result, integral abutments can be used only within a short span length range, because longer bridge spans would generally have excessive internal forces for which it could become very costly to withstand using member strength. Maximum span lengths, design considerations, and associated details of integral abutment spans should comply with recommendations outlined in the Federal Highway Administration (FHWA) Technical Advisory T 5140.13 (1980), except where substantial local experience indicates otherwise. *11.6.1.3*

To avoid water intrusion behind the abutment, the approach slab should be connected directly to the abutment (not to wingwalls), and appropriate provisions should be made to provide for drainage of any entrapped water.

Integral abutments should not be constructed on spread footings founded or keyed into rock unless one end of the span is free to displace longitudinally.

References

American Association of State and Highway Transportation Officials (AASHTO) (2012) *LRFD Bridge Design Specifications*, 6th ed., AASHTO, Washington, DC.

American Association of State and Highway Transportation Officials (AASHTO) (2011), *Manual for Bridge Evaluation*, 2nd ed., AASHTO, Washington, DC.

Federal Highway Administration (1980), "Integral, No-Joint Structures and Required Provisions of Movement," Technical Advisory T5140.13, U.S. Department of Transportation, Washington, DC.

Problems

6.1 Design an abutment for the bridge whose interior beams are as designed in Problem 4.6 and exterior beams as designed in Problem 4.7. Assume that this abutment is a partial-depth abutment and the abutment needs to be 30 ft high. The soil is 29 ft deep behind the abutment and 20 ft deep in front of the abutment. You may use other reasonable assumptions if needed. The assumptions should be consistent with what have been used in Problems 4.6, 4.7, and 3.3.

6.2 Design parallel wingwalls for the abutment in Problem 6.1 for the following conditions. The embankment is 15 ft wide from the edges of the abutment. Use other reasonable assumptions if needed. They should be consistent with what have been used in Problems 4.6, 4.7, and 3.3.

7 Highway Bridge Evaluation

7.1 Introduction

Many developed countries have a completed or almost completed highway infrastructure system that has significantly assisted in their economic development. The United States has such a system that includes about 600,000 bridges. Note that, according to the definition of the Federal Highway Administration (FHWA), any structure carrying a roadway with a span longer than 20 ft is inventoried as a bridge. About 40% of these bridges are legally owned by the states. The federal government and other private individuals own a small percentage. The rest, a majority of the total, is owned by counties, municipalities, townships, and so on. Realistically, local governments and private owners apply the specifications issued by the state of jurisdiction, which are largely consistent with the national specifications issued by the AASHTO and national guidelines published by the FHWA.

Maintaining the safety and normal operation of these bridges is the legal responsibility of the owners. The AASHTO has issued specifications for the evaluation of these bridges. The current AASHTO specifications for this purpose are included in the AASHTO *Manual for Bridge Evaluation* (2011). This set of specifications is referred to as the AASHTO manual or collectively with the AASHTO design specifications as the AASHTO specifications in this book. The articles from the AASHTO manual are quoted in this book with an *M* added as a prefix to the identification numbers. For example, *M1.2.3* refers to Article 1.2.3 in the AASHTO manual (2011). In contrast, *1.2.3* refers to Article 1.2.3 in the AASHTO *LRFD Bridge Design Specifications* (2012).

Currently, the main effort of U.S. bridge engineers is on maintaining the safety and normal operation of existing bridges in the highway infrastructure system. As a result, the goal of new bridge planning, design,

Figure 7.1-1
Steel bridge beam with corrosion and limited vertical clearance.

Figure 7.1-2
Cracked concrete bridge pier.

Figure 7.1-3
Reinforced concrete bridge pier with delaminated concrete marked and removed.

and construction is often to meet the needs of existing systems, such as replacing or rehabilitating aged structures. These decisions are made based on a number of factors, including, but not limited to, condition rating, load rating, functionality, traffic demand, and availability of funding. In this book, condition rating is briefly discussed in Section 7.2, load rating is introduced in Section 7.3, and fatigue evaluation of steel bridge components is presented in Section 7.4.

Condition rating is a result of inspection of the bridge in the field, performed largely based on visual observation and occasionally using other apparatus or equipment. Figures 7.1-1 through 7.1-3 show some example results of bridge inspection. Figures 7.1-1 and 7.1-2 are also often used to scope bridge rehabilitation when it is finally determined. Figure 7.1-3 is a result of inspection and the subsequent scoping, where deteriorated concrete in a bridge pier has been removed and new concrete is to be placed.

7.2 Inspection and Condition Rating

According to the requirement of the National Bridge Inspection Standards, every bridge (i.e., every structure on public roads over a depression or obstruction with a span 20 ft or longer) needs to be inspected at least once every 24 months. It is done to monitor the evolution or possible deterioration of the bridge's working condition and to allow timely decision for repair, rehabilitation, and/or emergency treatment such as strengthening or closure.

There are two major systems of condition rating being used in the United States: the FHWA national bridge inventory (NBI) system and the Pontis system. In some states there are also a number of rating systems that differ from the NBI. However, the FHWA requires that states report their bridge inspection results in the NBI format for convenience in data processing and subsequent reporting.

The FHWA NBI system uses a numerical rating framework to indicate a bridge's condition. The bridge component being rated can be an RC deck, a beam, a cable, and so on. The rating can be 0, 1, 2, ..., 9. Table 7.2-1 gives the definitions of these levels. As seen, these levels are defined verbally. Inevitably, subjective judgment is called for when using the definitions in Table 7.2-1. There are a few bridge component rating systems being used in some states with minor variation from the FHWA system. For example, New York State Department of Transportation has used a system between 1 and 7, as defined in Table 7.2-2.

It should be emphasized that the rating result is reached largely based on visual observation. Occasionally physical measurement using an apparatus may be performed if deemed necessary. For example, if a fatigue crack in a steel bridge beam is suspected, a simple dye penetrant test is a way to confirm or deny the observation. It can also identify the ends of the crack if indeed present and thus reach a better understanding as to how severe the crack is.

Bridge management system was one of the management systems mandated by the U.S. Congress to more quantitatively make decisions on bridge maintenance, repair, rehabilitation, and replacement. Over the years, the Pontis (a word derived from the Latin word *pons*, meaning "bridge")

Table 7.2-1
FHWA definitions for bridge component condition rating

Code	NBI Rating Definition
N	Not applicable
9	Excellent
8	Very good
7	Good
6	Satisfactory
5	Fair
4	Poor
3	Serious
2	Critical
1	Imminent failure
0	Failed

Table 7.2-2
Bridge component condition rating definition of New York State Department of Transportation

Rating	Description
9	Condition and/or existence unknown
8	Not applicable
7	New condition, no deterioration
6	Used to shade between ratings of 5 and 7
5	Minor deterioration but functioning as originally designed
4	Used to shade between ratings of 3 and 5
3	Serious deterioration or not functioning as originally designed
2	Used to shade between ratings of 1 and 3
1	Totally deteriorated or in failed condition

management system has gained momentum in the states. As of 2008 it was licensed by the AASHTO to 45 U.S. state transportation departments and other organizations in the United States and other countries. Therefore, the system's rating structure is briefly discussed here.

The Pontis system uses a rating scheme with up to five states (levels), depending on the bridge element. The minimum number of condition states in Pontis is 3. For example, element 314 (pot bearing) uses states 1 to 3 for rating; Element 103 (prestressed concrete box girder) uses 1 to 4; and element 12 (concrete deck and slab without overlay) uses 1 to 5. Usually, 1 is for the best/new condition as intended in the design and the maximum possible state (3, 4, or 5) refers to the worst condition.

While a state highway agency may use the Pontis system for bridge component rating and to predict future rating based on the Pontis probabilistic model, it is still required to report bridge condition in the NBI format (0 to 9) to the FHWA to contribute to the national bridge inventory. As a result, many states perform condition rating using both systems. Apparently, for the same bridge components, there must be a relation between the NBI and the Pontis systems, although it is certainly not linear (or there would be no need for the two systems).

Another major difference between the NBI system and the Pontis system is that the former is a database recording the condition evolution only, but the latter has an additional function of using the historical condition ratings recorded and cumulated to probabilistically predict what the future condition may be at the program level (e.g., at the level of state, city, and county).

The FHWA has issued guidelines on how to inspect and rate the condition of highway bridge components (FHWA, 2006). The bridge owners (largely the states or state transportation agencies) have also published various guidelines for such practice to be applied within the jurisdiction. Some are available online for free.

In addition to bridge component rating, states have developed comprehensive rating systems to aggregate the component ratings into a single rating for a subsystem or the entire bridge system. For example, many states have aggregated ratings for the superstructure system and the substructure system (including the bearings) of each bridge respectively using ratings for superstructure components and substructure components. They also have developed formulas to combine all component ratings into a single one for the entire bridge system. These formulas usually linearly combine critical component ratings with respective and different weights. For example, these components are considered more critical in a bridge system and thus have been given heavier weights: primary load-carrying members in the superstructure (beams, trusses, arches, etc.), abutments and piers, scour condition, and so on. These components are weighted more significantly because their failure may more likely lead to bridge system failure.

For the Pontis rating, some states also have developed guidelines/manuals on Pontis bridge inspection and rating particularly suitable within the jurisdiction. Some of these guidelines/manuals are also downloadable online for free. The interested reader may wish to consult them when needed. Many include photos for defining the ratings to educate new instructors and to make inspection results more consistent.

Note that elements refered to in bridge inspection and condition rating include also non-structural bridge components, such as paint systems for

Figure 7.2-1
Vacuum confinement for removing paint from
steel bridge beams to repaint.

steel members. Figure 7.2-1 shows operation of paint removal for a bridge in rehabilitation as a typical task of bridge condition upgrading.

There are also professional short courses offered to engineers to learn, strengthen, and update the needed knowledge for bridge inspection and condition rating. Further coverage of detailed inspection and rating is provided in these courses, which is beyond the scope of this introductory book on highway bridge design and evaluation.

7.3 Load Rating

Load rating is another important step in highway bridge evaluation. It is more quantitative than is condition rating. It addresses the safe load-carrying capability or quantitatively gives the load level the bridge is able to safely carry with consistent safety margin included. The load-rating process uses information about the strength and loading as in the design process, updated by field inspection of the bridge as available and/or appropriate.

The AASHTO manual specifies the following equation for load rating, indexed as a rating factor RF:

M6A4.2.1
$$RF = \frac{C - \gamma_{DC}DC - \gamma_{DW}DW \pm \gamma_P P}{\gamma_{LL}\,(LL \ + \ IM)}$$
(7.3-1)

For the Strength Limit States:

M6A.4.2.1
$$C = \phi_c \phi_s \phi_n \, R_n$$
(7.3-2)

where the following lower limit applies:

$$\phi_c \phi_s \geq 0.85$$

For the service limit states

M6A.4.2.1
$$C = f_R$$
(7.3-3)

where   RF = rating factor
       C = capacity
       f_R = allowable stress specified in AASHTO design specifications
       R_n = nominal member resistance as inspected
       DC = dead-load effect due to structural components and attachments
       DW = dead-load effect due to wearing surface and utilities
       P = permanent loads other than dead loads
       LL = live-load effect
       IM = dynamic load allowance

γ_{DC} = AASHTO design specification load factor for DC
γ_{DW} = AASHTO design specification load factor for DW
γ_p = AASHTO design specification load factor for permanent
         loads = 1
γ_{LL} = evaluation live-load factor (different from those in
         AASHTO design specifications)
ϕ_c = condition factor as shown in Table 7.3-1
ϕ_s = system factor as shown in Table 7.3-2
ϕ_n = AASHTO design specifications resistance factor

The numerator of the rating factor RF in Eq. 7.3-1 is the difference of the component's capacity less the dead-load effect in the component. In other words, the numerator is the component' available capacity for live load, with the resistance factor and load factors included. Capacity C can be updated based on inspection result. Figure 7.3-1 shows a case of Section loss of steel bridge component due to corrosion, which needs to be taken into account in estimating C. The denominator is the referenced live-load effect, including the dynamic effect and the live-load factor in Eq. 7.3-1. Note that all the items in Eq. 7.3-1 are the same as those defined in the AASHTO specifications for design, except the live-load factor. Namely, the rating factor defined in Eq. 7.3-1 is the ratio between the available live-load capacity and the required live-load effect, which may vary depending on the intended use of the bridge, to be discussed below in more details.

Table 7.3-1
Condition factor ϕ_c in AASHTO manual

Structural Condition of Member	φ_c
Good or satisfactory	1.00
Fair	0.95
Poor	0.85

Table 7.3-2
System factor ϕ_s in AASHTO manual

Superstructure Type	φ_s
Welded members in two-girder/truss/arch bridges	0.85
Riveted members in two-girder/truss/arch bridges	0.90
Multiple eyebar members in truss bridges	0.90
Three-girder bridges with girder spacing 6 ft	0.85
Four-girder bridges with girder spacing ≤ 4 ft	0.95
All other girder bridges and slab bridges	1.00
Floorbeams with spacing > 12 ft and noncontinuous stringers	0.85
Redundant stringer subsystems between floorbeams	1.00

Figure 7.3-1
Significant paint damage and corrosion that may cause section loss.

While the design of highway bridges is practiced in the United States using a set of specifications relatively uniform among the states, the load rating of existing bridges is performed under much more flexible specifications. The AASHTO manual allows much more jurisdiction-dependent practice, with consideration to cost implications, state practice history, and sometimes the site condition (e.g., the load requirement depending on the local load spectrum). This approach considers the fact that strengthening an existing bridge to increase its load-carrying capacity can be much more costly that adding capacity to a new bridge in the design stage. For example, research has found that increasing the live-load capacity by 25% will cost about 2 to 3% more for new bridges. Essentially, this increase in cost for new bridges' additional capacity is just for the additional material needed, with the labor cost almost unchanged. Nevertheless, strengthening an existing bridge's capacity can be as high as replacing the bridge, simply because there are no appropriate approaches available to reliably increase that capacity. One example is reinforced concrete members and prestressed concrete members. Adding more capacity to such members needs to address the issue of reliably anchoring new material to the existing members without damaging the existing member, which is difficult to do, if not impossible.

The AASHTO manual includes provisions for load rating using the concept of load and resistance factor rating (LRFR), parallel to allowable stress rating (ASR) and load factor rating (LFR). This book focuses on the LRFR approach, although the LFR and ASR are similar in format but are not calibrated based on the structural reliability concept. The LRFR approach has been calibrated based on the same probabilistic theory framework as that used in the calibration of the AASHTO design specifications, also taking into account cost effectiveness. The calibration concept has been discussed in Chapter 2.

The AASHTO manual includes provisions for three levels of load rating using respective trucks:

❏ Design load rating

❏ Legal load rating

❏ Permit load rating

7.3.1 Flexibility in Reference Truck and Resulting Level of Load Rating

This flexibility of several levels of rating offers options of acceptable bridge safe load-carrying capacity. Note that this capacity is different from the ultimate load-carrying capacity of the bridge, rather a level below it with a live-load factor already included. This is shown in Eq. 7.3-1.

Eq. 7.3-1 leads to a dimensionless factor that is to be used to find the load safely allowable to the bridge as

$$\text{Allowable load } = \text{RF} \times \text{LL} \qquad (7.3\text{-}4)$$

For example, for RF = 1.0 and LL being the HL93 load, the allowable load is then $1.0 \times$ HL93 = HL93. In a simplistic approach, which is being practiced in many states, this load is also referred to as 72 k or 36 t because the HL93 design truck weighs 72 k or 36 t. When the RF is below 1.0 (say 0.9), the allowable load is then proportionally reduced to the corresponding tonnage (64.8 k or 32.4 t) to simply and intuitively express the capacity of the bridge. This substandard tonnage also influences, if not determines, the posted load when such action is justified and/or warranted. Note that Eq. 7.3-1 refers to a cross section of a bridge component. Usually the lowest RF for a cross section controls the RF for the component and then the entire bridge.

Posting a substandard bridge notifies the traveling public of the current safe load-carrying capacity of the bridge. By local law, certain trucks are prohibited to cross these bridges. Such an action is a compromise measure for a bridge with substandard capacity before preparation and/or funds can be provided to enhance the capacity. Figure 7.3-2 shows an example of such posting for a highway bridge. As seen there, the tonnage of weight limit is dependent on the vehicle' configuration (axle arrangement), which is

Figure 7.3-2
A posted highway bridge for substandard load carrying capacity.

consistent with RF in Eq. 7.3-1. Equation 7.3-1 gives RF as a function of the referenced load LL. In other words, a specific LL gives a corresponding RF and tonnage. The three load-rating levels refer to three different reference loads LL. They are introduced individually below.

Examples 7.1 through 7.4 show how Eq. 7.3-1 is applied to the interior and exterior steel beams in Examples 4.5 to 4.8 respectively for the design load rating and the legal load rating. The two different ratings are further elaborated next.

Example 7.1 Design Load Rating (For Interior Beams in Examples 4.5 to 4.8)

❑ **Requirement**
> Perform load rating for the design load for the interior beams of the steel-rolled beam bridge in Examples 4.5, 4.6, 4.7, and 4.8.

❑ **Dead-Load Effects DC for Interior Beams**
According to Example 4.5

$$DC \text{ for moment } M_{DC} = M_{steel\_beam} + M_{deck} + M_{deck\_forms} + M_{miscellaneous} + M_{parapet}$$

$$= 38.8 + 192 + 24 + 4 + 36.6 = 295.4 \text{ kft}$$

$$DC \text{ for shear } V_{DC} = V_{steel\_beam} + V_{deck} + V_{deck\_forms} + V_{miscellaneous} + V_{parapet}$$

$$= 3.9 + 19.2 + 2.4 + 0.4 + 3.66 = 29.6 \text{ k}$$

❑ **Dead-Load Effect DW for Interior Beams**
According to Example 4.5

$$DW \text{ for moment } M_{DW} = 48 \text{ k}$$

$$DW \text{ for shear } V_{DW} = 4.8 \text{ k}$$

❑ **Design Live-Load Effect LL(1+IM) for Interior Beams**
According to Example 4.5

$$LL(1+IM) \text{ for moment } M_{LL(1+IM)} = 566.4 \text{ kft}$$

$$LL(1+IM) \text{ for shear } V_{LL(1+IM)} = 75 \text{ k}$$

❑ **Resistance of Interior Beams**
According to Example 4.6

$$\text{Resistance for moment } M_n = 2061 \text{ kft}$$

$$\text{Resistance for shear } V_n = 307 \text{ k}$$

□ **Design Load-Rating Factors for Interior Beams**

$$RF_{\text{interior moment design inventory}} = \frac{C - \gamma_{DC}DC - \gamma_{DW}DW \pm \gamma_P P}{\gamma_{LL}LL + IM}$$

$$= \frac{2061 - 1.25\,(295.4) - 1.5\,(48)}{1.75\,(566.4)} = 1.63$$

$$RF_{\text{interior shear design inventory}} = \frac{C - \gamma_{DC}DC - \gamma_{DW}DW \pm \gamma_P P}{\gamma_{LL}LL + IM}$$

$$= \frac{307 - 1.25\,(29.6) - 1.5\,(4.8)}{1.75\,(75)} = 2.00$$

$$RF_{\text{interior moment design operating}} = \frac{2061 - 1.25\,(295.4) - 1.5\,(48)}{1.35\,(566.4)} = 2.12$$

$$RF_{\text{interior shear design operating}} = \frac{307 - 1.25\,(29.6) - 1.5\,(4.8)}{1.35\,(75)} = 2.60$$

Example 7.2 Design Load Rating (for Exterior Beams in Examples 4.5 to 4.8)

□ **Requirement**

Perform load rating for the design load for the exterior beams of the steel-rolled beam bridge in Examples 4.5 to 4.8.

□ **Dead-Load Effect DC for Exterior Beams**

According to Example 4.5

DC for moment $M_{DC} = M_{\text{steel\_beam}} + M_{\text{deck}} + M_{\text{deck\_forms}} + M_{\text{miscellaneous}} + M_{\text{parapet}}$

$$= 38.8 + 182 + 12 + 4 + 36.6 = 273.4 \text{ kft}$$

DC for shear $V_{DC} = V_{\text{steel\_beam}} + V_{\text{deck}} + V_{\text{deck\_forms}} + V_{\text{miscellaneous}} + V_{\text{parapet}}$

$$= 3.9 + 18.2 + 1.2 + 0.4 + 3.66 = 27.4 \text{ k}$$

□ **Dead-Load Effect DW for Exterior Beams**

According to Example 4.5

DW for moment $M_{DW} = 48$ k

DW for shear $V_{DW} = 4.8$ k

☐ **Design Live-Load Effect LL(1+IM) for Exterior Beams**
According to Example 4.5

$$LL\,(1 + IM) \text{ for moment } M_{LL(1+IM)} = 624.5 \text{ kft}$$

$$LL\,(1 + IM) \text{ for shear } V_{LL(1+IM)} = 74.1 \text{ k}$$

☐ **Resistance of Exterior Beams**
According to Example 4.7

$$\text{Resistance for moment } M_n = 2051 \text{ kft}$$

$$\text{Resistance for shear } V_n = 307 \text{ k}$$

☐ **Design Load-Rating Factors for Exterior Beams**

$$RF_{\text{exterior moment design inventory}} = \frac{C - \gamma_{DC}DC - \gamma_{DW}DW \pm \gamma_P P}{\gamma_{LL}LL + IM}$$

$$= \frac{2051 - 1.25\,(273.4) - 1.5\,(48)}{1.75\,(624.5)}$$

$$= 1.50$$

$$RF_{\text{exterior shear design inventory}} = \frac{C - \gamma_{DC}DC - \gamma_{DW}DW \pm \gamma_P P}{\gamma_{LL}LL + IM}$$

$$= \frac{307 - 1.25\,(27.4) - 1.5\,(4.8)}{1.75\,(74.1)}$$

$$= 2.05$$

$$RF_{\text{exterior moment design operating}} = \frac{2051 - 1.25\,(273.4) - 1.5\,(48)}{1.35\,(624.5)}$$

$$= 1.94$$

$$RF_{\text{exterior shear design operating}} = \frac{307 - 1.25\,(27.4) - 1.5\,(4.8)}{1.35\,(74.1)}$$

$$= 2.65$$

For moment and shear in the interior and exterior beams, the case of moment in the exterior beams controls the rating factor for all the beams and both load effects of moment and shear.

Example 7.3 Legal Load Rating (for Interior Beams in Examples 4.5 to 4.8)

☐ **Requirement**

Perform load rating for the AASHTO legal load for the interior beams of the steel-rolled beam bridge in Examples 4.5 through 4.8.

☐ **Dead-Load Effect DC for Interior Beams**

According to Example 7.1

DC for moment $M_{DC} = M_{steel\_beam} + M_{deck} + M_{deck\_forms} + M_{miscellaneous} + M_{parapet}$

$$= 295.4 \text{ kft}$$

DC for shear $V_{DC} = V_{steel\_beam} + V_{deck} + V_{deck\_forms} + V_{miscellaneous} + V_{parapet}$

$$= 29.6 \text{ k}$$

☐ **Dead-Load Effect DW for Interior Beams**

According to Example 7.1

DW for moment $M_{DW} = 48 \text{ k}$

DW for shear $V_{DW} = 4.8 \text{ k}$

☐ **Legal Live-Load Effect LL(1+IM) for Interior Beams**

LL $(1 + IM)$ for moment $M_{LL(1+IM)} = 465 \text{ kft} (1.10) 0.78$

$$= 399 \text{ kft (Type 3 truck controls)}$$

LL $(1 + IM)$ for shear $V_{LL(1+IM)} = 40.7 \text{ k} (1.10) 0.87$

$$= 38.9 \text{ k (Type 3 truck controls)}$$

MC6A4.4.3   *Smooth road surface. IM* $= 1.10$.

Table Ex7.3-1

Maximum moment in simple spans induced by AASHTO legal load

Span, ft	AASHTO Legal Loads		
	Type 3	Type 3-S2	Type 3-3
36	399.0	376.9	327.4
38	432.0	404.1	356.4
40	465.0	431.5	385.2
42	498.0	458.6	414.2
44	531.0	486.0	443.2

❑ **Resistance of Interior Beams**
According to Example 4.5

Resistance for moment $M_n = 2061$ kft

Resistance for shear $V_n = 307$ k

❑ **Legal Load-Rating Factors for Interior Beams**

$$RF_{\text{interior moment legal}} = \frac{C - \gamma_{DC}DC - \gamma_{DW}DW \pm \gamma_P P}{\gamma_{LL}LL + IM}$$

$$= \frac{2061 - 1.25\,(295.4) - 1.5\,(48)}{1.8\,(399)} = 2.25$$

$$RF_{\text{interior shear legal}} = \frac{C - \gamma_{DC}DC - \gamma_{DW}DW \pm \gamma_P P}{\gamma_{LL}LL + IM}$$

$$= \frac{307 - 1.25\,(29.6) - 1.5\,(4.8)}{1.8\,(38.9)} = 3.75$$

Example 7.4 Legal Load Rating (for Exterior Beams in Examples 4.5 to 4.8)

❑ **Requirement**
Perform load rating for the AASHTO legal load for the exterior beams of the steel-rolled beam bridge in Examples 4.5, 4.6, 4.7, and 4.8

❑ **Dead-Load Effect DC for Exterior Beams**
According to Example 7.2

DC for moment $M_{DC} = M_{\text{steel\_beam}} + M_{\text{deck}} + M_{\text{deck\_forms}} + M_{\text{miscellaneous}} + M_{\text{parapet}}$

$= 273.4$ kft

DC for shear $V_{DC} = V_{\text{steel\_beam}} + V_{\text{deck}} + V_{\text{deck\_forms}} + V_{\text{miscellaneous}} + V_{\text{parapet}}$

$= 27.4$ k

❑ **Dead-Load Effect DW for Exterior Beams**
According to Example 7.2

DW for moment $M_{DW} = 48$ k

DW for shear $V_{DW} = 4.8$ k

☐ **AASHTO Legal Live-Load Effect LL(1+IM) for Exterior Beams**
Using the distribution factor from Example 4.5

$$LL\,(1+IM)\text{ for moment } M_{LL(1+IM)} = 465\text{ kft }(1.10)\,0.86 = 440\text{ kft}$$

(Type 3 truck controls, see Table Ex7.4-1)

$$LL\,(1+IM)\text{ for shear } V_{LL(1+IM)} = 40.7\text{ k }(1.10)\,0.86 = 38.5\text{ k}$$

(Type 3 truck controls)

> MC6A4.4.3   *Smooth road surface. IM = 1.10.*

Table Ex7.4-1
Maximum moment in simple spans induced by AASHTO legal load

Span, ft	AASHTO Legal Loads		
	Type 3	Type 3-S2	Type 3-3
36	399.0	376.9	327.4
38	432.0	404.1	356.4
40	465.0	431.5	385.2
42	498.0	458.6	414.2
44	531.0	486.0	443.2

☐ **Resistance of Exterior Beams**
According to Example 4.6

$$\text{Resistance for moment } M_n = 2051 \text{ kft}$$

$$\text{Resistance for shear } V_n = 307 \text{ k}$$

☐ **Legal Load-Rating Factors for Exterior Beams**

$$RF_{\text{exterior moment legal}} = \frac{C - \gamma_{DC}DC - \gamma_{DW}DW \pm \gamma_P P}{\gamma_{LL}LL + IM}$$

$$= \frac{2051 - 1.25\,(273.4) - 1.5\,(48)}{1.8\,(440)} = 2.07$$

$$RF_{\text{exterior shear legal}} = \frac{C - \gamma_{DC}DC - \gamma_{DW}DW \pm \gamma_P P}{\gamma_{LL}LL + IM}$$

$$= \frac{307 - 1.25\,(27.4) - 1.5\,(4.8)}{1.8\,(38.5)} = 3.83$$

M6A4.4.2.3 Live-load factor = 1.80 for ADTT = 5000.

DESIGN LOAD RATING

The design load rating assesses the performance of existing bridges with reference to the AASHTO design specification HL-93 load, hence the name of rating. The design load rating of bridges may be performed at two different levels: (1) the inventory level, which is the same as the design level using $\gamma_{LL} = 1.75$, and (2) the operating level with $\gamma_{LL} = 1.35$. Apparently, these two live-load factors correspond to different structural reliability levels for the component being load rated. The former is the same as that for new bridges designed according to the AASHTO design specifications, which specify $\gamma_{LL} = 1.75$. The latter represents a lower reliability level comparable to the operating level in past load-rating practice, according to previous AASHTO specifications for bridge evaluation and load rating.

Therefore, the rating factor for design load rating in Eq. 7.3-1 uses the HL93 load effect in the focused component as the reference in the denominator (e.g., if a beam is being load rated, then the corresponding load distribution is included in the live-load effect LL in Eq. 7.3-1). The rating factor in Eq. 7.3-1 can be read as "how many times" the reference load LL can be safely carried by the focused component as part of the structural system. This level of load rating represents the highest requirement at the same level as the design requirement. For comparison, the HL93 load is given again in Figure 7.3-3.

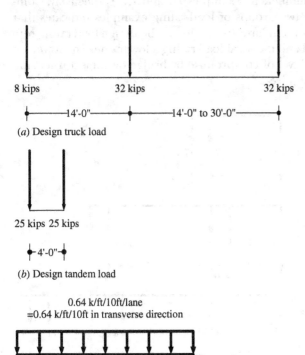

8 kips   32 kips   32 kips

|———14'-0"———|———14'-0" to 30'-0"———|

(*a*) Design truck load

25 kips 25 kips

|—4'-0"—|

(*b*) Design tandem load

0.64 k/ft/10ft/lane
=0.64 k/ft/10ft in transverse direction

(*c*) Design lane load

Figure 7.3-3
AASHTO HL93 live load: maximum of (truck load, tandem load) + lane load.

Examples 7.1 and 7.2 illustrate the application of design load rating for the primary superstructure members for the bridge designed in Examples 4.5 through 4.8. Note that some state transportation agencies require load rating to be performed whenever the design is completed and the rating results to be documented in the design plans. Examples 7.1 and 7.2 show that most, if not all, of the information needed to perform load rating is the same as that used in the design. When the bridge ages, strength, resistance, or capacity updating may be required depending on how severe the deterioration has been. This subject is discussed below.

LEGAL LOAD RATING

The legal load rating uses the AASHTO legal truck loads or rating trucks as shown in Figure 7.3-4 in the denominator of Eq. 7.3-1. These trucks induce lower load effects than the HL93 load. However, the AASHTO legal truck loads are legal and thus more realistic than the HL93 load. In other words, for the same bridge component, if a rating factor of 1.0 or higher results using Eq. 7.3-1 satisfying the design load-rating requirement, a higher-than-1.0 rating factor will result when it is rated against the AASHTO legal loads. Thus, obviously, the legal load rating is less conservative than the HL93 design load rating.

Examples 7.3 and 7.4 present applications of legal load rating for the same superstructure members in Examples 7.1 and 7.2, respectively. Comparison between these two groups of load-rating examples indicates that they merely use different standards for live load. The design load rating represents a high standard and the legal load rating a lower one. The lower set of standards reflects a level of compromise by bridge owners, considering the cost-effectiveness required.

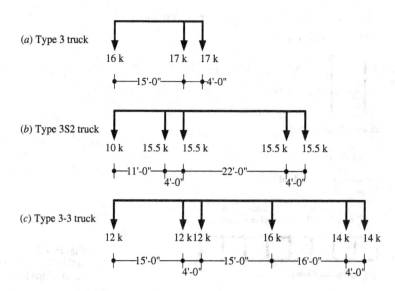

Figure 7.3-4
AASHTO legal loads.

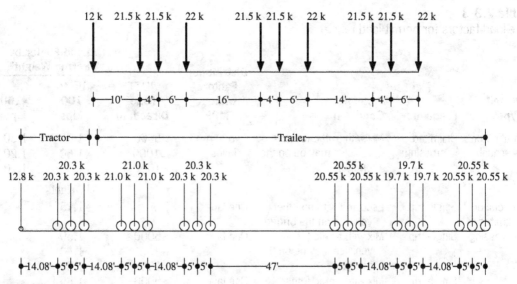

Figure 7.3-5
Examples of permit loads in (top) Washington State and (bottom) Connecticut State.

PERMIT LOAD RATING

In general, permit loads are those that exceed the truck weight limit of the jurisdiction. Namely they are overweight loads. They are thus required to have a permit to travel. The AASHTO manual does not specify the truck load to be used for this level of load rating as it does for the Strength II limit state where the jurisdiction-specific load needs to be used. Figure 7.3-5 shows two more examples of permit truck loads in Washington State and Connecticut State. See also those in Figure 3.3-7.

Permit load rating is a new load rating in the AASHTO manual with the LRFR approach, although it was practiced with implicit provisions in the previous versions of the manual. The main difference of the new provisions is in the prescribed flexible live-load factors. The live-load factors are given based on consideration of the likelihood of the critical loading on the bridge span. Table 7.3-3 displays the live-load factors given in the current AASHTO manual. They are seen to decrease with increases in truck weight. This may appear to be contradicting the conventional thinking that higher loads require higher load factors. The justification behind this approach of permit load rating is that the heavier the permit load, the more certain that another truck of similar weight will not be present in other lanes or the same lane. Because the live-load factor mainly covers the uncertainty of the live load, a lower live-load factor is given in the AASHTO manual. This structure of permit load checking with the live-load factor decreasing with increasing permit load also reduces the pressure on the transportation agency when a very heavy load is required to travel due to economic development.

Table 7.3-3
Live-load factors for permit load rating

Permit Type	Frequency	Loading Condition	Distribution Factors without MPF	ADTT (One Direction)	Load Factor by Permit Weight[a]	
					Up to 100 kips	≥150 kips
Routine or annual	Unlimited crossings	Mix with traffic (other vehicles may be on the bridge)	Two or more lanes	>5000	1.80	1.30
				=1000	1.60	1.20
				<100	1.40	1.10
					All Weights	
Special or limited crossing	Single trip	Escorted with no other vehicles on the bridge	One lane	N/A	1.15	
	Single trip	Mix with traffic (other vehicles may be on the bridge)	One lane	>5000	1.50	
				=1000	1.40	
				<100	1.35	
	Single trip	Mix with traffic (other vehicles may be on the bridge)	One lane	>5000	1.85	
				=1000	1.75	
				<100	1.55	

[a]For routine permits between 100 and 150 kips, interpolate the load factor by weight and ADTT value. Use only axle weights on the bridge

7.3.2 Resistance Updating in Load Rating

Another important feature of load rating for highway bridges, different from design, is that the resistance or capacity of the member needs to be updated when needed information becomes available. For example, for the steel beam bridge shown in Figure 7.3-6, inspection observed corrosion

Figure 7.3-6
Corroded steel beam and deteriorated concrete deck in highway bridge.

needs to be taken into account in load rating in terms of resistance reduction resulting from partial cross-sectional loss. The bridge load-rating engineer needs to exercise judgment as to how much reduction would reflect reality. Conservative estimation is advised, but overconservative reduction of the cross section is not favored, which could impose an excessively high requirement on the component and possibly waste valuable resources unnecessarily.

Figure 7.3-7 shows another example of component deterioration, a reinforced concrete deck this time. It appears that some loss of concrete has

Figure 7.3-7
Concrete deck spalling (top) untreated and (bottom) patched.

occurred due to the observed cracking and possibly delamination in concrete. For load rating bridge decks, such loss of cross section should be accounted for.

On the other hand, some states do not treat the deck as a critical member in the bridge system for load rating. Concrete decks are considered highly redundant structural members typically being supported on several beams and further loss of concrete may not jeopardize the safety of the entire bridge system.

It should be noted that it is commonly practiced to use the lowest load rating of the components as the rating for the entire bridge, although some states ignore the lowest deck rating and use another nondeck lowest rating to represent the bridge's load rating. The rating is required to be reported to the FHWA to be recorded in the NBI.

7.4 Fatigue Evaluation for Steel Components

Besides strength-concerned load rating, existing steel bridge components are also evaluated for the remaining life according to Section 7 of the AASHTO manual. Similar to steel component fatigue design, there are two categories of remaining fatigue life offered in the AASHTO manual for evaluation: infinite and finite remaining fatigue lives.

7.4.1 Infinite Remaining Fatigue Life

As presented in Section 4.7.8, when the stress range is low enough, the steel detail can be considered immunized for fatigue failure, that is, fatigue cracking. Therefore, the detail is thought to have an infinite fatigue life. For existing steel bridge components, this may be the case so that fatigue failure is of no concern. The AASHTO manual requires the following criterion to determine whether this is the case for a particular steel bridge detail:

$M7.2.4$
$$(\Delta f)_{max} \leq (\Delta F)_{TH} \qquad (7.4\text{-}1)$$

where $(\Delta f)_{max}$ = maximum stress range expected at the fatigue-prone detail, which may be taken as $2(\Delta f)_{eff}$, where $(\Delta f)_{eff}$ is the effective stress range as defined in Eq. 7.4-2 for steel component fatigue limit states design

$(\Delta F)_{TH}$ = constant-amplitude fatigue thresholds (CAFT) for infinite life in Table 4.7-3

If Eq. 7.4-1 is not satisfied, the focused detail then is deemed to have a finite life, to be estimated as discussed in the next section, according to the AASHTO manual. Note also that the evaluation using Eq. 7.4-1 corresponds to steel component design for the Fatigue I limit state discussed in Section 3.4.5.

The AASHTO manual gives the following formula to find the total fatigue life of the detail in year Y:

7.4.2 Finite Remaining Fatigue Life

M7.2.5

$$Y = \frac{R_R A}{365n \ \text{ADTT}_{\text{SL}} \left[(\Delta f)_{\text{eff}} \right]^3} \tag{7.4-2}$$

Most symbols used in Eq. 7.4-2 have been used in steel member design for the Fatigue II limit state, presented in Chapters 3 and 4. The definitions of the symbols used here are the same as Eq. 4.7-34 used in steel fatigue limit state design in Chapter 4. They are repeated here for the reader's convenience.

Y = total fatigue life in year, which can be mean life, evaluation life, and minimum life depending on R_R

R_R = resistance factor specified for evaluation, minimum, or mean fatigue life as given in Table 7.4-1 *7.2.5.2*

A = fatigue strength coefficient given in Table 4.7-4

a = present age of detail in year

n = number of stress cycles per truck crossing, given in Table 4.7-5

$(\Delta f)_{\text{eff}}$ = effective stress range at the detail = $R_s \ \Delta f$

Δf = estimated stress range. A number of approaches may be used for this estimation. For the load, the fatigue truck model introduced in Chapter 3 with a load factor of 0.75 and field measured truck weights (weigh-in-motion data) are two options. The latter requires more effort and equipment and certainly is more expensive. For structural analysis, a simplified analysis as used in design and a refined analysis using numerical computation methods (such as the finite element analysis methods) are two general options. The latter requires more effort and sophisticated computer programs and thus is usually more expensive.

Table 7.4-1
Resistance factor R_R *M7.2.5.2*

Detail Category	R_R Evaluation Life	Minimum Life	Mean Life
A	1.7	1.0	2.8
B	1.4	1.0	2.0
B′	1.5	1.0	2.4
C	1.2	1.0	1.3
C′	1.2	1.0	1.3
D	1.3	1.0	1.6
E	1.3	1.0	1.6
E′	1.6	1.0	2.5

Table 7.4-2
Partial live-load factor $R_s = R_{sa} R_{st}$ M7.2.2.1

Method to estimate stress range	R_{sn}	R_{st}	R_s
For evaluation or minimum fatigue life			
Simplified analysis and AASHTO fatigue truck	1.00	1.00	1.00
Simplified analysis and AASHTO weigh-in-motion data	1.00	0.95	0.95
Refined analysis and AASHTO weigh-in-motion data	0.95	1.00	0.95
Refined analysis and weigh-in-motion data	0.95	0.95	0.90
Stress range by field-measured strains	N/A	N/A	0.85
For mean fatigue life			
All methods	N/A	N/A	1.00

Source: AASHTO.

$R_s = R_{sa} R_{st}$ = reliability factor given in Table 7.4-2 according to the AASHTO manual. These parameters are similar to live-load factors to cover uncertainty involved in the estimation processes. The less uncertainty, the smaller the factors R_{sa}, R_{st}, and R_s should be.

ADTT_{SL} = average daily truck traffic in shoulder lane, averaged over the fatigue life

Based on present ADTT, ADTT_{SL} can be approximated using the ratio of $(\text{ADTT})_{SL} / [(\text{ADTT})_{SL}]_{present}$ in Figure 7.4-1 provided in the AASHTO

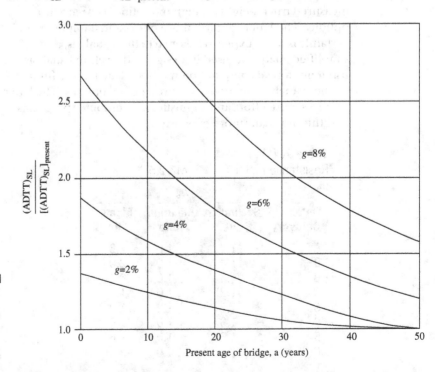

Figure 7.4-1
Approximate method to find ratio of life average ADTT and current ADTT M7.2.5.1. From AASHTO Manual for Bridge Evaluation, 2011, used by permission.

Table 7.4-3
Fraction of trucks in traffic   C3.6.1.4.2.1

Class of Highway	Fraction of Trucks in Traffic
Rural interstate	0.20
Urban interstate	0.15
Other rural	0.15
Other urban	0.10

manual. This ratio is given as a function of present age and annual growth rate g in Figure 7.4-1.

A new accurate formula has been developed and proposed to replace Eq. 7.4-2 and Figure 7.4-1:

$$Y = \frac{\log\left[\dfrac{R_R A}{365n\left[(\text{ADTT})_{\text{SL}}\right]_{\text{PRESENT}}\left((\Delta f)_{\text{eff}}\right)^3}g\left(1+g\right)^{a-1} + 1\right]}{\log\left(1+g\right)} \tag{7.4-3}$$

Sometimes estimation of the truck volume is difficult because ADTT data are not available. Table 7.4-3 is provided in the AASHTO design specifications (C3.6.1.4.2.1) to help this estimation based on average daily traffic (ADT). ADTT can then be estimated as the product of ADT and the fraction value taken from Table 7.4-3.

References

American Association of State and Highway Transportation Officials (AASHTO) (2012), *LRFD Bridge Design Specifications*, 6th ed., AASHTO, Washington, DC.

American Association of State and Highway Transportation Officials (AASHTO) (2011), *Manual for Bridge Evaluation*, 2nd ed., AASHTO, Washington, DC.

Federal Highway Administration (FHWA), (2006, Dec.), *Bridge Inspector's Reference Manual*, FHWA, Washington, DC.

Problems

7.1   Many bridge owners in the United States require new bridges to be load rated to start a record of the service history. Based on this concept, please load rate the superstructure interior and exterior beams designed in Problems 4.4 and 4.5 for the design load.

7.2   Load rate the superstructure interior and exterior beams designed in Problems 4.4 and 4.5 for the AASHTO legal loads. Compare and comment on the results of this problem and Problem 7.1.

7.3 Load rate the superstructure exterior steel beam in Problem 4.5 for the design load, assuming that the bottom flange of the beam is corroded and has lost 20% of its original capacity.

7.4 Load rate the superstructure exterior steel beam in Problem 4.5 for the AASHTO legal loads, assuming that the bottom flange of the beam is corroded and has lost 20% of its original capacity.

7.5 Search the Internet to find two sets of guidelines/manuals/instructions for highway bridge inspection issued by a state or federal agency. Review the contents. Comment on the differences between the two sets identified and acquired.

Index